Performance Engineering of Computer and Telecommunications Systems

Springer

London
Berlin
Heidelberg
New York
Barcelona
Budapest
Hong Kong
Milan
Paris
Santa Clara
Singapore
Tokyo

Madjid Merabti, Michael Carew and Frank Ball (Eds)

Performance Engineering of Computer and Telecommunications Systems

**Proceedings of UKPEW'95,
Liverpool John Moores University, UK.
5-6 September 1995**

 Springer

Madjid Merabti, BSc, MA, PhD
Michael Carew, BEng (Hons)
School of Computing and Mathematical Sciences
Liverpool John Moores University
Byrom Street, Liverpool L3 3AF, UK

Frank Ball, BSc (Hons)
Department of Computing Sciences
Lancaster University
Bailrigg, Lancaster LA1 4YR, UK

ISBN 3-540-76008-3 Springer-Verlag Berlin Heidelberg New York

British Library Cataloguing in Publication Data
Performance engineering of computer and telecommunications systems : proceedings of UKPEN '95, Liverpool John Moores University, 5-6 September 1995
   1. Electronic digital computers 2. Telecommunication systems 3. Systems engineering - congresses
I.Merabti, Madjid II. Carew,Michael III.Ball, Frank
621.3'82
ISBN 3540760083

Library of Congress Cataloging-in-Publication Data
A Catalog record for this book is available from the Library of Congress

© Springer-Verlag London Limited 1996
Printed in Great Britain

Typesetting: Camera ready by contributors
Printed and bound at the Athenæum Press Ltd., Gateshead, Tyne and Wear
34/3830-543210 Printed on acid-free paper

Preface

This book is the proceedings of the Workshop on the Performance Engineering of Computer and Telecommunications Systems. The workshop was held at Liverpool John Moores University, England on the 5th and 6th September 1995. The workshop follows a series organised by the British Computer Society (BCS) Special Interest Group on Performance Engineering.

The workshop addressed most techniques and experiences in the Engineering of Computer and Telecommunications Systems that provide a guaranteed quality of service. Techniques such as measurements, simulation, and analytical models and their applications to ATM networks, Multimedia Systems, Distributed Systems, Access and Wide Area Networks were presented. In addition a number of papers dealt with advances in the development of analytical models, simulation architectures and the application of formal methods, such as Process Algebra, to the specification and building of performance biased computer systems.

The book is suitable for systems designers, engineers, researchers and postgraduate students interested in the design and implementation of Computer Systems, Networks and Telecommunications.

Many people assisted in the arrangements and success of this workshop. I would like to thank them all and in particular the reviewers.

I would also like to particularly thank our industrial sponsors GPT Public Networks Group, Liverpool and BICC Cables, Chester, England for their generous financial and material support.

Madjid Merabti, Workshop Chair,
Liverpool, September 1995

Contents

Formal Methods for Performance

Access Networks

Distributed Systems

Simulation

Wide Area Networks

Invited Paper

Cost Effective Methodologies for Queueing Models of ATM Switched Networks[*]

DEMETRES KOUVATSOS
Computer Systems Modelling Research Group
University of Bradford
Bradford, BD7 1DP, West Yorkshire
England

Abstract

Discrete-time queueing models are naturally recognised as powerful and realistic tools for the performance evaluation of Asynchronous Transfer Mode (ATM) switch architectures. However, analytic solutions of such models are often hindered by the generation of large state spaces requiring further approximations and a considerable (or even prohibitive) amount of computation. This paper reviews two cost-effective methodologies for the exact and approximate analysis of discrete-time queueing models of multi-buffered and shared buffer ATM switch architectures with bursty and/or correlated traffic. The methodologies are based on the principle of entropy maximisation, queueing theoretic concepts and batch renewal processes. Comments on future research work are included.

1   Introduction

Information highways to support B-ISDNs (Broadband Integrated Services Digital Networks), as well as local and wide area networks will be based on the Asynchronous Transfer Mode (ATM) transmission protocols. B-ISDN is the medium that will support vast volumes of traffic serving multi-media applications in business, scientific, medical and entertainment sectors. The performance prediction and quantitative analysis of ATM systems are therefore extremely important in view of their ever expanding usage and the multiplicity of their component parts together with the complexity of their functioning. The principal target is a constant search for new evaluation tools for actual system enhancement, identifying the most cost-effective means for better resource utilisation with the maximum of economy and efficiency. However, such predictions are even more valuable during the design and development stages of ATM networks, where experimentation is no longer an option and careful performance modelling and analysis provide crucial guidance to the network designer.

ATM is designed to integrate existing and future voice, audio, image and data services and to support transmission rates above 150 Mbps. Moreover, ATM aims to minimise the complexity of switching and buffer management, to optimise intermediate node processing and buffering and to bound transmission delays.

[*] This work is supported by the Engineering and Physical Sciences Research Council (EPSRC), UK, under grants GR/H/18609 and GR/K/67809.

These design objectives are met at high transmission speeds by keeping the basic unit of ATM transmission - the ATM cell - short and of fixed length. However, to support such a diverse range of services on one integrated communication platform, it is necessary to provide a most careful network engineering in order to achieve a fruitful balance amongst the conflicting requirements of different quality of service constraints, ensuring one service does not have adverse implications on another.

Over recent years a good deal of progress has been made in the research and development of ATM technology, but there are still many fundamental performance problems to resolve, such as: ATM traffic modelling and characterisation; traffic control, routing and optimisation; ATM switching techniques and provision of a specified quality of service (QoS).

Emerging architectural designs for ATM networks have R input and R output interconnected ports. Of particular importance are multi-buffered [1-4] and shared buffer [5-10] ATM switch architectures. Multi-buffered ATM switch architectures usually involve First-Come-First-Served (FCFS) buffering at either the input or output ports of each switching element. In such cases an arriving cell will be blocked (lost) if it finds a full buffer. Shared buffer ATM switch architectures incorporate a single memory of fixed size which is shared by all output ports. An incoming cell is stored in a shared buffer of finite capacity while its address is kept in the address buffer. Cells destined for the same output port can be linked by an address chain pointer [6] or their addresses can be stored into a FCFS buffer which relates to a particular output port. A cell will be lost if on arrival it finds either the shared buffer or the address buffer full. An example of such a switch architecture is the Prelude architecture proposed by CNET, France (cf. [5]). For other examples see [7, 8]. Moreover, buffered space division ATM switches, based on multistage interconnection networks, consist of switching components of the shared buffer type [11].

Traffic in B-ISDN is essentially discrete and basic operational parameters are known via measurements obtained at discretised points of time. Thus, discrete-time models, such as those based on queueing networks (QNs) and Stochastic Petri Nets (SPNs), offer powerful and realistic tools for representing B-ISBN and optimising their performance. Under this framework the time axis is segmented into a sequence of time intervals (or slots) of unit duration corresponding to the elementary unit of time in the system. In this context, arrivals and departures are allowed to occur at the boundary epochs of a slot, while during a slot no cells enter or leave the system.

The problem of deciding on an ATM traffic admission scheme is a complex one due to i) difficulties in characterising a bursty and correlated arrival process at the edges of the network [12-16] ii) current lack of understanding as to how the superposition of arrival processes is shaped deep in the network [17-21] and iii) bandwidth allocation decisions which are very difficult and CPU intensive and cannot be done on the fly [22].

It is widely accepted that there is a need for a unified approach for dealing with both traffic and network characteristics. SPNs [23-25] attempt to achieve this objective, but their notation is cumbersome and can obscure essential system structure making analytical solution difficult. Recently proposed Stochastic Timed Process Algebra (STPA) formalisms offer compositionality and a promising approach for analysis [26], although cost-effective solutions are not easy to obtain. Traditionally, Queueing network models (QNMs) are widely recognised as natural and valuable performance evaluation tools in the field. However, there are inherent difficulties and open issues associated with the study of these models, especially in the discrete time domain, due to the simultaneous occurrence of events, including bulk arrivals and departures, at the boundary epochs of slot and also the complexity of arbitrary queueing networks with various blocking mechanisms, flow correlation and traffic variability constraints.

Recently proposed queueing models for ATM switches and networks are not analytically tractable except in special cases (e.g. [11,27,28]). Usually it is necessary to resort to either simulation or numerical methods: simulation is time consuming, and cannot easily yield the great precision needed for some rare events such as cell loss, while numerical methods are severely limited in scope as the system size increases. Alternatively, existing fluid flow and diffusion approximations are based on the unrealistic assumption that the cross traffic is Poisson [29]. Thus, there is a great need to consider alternative methodologies leading to both credible and cost-effective approximations for the performance prediction and optimisation of ATM switches and networks.

This paper reviews two cost-effective methodologies which have recently been applied towards the exact and approximate analysis of discrete-time finite capacity queues and QNMs of multi-buffered and shared buffer ATM switch architectures [30-32]. These methodologies are based on the principle of maximum entropy (ME), queueing theoretic concepts and batch renewal processes.

The paper is divided into three sections. Section 2 introduces the ME formalism and presents cost-effective solutions for discrete-time queueing models of shared buffer ATM switch architectures and networks with bursty traffic and departures first (DF) buffer management policy. Section 3 describes batch renewal processes and reviews exact analytic solutions for a discrete-time queueing model under DF policy. This model represents a multibuffered ATM switch architecture decomposed into individual finite capacity output port queues with correlated and bursty traffic. Section 4 comments on future research work in the field.

Remarks

Buffer management policies for discrete time queues stipulate how a buffer is filled or emptied in the case of simultaneous arrivals and departures at a boundary epoch of a slot. In such cases, according to DF policy, departures take precedence over arrivals while under arrivals first (AF) policy the opposite effect is observed (see figure 1). Buffer management policies may play a significant rôle in the determination of blocking probabilities in discrete time finite capacity queues [33].

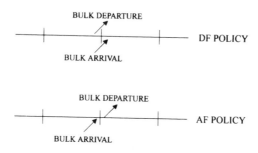

Figure 1. Effects of AF and DF Buffer management policies at slot boundary epoch.

2. Entropy Maximimisation and Shared Buffer ATM Switch Architectures

The principle of Maximum Entropy (ME), a probability inference method [34], has been applied successfully, in conjunction with queueing theoretic mean value constraints, to approximate analysis of both continuous time and discrete time QNMs of arbitrary configurations of single general queues with finite or infinite capacity [35-39]. The principle has been recently used in the study of general multibuffered and shared buffer queueing models of ATM switches and networks and closed form expressions have been obtained for queue length distribution, cell loss probability and mean delay [30, 31, 40, 41]. In the aforementioned studies the arrival process at each queue has been assumed to be bursty and was modelled by a compound Bernoulli process (CBP) with geometrically distributed bulk sizes [39]. In this context, the burstiness of the arrival process is captured by the squared coefficient of variation of the arrival process and, subsequently, the size of the incoming bulk. This particular CBP implies a generalised geometric (GGeo) interarrival time distribution, of which the pseudo-memoryless properties facilitate analysis of complex discrete time queues and networks [30, 31, 41].

This section introduces the ME formalism (Section 2.1) and the GGeo distributions (Section 2.2) and reviews their applicability towards the analysis of shared buffer ATM switches (Section 2.3) and networks (Section 2.4).

2.1 Maximum Entropy Formalism

Consider a system Q which has a set S of possible discrete states $\{S_0, S_1, S_2, ..\}$ which may be finite or countably infinite and state S_n, $n = 0,1,2,...$ may be specified arbitrarily. Suppose that the available information about Q places a number of constraints on $p(S_n)$, the probability distribution that the system Q is in state S_n. Without loss of generality, it is assumed that these constraints take the form of mean values of suitable functions $\{f_1(S_n), f_2(S_n), ..., f_m(S_n)\}$, where m is less than the number of possible states. The principle of maximum entropy [34] states that, of all distributions which satisfy the constraints, the minimally biased distribution is the one which maximises the system's entropy function

$$H(p) = -\sum_{S_n \in S} p(S_n) \ln p(S_n) \tag{2.1.1}$$

subject to the constraints

$$\sum_{S_n \in S} p(S_n) = 1 \tag{2.1.2}$$

$$\sum_{S_n \in S} f_k(S_n) p(S_n) = \langle f_k \rangle, \quad k = 1, 2, \ldots, m \tag{2.1.3}$$

where $\{\langle f_k \rangle\}$ are the prescribed mean values defined on the set of functions $\{f_k(S_n)\}$, $k=1,2,...,m$. The maximisation of (2.1.1), subject to the constraints (2.1.2) and (2.1.3), can be carried out using Lagrange's method of undermined multipliers and leads to the solution

$$p(S_n) = \frac{1}{Z} \exp\left(-\sum_{k=1}^{m} \beta_k f_k(S_n)\right) \tag{2.1.4}$$

where $\{\beta_k\}$, $k=1,2,...,m$, are the Lagrangian multipliers determined from the set of constraints (2.1.3) and Z, known in statistical physics as the "partition function", is given by

$$Z = \exp(\beta_0) = \sum_{S_n \in S} \exp\left(-\sum_{k=1}^{m} \beta_k f_k(S_n)\right) \tag{2.1.5}$$

where $\{\beta_0\}$ is the Lagrangian multiplier determined by the normalisation constraint (2.1.2).

Jaynes [42] has shown that, if the prior information includes all constraints actually operative during a random experiment, the distribution predicted by entropy maximum can be realised in overwhelmingly more ways than by any other distribution. The principle of maximum entropy has also been shown, by Shore and Johnson [43], to provide a "uniquely correct self-consistent method of inference" for estimating probability distributions based on the available information.

Maximum entropy formalism can be used in the performance analysis of queueing systems because expected values of various distributions of interest are usually known in terms of moments of the interarrival and service time distributions.

2.2. The GGeo-Type Distribution

Consider a discrete time random variable (rv) W representing the interarrival time or the service time of a stable single server queue. Let $E[W] = 1/v$ be the mean and C_W^2 be the SCV of W. The GGeo discrete time distribution [39] is defined by

$$\Pr[W = n] = \begin{cases} 1 - \tau & n = 0 \\ \tau\sigma(1-\sigma)^{n-1} & n \geq 1 \end{cases} \tag{2.2.1}$$

where

$$\tau = \frac{2}{C_W^2 + 1 + v}, \quad 0 < \tau \leq 1, \tag{2.2.2}$$

$$\sigma = \tau v, \quad\quad\quad 0 < \sigma < 1, \tag{2.2.3}$$

and $|1 - v| \leq C_W^2$ (see Figure 2).

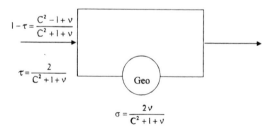

Figure 2. The GGeo distribution with parameters τ and σ $(0 < \tau, \sigma \leq 1)$.

Since rv W can realise a value of zero, it is implied that the GGeo distribution accommodates bulk arrivals or departures. In essence, the GGeo distribution corresponds to a CBP with rate σ, whilst the number of events (i.e. arrivals or departures) in a slot (i.e. the bulk size) is distributed geometrically with parameter τ. Thus, the GGeo pattern comprises a sequence of bulks with independent and identically distributed non-negative integer valued rv's $\{W_k\}$, where W_k, k=1,2,... is the number of events occurring at the kth slot, with a fixed probability distribution given by

$$\Pr\left[W_k = \ell\right] = \begin{cases} 1-\sigma & e = 0 \\ \sigma\tau(1-\tau)^{\ell-1} & e \geq 1 \end{cases} \qquad (2.2.4)$$

and $v = \sigma/\tau$ events per slot is the mean event (arrival or departure) rate and $C_w^2 \geq |1 - v|$. The $GGeo(\tau,\sigma)$ distribution is versatile, possessing the "pseudo memoryless" attribute (i.e. the remaining interevent time of a $GGeo(\tau,\sigma)$ distribution is geometric with parameter σ) and other interesting properties [39].

In the context of an ATM switch architecture, due to the fixed cell size and the nature of the associated output links, transmission times are assumed to be deterministic, with SCV=0. It is interesting to note in this case that, if $v = 1$, then $\tau = 1$ and $\sigma = 1$ and the $GGeo(\tau,\sigma)$ distribution reduces to a true deterministic (D) distribution.

2.3 ME Application to a Shared Buffer Queue

2.3.1. Model Formulation

Consider a general queueing model of a shared buffer switch with bursty arrivals and DF buffer management policy depicted in Figure 3. The queueing model consists of R parallel single server queues, where R is the number of output ports. Each server represents an output port and each queue corresponds to the address queue for the output port. There are R bursty and heterogeneous arrival streams of cells, one stream to each of R input ports. Each stream is described by a $GGeo\left(\tau_{ai},\sigma_{ai}\right)$ distribution with mean arrival rate Λ_i cells per slot and SCV of interarrival time C_{ai}^2 for stream i, i=1,2,...,R. Similarly, the transmission (or service) time of a cell at queue i follows a $GGeo\left(\tau_{si},\sigma_{si}\right)$ distribution with mean rate μ_i cells per slot and SCV of interdeparture time C_{si}^2 for stream i, i=1,2,...,R. Let N be the size of the total shared buffer. A cell is lost if it arrives at a time when there are a total of N cells in the R queues. Without loss of generality, it is assumed that any of the R queues may attain the maximum size N.

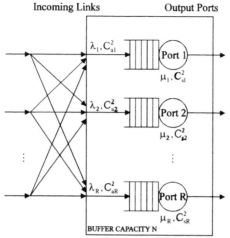

Incoming Links   Output Ports

The $S_{RxR}(G/G/1)/N$ building block

Figure 3. The $S_{R \times R}$(GGeo/GGeo/1)/N queueing model of a shared buffer switch.

The queueing model of the shared buffer switch is denoted by $S_{R \times R}$ (GGeo/GGeo/1)/N, such that

i)   the overall (merged) interarrival times and service times at a $R \times R$ switch queue are heterogeneous and GGeo distributed,

ii)   each output port has a single server,

iii) the total buffer capacity of the switch is N.

Finally, let the state of the system at any time be represented by a vector $\mathbf{n} = (n_1, n_2, \ldots, n_R)$, where n_i is the number of cells in queue i, $i=1,2,\ldots,R$.

2.3.2. A ME Solution for a $S_{R \times R}$(GGeo/GGeo/1)/N Queue

The form of the ME solution $p(\mathbf{n})$,

$$\mathbf{n} \in S(N, R) = \left\{ \mathbf{n} = (n_1, n_2, \ldots, n_R): \sum_{i=1}^{R} n_i \leq N, 0 \leq n_i \leq N, i = 1, \ldots, R \right\}$$

of a $S_{R \times R}$(GGeo/GGeo/1)/N queueing model has been characterised in [30, 40], subject to normalisation and the constraints: server utilisation U_i, $0<U_i<1$; mean queue length $L_i, U_i \leq L_i < N$; aggregate probability φ_i of full buffer with $n_i > 0$,

$0<\varphi_i<1$, $i=1,2,...,R$, and is given - by the method of Lagrange's undetermined multipliers - as

$$p(\mathbf{n}) = \frac{1}{Z} \prod_{i=1}^{R} g_i^{s_i(\mathbf{n})} x_i^{n_i} y_i^{f_i(\mathbf{n})}, \quad \forall \mathbf{n} \in S(N,R) \tag{2.3.1}$$

where Z is the normalising constant

$$Z = \sum_{\mathbf{n} \in S(N,R)} \prod_{i=1}^{R} g_i^{s_i(\mathbf{n})} x_i^{n_i} y_i^{f_i(\mathbf{n})}, \tag{2.3.2}$$

$s_i(\mathbf{n})$ and $f_i(\mathbf{n})$ are suitable auxiliary functions and $\{g_i, x_i, y_i : i = 1,2,...,R\}$ are the GGeo type Lagrangian coefficients corresponding to the constraints $\{U_i, L_i, \varphi_i : i = 1,2,...,R\}$, respectively.

The GGeo Lagrangian coefficients $\{g_i, x_i : i = 1,2,...,R\}$ are obtained by making asymptotic connections with a stable GGeo/GGeo/1 infinite capacity queue, namely

$$g_i = \frac{\rho_i(1-x_i)}{x_i(1-\rho_i)}, \qquad x_i = \frac{\tilde{L}_i - \rho_i}{\tilde{L}_i}, \qquad i = 1,2,...,R, \tag{2.3.3}$$

where $\rho_i = \Lambda_i / \mu_i$ and $\tilde{L}_i$ is the mean queue length of the GGeo/GGeo/1 queue [39].

The normalising constant can be determined by applying the generating function approach [44] and can be computed by

$$Z = \sum_{v=0}^{N-1} C_1(v) + C_2(N) \tag{2.3.4}$$

where $\{C_1(v) : v = 0,1,...,N-1\}$ and $\{C_2(N)\}$ are determined via appropriate recursive formulae [30, 40].

Similarly the utilisation U_i can be expressed as

$$U_i = \frac{1}{Z} \left(\sum_{v=1}^{N-1} C_1^{(i)}(v) + C_2^{(i)}(N) \right), \quad i = 1,2,...,R \tag{2.3.5}$$

where

$$C_k^{(i)}(v) = (1-B_i)x_i C_k^{(i)}(v-1) + B_i x_i C_k(v-1), \quad v=2,..,N-2+k, \quad k=1,2, \quad i=1,...,R,$$

with initial conditions:

$$C_k^{(i)}(1) = B_i x_i, \quad B_1 = g_1 \text{ and } B_2 = g_2 y_2.$$

The marginal state probabilities can be determined by using ME solution (2.3.1) and the recursive expressions for $C_k^{(i)}(v)$. Let $n(i)$ be the random variable for the number of cells at queue i, $i = 1, 2, \ldots, R$. Then the marginal state probabilities are given by [30]

$$\Pr[n(i) = l_i] = \Pr[n(i) \geq l_i] - \Pr[n(i) \geq l_i + 1] \tag{2.3.6}$$

where

$$\Pr[n(i) \geq l_i] = \frac{x_i^{l_i-1}}{Z}\left(\sum_{v=l_i}^{N-1} C_1^{(i)}(v - l_i + 1) + C_2^{(i)}(N - l_i + 1)\right)$$

$i = 1, 2, \ldots, R, \, l_i = 1, 2, \ldots, N$.

Moreover, by focusing on a stable $S_{R\times R}(GGeo/GGeo/1)/N$ queue under DF policy and by using GGeo type probabilistic arguments, the marginal cell-loss probabilities are seen to be

$$\pi_i = \frac{1}{Z}\left(F_i(N) + C_2(N) - \mu_i \tau_{ai} \delta_i C_2^{(i)}(N)\right) \tag{2.3.7}$$

where

$$\delta_i = \frac{\tau_{si}}{\tau_{si}(1 - \tau_{ai}) + \tau_{ai}} \quad \text{and} \quad F_i(N), i = 1, 2, \ldots, R \quad \text{are suitable functions}$$

involving input data and recursive formulae $C_1(v)$ and $C_1^{(i)}(v), i = 1, 2, \ldots, R$ [30].

The GGeo type Lagrangian coefficients $\{y_i : i = 1, 2, \ldots, R\}$ can be determined numerically by substituting U_i of (2.3.5) and π_i of (2.3.7) in the flow balance conditions

$$\Lambda_i(1 - \pi_i) = \mu_i \upsilon_i, \quad i = 1, 2, .., R \tag{2.3.8}$$

and solving the resulting system of R non-linear equations with R unknowns $\{y_i : i = 1, 2, \ldots, R\}$, namely

$$\left(1-\mu_i\tau_{ai}\delta_i\rho_i\right)C_2^{(i)}(N) = \rho_i\left(\sum_{v=1}^{N-1}C_1(v)-F_i(N)\right)-\sum_{v=1}^{N-1}C_1^{(i)}(v) \qquad (2.3.9)$$

for all $i = 1, 2, \ldots, R$ and $N \geq 2$.

System (2.3.9) can be solved by applying the numerical algorithm of Newton Raphson in which the partial derivatives of the Jacobian matrix are calculated by applying an efficient recursive scheme (cf[40]). Thus, the $S_{R\times R}(\text{GGeo}/\text{GGeo}/1)/N$ queueing model can be used as an effective building block in the analysis of large discrete time queueing networks of shared buffer ATM switches. An outline of such computational procedure is presented in the following section.

2.4 ME Application to Arbitrary Open Queueing Networks of Shared Buffer Switches

2.4.1 Model Formulation

Consider an arbitrary discrete time open queueing network at equilibrium consisting of M nodes under DF buffer management policy, as depicted in Figure 4. Each node i, $i = 1, 2, \ldots, M$, is a $R_i \times R_i$ shared buffer queueing model with finite capacity N_i (see Figure 3). At any given time, the joint state of the network is denoted by $\mathbf{n} = (\mathbf{n}_1, \mathbf{n}_2, \ldots, \mathbf{n}_M)$ where $\mathbf{n}_i = (n_{i1}, n_{i2}, \ldots, n_{iR_i})$ is the joint state of shared buffer queueing system i and n_{ij} is the number of cells queueing for output port j, $j = 1, 2, \ldots, R_i$. Moreover, let $p(\mathbf{n})$ be the joint state probability of the network.

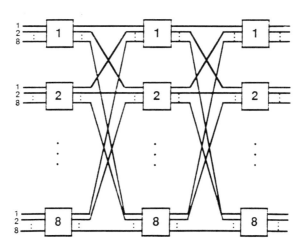

Figure 4. A network configuration of shared buffer switches.

2.4.2 A ME Solution for a Network of $S_{RxR}(GGeo/GGeo/1)/N$ Switches

The form of the ME solution, $p(\mathbf{n})$, for a FCFS GGeo type open queueing network, subject to normalisation and the marginal constraints of shared buffer queueing systems used in Section 4, namely utilisation, U_{ij}, $0<U_{ij}<1$, mean queue length, L_{ij}, $U_{ij}<L_{ij}<N_i$, and aggregate full buffer probability with $n_{ij}>0$, φ_{ij}, $0<\varphi_{ij}<1$, $j=1,2,\ldots,R_i$, $i=1,2,\ldots,M$, is given — by the method of Lagrange's undetermined multipliers — as

$$p(\mathbf{n}) = \frac{1}{Z} \prod_{i=1}^{M} \prod_{j=1}^{R_i} g_{ij}^{s_{ij}(\mathbf{n}_i)} x_{ij}^{n_{ij}} y_{ij}^{f_{ij}(\mathbf{n}_i)} \tag{2.4.1}$$

where Z is the normalising constant and $\{g_{ij}, x_{ij}, y_{ij}\}$ are the Lagrangian coefficients corresponding to constraints $\{U_{ij}, L_{ij}, \varphi_{ij}\}$, respectively. The form of ME solution (2.4.1) clearly suggests a product form approximation, namely

$$p(\mathbf{n}) = \prod_{i=1}^{M} p(\mathbf{n}_i) \tag{2.4.2}$$

where $p(\mathbf{n}_i)$ is determined by the ME solution (2.3.1) of each shared buffer $S_{RxR}(GGeo/GGeo/1)/N_i$ queueing model, $i=1,2,\ldots,M$.

The ME solution (2.4.1) can be implemented computationally by decomposing the network into individual building blocks of $S_{Ri \times Ri}(GGeo/GGeo/1)/N_i$ queues, $i=1,2,\ldots,M$, each of which can be solved in isolation be determining iteratively the first two moments of the overall flows in the network.

2.4.3 Flows in the Network

In the sequel, routing of cells through the network is based upon the notion of the virtual circuit (VC). A VC has a fixed path through the network. All cells that belong to a particular VC flow along its path. A number of different VCs will exist across the network. It is assumed that the first two moments of the external flow of each VC as it arrives at the network is known. To this end, GGeo type ME solution and GGeo flow formulae can be applied in a similar fashion to that established in [41]. However, the flow of cells belonging to VCs must be converted to flows through each switch/port, and from one switch/port to another. Due to finite buffer sizes cell loss will occur at switches and thus within a VC the flow of cells will reduce at each link composing its path. Because cell flows are attenuated, it is not possible to calculate apriori the flows required in the GGeo flow formulae.

The path of a VC can be represented as an ordered and finite list of switch/port pairs;

$$(0,0) \rightarrow (i,a) \rightarrow, .., \rightarrow (j,b) \rightarrow (k,c) \rightarrow, .., \rightarrow (0,0)$$

where $(0,0)$ represents the outside world. Figure 5 shows two VCs, namely

$$(0,0) \rightarrow (1,3) \rightarrow (2,1) \rightarrow (0,0),$$

and $(0,0) \rightarrow (1,1) \rightarrow (2,2) \rightarrow (3,2) \rightarrow (0,0)$ following predefined routes through three 3×3 switches.

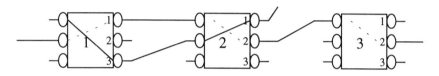

Figure 5. Routes of VCs across a Network of ATM Switches.

The mean rate of cells flowing on a link, say $(j,b),(k,c)$ of the path of VCl is given by

$$\Lambda^{vc\,\ell}_{(j,b),(k,c)} = \left(1 - \pi_{(i,a),(j,b)}\right)\Lambda^{vc\,\ell}_{(i,a),(j,b)} \qquad (2.4.3)$$

where $\Lambda^{vc\,\ell}_{(i,a),(j,b)}$ is the mean overall arrival rate of cells on VCl on the link $(i,a),(j,b)$ and $\pi_{(i,a),(j,b)}$ is the cell loss or blocking probability that cells from switch/port (i,a) are lost on arrival to switch/port (j,b). It can be shown, by using GGeo type probabilistic arguments (as in the case of a network with ordinary finite capacity queues [41]), that probability $\pi_{(i,a),(j,b)}$ can be obtained by equation (2.3.7) with the subscripts of parameters change to $(i,a),(j,b)$ to reflect that arrivals to switch/port (j,b) are considered from one stream emanating from switch/port (i,a) instead of the merged stream of all arrivals. Once all the blocking probabilities are known the flow on each link of each VC can be calculated (starting with $\Lambda^{vc\,\ell}_{(0,0),(i,a)}$, the known external arrival rate of VCl) by repeated application of equation (2.4.3).

The mean flow of cells generated by switch/port (i,a) upon VCl and entering switch/port (j,b) is referred to as the mean effective rate $\lambda^{vc\,\ell}_{(i,a),(j,b)}$. This is the same as the mean rate of cells of VCl that leave switch/port (j,b). As this is the overall mean rate on the next link (k,c) of the path of VCl, it follows that

$$\lambda^{vc\,\ell}_{(i,a),(j,b)} = \Lambda^{vc\,\ell}_{(j,b),(k,c)} \tag{2.4.4}$$

If VCl enters the network at switch/port (i,a), its mean overall external arrival rate is denoted by

$\Lambda^{vc\,\ell}_{(0,0),(i,a)}$, and its SCV by $C^{vc\,\ell}_{(0,0),(i,a)}$. The mean effective arrival rate at switch/port (i,a), obtained via equation (2.4.3), is $\Lambda^{vc\,\ell}_{(i,a),(j,b)}$. By using the GGeo split flow formulae [39], the SCV of the effective external interarrival time at switch/port (i,a) is clearly given by

$$\tilde{C}^{vcl}_{(0,0),(i,a)} = \pi_{(0,0),(i,a)} + C^{vcl}_{(0,0),(i,a)} \left(1 - \pi_{(0,0),(i,a)}\right) \tag{2.4.5}$$

The overall mean rate of cells that flow between upstream switch/port (r,s) and switch/port (k,j) is given by

$$\Lambda_{(r,s),(k,j)} = \sum_{l \in S_{(r,s),(k,j)}} \Lambda^{vc\,l}_{(r,s),(k,j)} \tag{2.4.6}$$

where $S_{(r,s),(k,j)}$ is the set of all VCs {l}having (r,s),(k,j) as a link on their path. Moreover, the overall mean arrival rate of cells at switch/port (k,j) is given by

$$\Lambda_{(k,j)} = \sum_{(r,s) \in S_{(k,j)}} \Lambda_{(r,s),(k,j)} \tag{2.4.7}$$

where $S_{(k,j)}$ is the set of all upstream switch/ports {(r,s)} linking to switch/port (k,j).

To simplify notation, in what follows the suffix (k,j) for port j of switch k is represented by i and the suffix (r,s) for port s of switch r is represented by j.

The SCV of effective interarrival time at output port i is given by the merging of GGeo type streams [39]

$$\tilde{C}^2_{ai} = -1 - \lambda_i + \left(\sum_{l \in S^a_{li}} \frac{\lambda^{vcl}_{0i}}{\lambda_i \left(\tilde{C}^{vcl}_{0i} + \lambda^{vcl}_{0i} + 1 \right)} + \sum_{j \in A(i)} \frac{\lambda_{ji}}{\lambda_i \left(\tilde{C}a^2_{ji} + \lambda_{ji} + 1 \right)} \right)^{-1}, i = 1,2...,M \tag{2.4.8}$$

where S^a_{li} is the set of VCs that enter the network at port i, $A(i)$ is the set of all upstream ports immediately connected to port i, λ^{vcl}_{0i} and $\tilde{C}^{vcl}_{0i}$ is the effective mean rate and SCV, respectively, of the interarrival times of cells entering the network at port i upon $VC\ell$ (cf, (2.4.4), (2.4.5)) and $\tilde{C}a_{ji}^2$ is the SCV of the effective stream of cells that leave port j and enter port i. Clearly,

$$\lambda_{ji} = \left(1 - \pi_{ji}\right)\Lambda_{ji}, \; j = 0,1,..,m, \quad i = 1,2,..,M \tag{2.4.9}$$

where π_{ji} is the blocking probability determined by (2.3.7) with parameters Λ_{ji} and C_{dji}^2 the latter being the SCV of the interdeparture time of the overall stream of cells from port j to port i.

The SCV of the overall interarrival times at output port i is clearly given by the merging of overall GGeo type streams [39], namely

$$C_{ai}^2 = -1 - \Lambda_i + \left(\sum_{l \in S_{li}^a} \frac{\Lambda_{0i}^{vcl}}{\Lambda_i\left(C_{0i}^{vcl} + \Lambda_{0i}^{vcl} + 1\right)} + \sum_{j \in A(i)} \frac{\Lambda_{ji}}{\Lambda_i(C_{dji}^2 + \Lambda_{ji} + 1)}\right)^{-1}, j = 0,1,..,M; \; i = 1,2,..,M \tag{2.4.10}$$

The departure rate from (or the effective arrival rate to) an output port j is a non-linear function of the switching node traffic characteristics and is given by the sum of the rates from that port to all downstream destinations

$$\lambda_j = \sum_{l \in S_{lj}^a} \Lambda_{j0}^{vcl} + \sum_{i \in D(j)} \Lambda_{ji}, \; j = 1,2,..,M; \; i = 1,2,..,M \tag{2.4.11}$$

where s_{lj}^a is the set of VCs that leave the network at port j and Λ_{j0}^{vcl} is the rate at which cells leave the network from port j upon VCl and $D(j)$ is the set of all downstream ports immediately connected to port j. Moreover the SCV of the interdeparture time C_{di}^2 can be approximated by analysing a stable the departure process of a GGeo/GGeo/1 queue with infinite capacity and effective revised arrival rate, and is given by [39]

$$C_{di}^2 = \tilde{\rho}_i^2(C_{si}^2 + \mu_i - 1) - \tilde{\rho}_i(\tilde{C}_{ai}^2 + \lambda_i - 1) + \tilde{C}_{ai}^2, \quad i = 1,2,...,M \tag{2.4.12}$$

where $\tilde{\rho}_i = \lambda_i / \mu_i$.

The splitting of the interdeparture times at each output port j with parameters $(\Lambda_{ji}, C_{dji}^2)$ can be made dynamically via computational iteration which necessitates the estimation of transition probabilities $\{\alpha_{ji}\}$ that a departing cell from port j will go to port i. Probabilities $\{\alpha_{ji}\}$ can be determined by observing that each output port i experiences a virtual arrival process of cell complying with a random routing policy and is clearly given by

$$\alpha_{ji} = \frac{\Lambda_{ji}}{\lambda_j} \qquad\qquad (2.4.13)$$

The SCV of interdeparture time for the overall flow is given by

$$C_{d_{ji}}^2 = 1 + \alpha_{ji}\left(C_{d_j}^2 - 1\right), \ j = 1,2,..,M; \ i = 0,1,..,M \qquad\qquad (2.4.14)$$

whilst, the SCV of effective interarrival time at port i from port j can be expressed by

$$\tilde{C}_{a_{ji}}^2 = \pi_{ji} + C_{d_{ji}}^2\left(1 - \pi_{ji}\right), \ j = 1,2,..,M; \ i = 1,2,..,M \qquad\qquad (2.4.15)$$

2.4.4 A ME Algorithm

The steps of an iterative ME algorithm broadly follow those established by the ME methodology as applied to simpler types of networks of ordinary finite capacity queues [38, 41] and are based on the following building blocks

i)   the ME solution of the $S_{R \times R}(GGeo / GGeo / 1) / N$ shared buffer queuing model,

ii) the GGeo type flow approximation formulae for calculation of the mean and SCV of the interarrival and interdeparture times at each output port and

iii) the mean rate of VCs on each link of their paths .

A sketch of the ME algorithm is presented below:

Step 1. Read in input parameters and routes of VCs and service time parameters of output port queue and initialise all cell loss probabilities to zero. Set SCV of interdeparture times to 1.

Step 2. Compute the mean flows rates for each VC on each link in their paths using equation (2.4.3) and calculate effective flows of each VC as it enters the network using equations (2.4.3) and (2.4.5).

Step 3. Calculate the overall and effective mean arrival rate to each output port from all other output ports and the overall mean departure rate from each output port to all other output ports using equations (2.4.4), (2.4.6), (2.4.7), (2.4.9) and (2.4.11).

Step 4. Obtain the routing probabilities via equation (2.4.13).

Step 5. Utilise the routing probabilities to obtain the SCV of interdeparture times from each output port to other output ports via equations (2.4.14).

Step 6. Calculate the SCV of the merging streams (overall and effective) at each output port using equations (2.4.8), (2.4.10) and (2.4.15).

Step 7. Obtain the queue length distribution of the $S_{R \times R}$(GGeo/GGeo/1)/N as described in Section 4 (cf equations (2.3.1, 2.3.8)) with the overall interarrival time parameters obtained in steps 3 and 6.

Step 8. Compute the blocking probabilities using equation (2.3.7).

Step 9. Estimate the overall SCV of the interdeparture time of each output port using equation (2.4.12).

Repeat steps 2–9 until convergence of the calculated value of the SCV of the interdeparture time.

Step 10. Obtain cell loss probability and mean delay for each output port in the network.

Remarks

The main computational cost of the ME algorithm is the calculation of cell loss probabilities at the output ports of the shared buffer switch, which must be obtained at each iteration (step 8). However, these computations are based on the effective building block $S_{RXR}(GGeo/GGeo/1)/N$ queue and can be performed in a few minutes on a SUN workstation.

The existence and uniqueness of the solution of the system of non-linear equations (2.3.9) cannot be proved analytically due to the complexity of the expressions involved. Moreover, no rigorous mathematical justification can be given for convergence of the C_{di}^2's. Nevertheless, there has been no incidence of numerical instability and no lack of convergence in the many experiments which have been performed.

More details and numerical results on the ME analysis of shared buffer ATM switch architectures can be seen in [30, 31, 40].

3. Batch Renewal Processes and Multi-buffered ATM Switch Architectures

In the field of ATM technology, the indices of dispersion have been proposed as appropriate characterisation of bursty, correlated traffic and Markov modulated processes as models of sources of bursty traffic with correlation. Sriram and Whitt [12] described superposition of bursty sources (modelled by renewal processes) in terms of the indices of dispersion for intervals (IDI). Heffes and Lucantoni [13] model superposition of bursty renewal processes approximately by a 2-phase Markov modulated Poisson process (MMPP) matched on three features of the indices of dispersion for counts (IDC) and mean arrival rate. Gusella [14] estimated indices of dispersion for measured LAN traffic and modelled the traffic approximately by a 2-phase MMPP matched on three features of the IDC and the SCV of one interarrival time. Fowler and Leland [45] have reported LAN traffic with unbounded IDC. However, it is to be expected that performance of restricted buffer systems with deterministic service (as in ATM switches) would not be affected by the magnitude of IDC for long intervals. Andrade and Martinez-Pascua [46] have shown that queue length distribution, and other performance measures are affected by IDC only up to a certain size of interval (determined by the buffer size) and "the value of the IDC at infinity has little importance". Recently , Kouvatsos and Fretwell [32, 47, 48] have shown that if the indices of dispersion are all that is known about certain traffic (as might result from measurements of real traffic) then a batch renewal process may be constructed which incorporates all that information and no other: in that sense, batch renewal processes provide a description of the traffic which is both complete and least biased - which models, such as MMPP, with limited parameterisation do not.

This section reviews an exact technique, based on batch renewal processes, for the performance analysis of simple multi-buffered ATM switch architectures. Such architectures can be decomposed into individual censored GI^G/D/1/N queues each of which can be analysed in isolation.

Section 3.1 defines batch renewal processes, lists their properties and presents the relationships between the component distributions of the batch renewal arrivals process and the queue the length distribution, waiting time and blocking probability in a finite buffer queue with deterministic service and censored batch renewal arrivals process. Such a queueing system is an appropriate model for an ATM multiplexer or partitioned buffer switch. The analysis is specialised in Section 3.2 to batch renewal processes in which the component distributions are shifted generalised geometric (shifted GGeo). This form of process appears to be appropriate to measured traffic, especially where the usable data set be limited by (say) the time for which the actual process may be regarded as being stationary. Closed form expressions for queue length distribution, waiting time and blocking probability are given.

3.1 Batch Renewal Processes and Censored GI ^{G}/1/N/Queue

A batch renewal arrival process is a process in which there are batches of simultaneous arrivals such that

- the number of arrivals in batches are independent identically distributed random variables,

- the intervals between batches are independent identically distributed random variables,

- the batch sizes are independent of the intervals between batches.

It has been shown [32, 47] that a discrete time batch renewal arrival process may be constructed to give any degree of correlation between numbers of arrivals at different epochs and, simultaneously, any degree of correlation between interarrivals times at arbitrary lags. Indeed, there is a *one-to-one* correspondence between arbitrary set of indices of dispersion (or, equivalently, of correlation functions or covariances) and a batch renewal process. Furthermore, the corresponding batch renewal process is, in the information theoretic sense, the *least biased choice* given only a set of indices of dispersion or of correlation functions [32].

Consider a discrete time batch renewal process in which

- the distribution of batch size is given by the probability mass function (*pmf*) $b(n)$, $n = 1, 2,...$, with mean b, square coefficient of variation (SCV) C_b^2 and probability generating function (*pgf*) $B(z) = \sum_{n=0}^{\infty} b(n) z^n$, and

- the distribution of intervals between batches is given by the *pmf*

  $a(t)$, $t = 1, 2,....$, with mean a, SCV C_a^2 and *pgf* $A(w) = \sum_{t=0}^{\infty} a(t) w^t$.

Observe that no loss in generality ensues from the assumption that

$a(0) = 0$, $b(0) = 0$.

The correlation functions (covariances) at lag l, $l = 1, 2,...$, are related to the generating functions of the component distributions of the batch renewal process by

$$K(w) \overset{\Delta}{=} \frac{1}{\lambda} \left(Var[N] + 2 \sum_{l=1}^{\infty} Cov[N(t), N(t+l)] w^l \right)$$

$$= bC_b^2 + b \frac{1 + A(w)}{1 - A(w)} - \lambda \frac{1 + w}{1 - w} \tag{3.1.1}$$

$$L(z) = \lambda^2 \left(Var[X] + 2 \sum_{l=1}^{\infty} Cov[X(t), X(t+l)] z^l \right)$$

$$= bC_a^2 + b \frac{1 + B(z)}{1 - B(z)} - \frac{1 + z}{1 - z} \tag{3.1.2}$$

where $\lambda = b/a$ is the mean arrival rate.

Remarks

i) $\lambda K(z)$ and $L(w)/\lambda^2$ are *pgf* analogs of the spectral density functions for the random sequence of interarrival times, respectively.

ii) The relationship between correlation functions (covariances) and indices of dispersion may be described by

$$K(w) = (1-w)^2 I'(w) \text{ and } L(z) = (1-z)^2 J'(z)$$

where $I(w)$ is the *pgf* of the indices of dispersion I_t for counts, $J(z)$ is the *pgf* of the indices of dispersion J_n for intervals and where the prime (') indicates the derivative.

Consider a $GI^G/D/1/N$ queue in discrete time in which arrivals to a full system are turned away and simply lost (i.e. censored arrivals). Events (arrivals and departures) occur at discrete points in time (epochs) only. The intervals between (epochs) are called *slots* and, without loss of generality, may be regarded as being of constant duration. At an epoch at which both arrivals and departures occur, the departing customers release the places, which they had been occupying, to be available to arriving customers (DF buffer management policy). The service time for a customer is one slot and the first customer arriving to an empty system (after any departures) receives service and departs at the end of the slot in which it arrived (*immediate service* policy). By GI^G arrivals process is meant the intervals between batches are independent and of general distribution and the batch size distribution is general (batch renewal process).

Consider further two processes embedded at points immediately before and immediately after each batch of arrivals. Each process may be described independently by a Markov chain but the processes are mutually dependent. Let

$P_N(n)$ be the steady state probability that there be $n = 0, 1,...., N$ customers in the system (either queueing or receiving service) during a slot (i.e. $\{ p_N(n) : n = 0,...,N \}$ is random observer's distribution),

$P_N^A(n)$ be the steady state probability that a batch of arrivals 'see' $n = 0,...,N - 1$ customers in the system (i.e. $\{ p_N^A(n) : = 0,...,N \}$ is the stationary distribution of the Markov chain embedded immediately before batch arrivals),

$p_N^D(n)$ be the steady state probability that a batch of arrivals 'create' $n = 1,...,N$ customers in the system (i.e. $\{ p_N^D(n) : n = 1,...,N \}$ is the stationary distribution of the Markov chain embedded immediately after batch arrivals).

Solution of a discrete time $GI^G/D/1/N$ queue is achieved by exploiting the following relationships [32]:

$$p_N^A(n) = \begin{cases} \sum\limits_{k=1}^{N} p_N^D(k) \sum\limits_{t=k}^{\infty} a(t) & n=0 \\ \sum\limits_{k=n+1}^{N} p_N^D(k) a(k-n) & n=1,...,N-1 \end{cases} \qquad (3.1.3)$$

$$p_N^D(n) = \begin{cases} \sum\limits_{k=0}^{n-1} p_N^A(k) b(n-k) & n=1,...,N-1 \\ \sum\limits_{k=0}^{N} p_N^A(k) \sum\limits_{r=N-k}^{\infty} b(r) & n=N \end{cases} \qquad (3.1.4)$$

Random Observer's Probabilities, $\{P_N(n) := 0,...N\}$

$$P_N(n) = \begin{cases} \dfrac{1}{a} \sum\limits_{k=1}^{N} p_N^D(k) \sum\limits_{t=k+1}^{\infty} (t-k) a(t) & n=0 \\ \dfrac{1}{a} \sum\limits_{k=n}^{N} p_N^D(k) \sum\limits_{t=k-n+1}^{\infty} a(t) & n=1,...,N \end{cases} \qquad (3.1.5)$$

Blocking Probability, π_N^B

$$\pi_N^B = \sum_{k=0}^{N-1} p_N^A(k) \sum_{r=1}^{\infty} \frac{r}{b} b(N-k+r) \qquad (3.1.6)$$

Waiting time distribution, $\{w(t) : t = 1,...,N\}$

$$w(t) = \sum_{k=0}^{t-1} p_N^A(k) \frac{1}{b_{N-k}} \sum_{r=t-k}^{N-k} {}^b N-k \,(r) = \sum_{k=0}^{t-1} P_N^A(k) \frac{1}{b_{N-k}} \sum_{r=t-k}^{\infty} b(r) \qquad (3.1.7)$$

where

$$b_k = (k>1) \sum_{n=1}^{k-1} n b(n) + k \sum_{n=k}^{\infty} b(n) \qquad (3.1.8)$$

and b_k is the mean effective batch size given that an arriving batch see k,

$k=1,...,N-1$, places available in the buffer.

3.2 Shifted GGeo Distributions of Batch Size and Intervals

This section presents particular forms of batch renewal arrivals process which appear to be especially appropriate to models of traffic where there are relatively few measurements from which the correlation functions (covariances) may be estimated. In such cases it is natural to plot the logarithms of covariances against lags and fit a straight line to the plot. Then, if

$$\log Cov\left[N(t), X(t+\ell)\right] \cong -C - m\ell$$

(for some constants C and m), equation (3.1.2) implies that the corresponding batch renewal process has batch size distribution of the form

$$b(n) = \begin{cases} 1-\eta & n=1 \\ \eta v(1-v)^{n-2} & n=2,... \end{cases} \tag{3.2.1}$$

in which

$$\eta = \frac{\lambda^2 e^{-C}}{1+\lambda^2 e^{-C}}\left(1-e^{-m}\right) \quad \text{and} \quad v = \frac{1}{1+\lambda^2 e^{-C}}\left(1-e^{-m}\right).$$

Distributions of form (3.2.1) are known as shifted generalised geometric (shifted GGeo) with parameters η and v.

Similarly, if

$$\log Cov\left[N(t), N(t+\ell)\right] \cong -C - m\ell$$

(for some constants C and m), equation 1 implies that the corresponding batch renewal process has intervals between batches distributed as a shifted GGeo with parameters σ and τ, where

$$\sigma = \frac{e^{-C}}{\lambda + e^{-C}}\left(1-e^{-m}\right) \quad \text{and} \quad \tau = \frac{\lambda}{\lambda + e^{-C}}\left(1-e^{-m}\right).$$

When both batch sizes and intervals between batches are distributed as shifted GGeo, equations (3.1.3)-(3.1.8) lead to the following closed form expressions for queue length distribution, waiting time distribution and blocking probability:

Queue Length Distribution is $\{P_N(n); n = 0, \dots, N\}$

$$P_N(n) = \begin{cases} \dfrac{1}{Z}\dfrac{1}{\sigma+\tau}\dfrac{1}{v}(\sigma v - \tau\eta) & n = 0 \\[2ex] \dfrac{1}{Z}\dfrac{\tau}{\sigma+\tau}\dfrac{1}{v}\dfrac{\sigma v - \tau\eta}{\sigma+(1-\sigma-\tau)\eta} & n = 1 \\[2ex] \dfrac{1}{Z}\dfrac{\tau}{\sigma+\tau}\dfrac{\eta}{v}\dfrac{\sigma v - \tau\eta}{\sigma+(1-\sigma-\tau)\eta}\dfrac{1-(1-\sigma-\tau)(1-\eta-v)}{\sigma+(1-\sigma-\tau)\eta} & \\[2ex] \qquad\qquad\times\left(\dfrac{\sigma(1-\eta-v)+\eta}{\sigma+(1-\sigma-\tau)\eta}\right)^{n-2} & n = 2,\dots,N-1 \\[3ex] \dfrac{1}{Z}\dfrac{\tau}{\sigma+\tau}\dfrac{\eta}{v}\dfrac{\sigma v - \tau\eta}{\sigma+(1-\sigma-\tau)\eta}\dfrac{1}{\sigma}\left(\dfrac{\sigma(1-\eta-v)+\eta}{\sigma+(1-\sigma-\tau)\eta}\right)^{N-2} & n = N \end{cases}$$

(3.2.2)

where Z is the normalising constant and is given by

$$Z = 1 - \frac{\eta}{v}\frac{\tau}{\sigma}\left(\frac{\sigma(1-\eta-v)+\eta}{\sigma+(1-\sigma-\tau)\eta}\right)^{N-1}$$

(3.2.3)

Blocking Probability, π_N^B

$$\pi_N^B = \frac{1}{Z}\frac{\eta}{\eta+v}\frac{\sigma v - \tau\eta}{\sigma v}\left(\frac{\sigma(1-\eta-v)+\eta}{\sigma+(1-\sigma-\tau)\eta}\right)^{N-1} = \frac{1-Z}{Z}\frac{1-\lambda}{\lambda}$$

(3.2.4)

Hence

$$\frac{\pi_{N+1}^B}{\pi_N^B} \rightarrow \frac{\sigma(1-\eta-v)+\eta}{\sigma+(1-\sigma-\tau)\eta} \quad \text{as } N \rightarrow \infty$$

which illustrates the typical log-linear relationship between blocking probability π_N^B and buffer size N.

Waiting Time Distribution, $\{w(t): t = 1, \dots N\}$

$$w(t) = (t>1)\sum_{k=0}^{t-2} P_N^A(k)\frac{\eta(1-v)^{t-k-2}}{1+\frac{\eta}{v}\left(1-(1-v)^{N-k-1}\right)} + P_N^A(t-1)\frac{1}{1+\frac{\eta}{v}\left(1-(1-v)^{N-t}\right)}, t = 1,\dots,N.$$

(3.2.5)

An investigation into the effects of correlation on basic performance metrics and burst structure of a multi-buffered ATM switch decomposed into individual finite capacity $GGeo^{GGeo}$/D/1/N queues can be seen in [32, 48].

4. Further Research Work

In the previous sections a review of two cost-effective methodologies, based on the principle of ME, queueing theoretic concepts and batch renewal processes, is presented for the performance analysis of discrete time queueing models of shared buffer and multi-buffered ATM switch architectures with bursty and/or correlated traffic. These methodologies provide relatively simple but effective means for accounting for traffic variability and correlation and also simultaneity of arrivals and departures in ATM environments. However, further research work is needed for the derivation of credible flow (merging, splitting, departing) formulae to capture the behaviour of correlated arrival processes in the interior of an ATM network and also analyse arbitrary discrete-time QNMs of various types of ATM switch architectures.

Work of this nature is described below. It can be based on the principles of ME and minimum relative entropy (MRE), a generalisation, batch renewal processes and an asymptotic approximation technique.

4.1 The Principles of ME and MRE

The principle of MRE, a generalisation of the principle of ME, is a self consistent method of probabilistic inference of estimating uniquely an unknown but true probability distribution, based on information expressed in terms of a prior distribution and known mean value constraints. Both ME and MRE solutions are considered to be minimally biased in information theoretic terms and can be applied, in conjunction with classical queueing theory, in order to characterise extended product-form approximations and cost-effective algorithms for discrete-time QNMs of various ATM switch architectures with bursty and/or correlated arrival processes, space priorities and single/multiple classes of cells. These include shared buffer and shared medium switch architectures, polling and tree-like ATM switch configurations and buffered space division switches with blocking. Performance results obtained in the discrete-time domain under different buffer management simultaneity schemes can be compared with corresponding continuous-time results which are expected to provide good approximations and may be more appropriate in cases with large numbers of cells in each slot (c.f. fluid flow approach) [29, 49].

4.2 Batch Renewal Processes

The review of Section 3 demonstrated that batch renewal processes provide the means of defining the effects of correlated traffic in queueing systems and its behaviour as it traverses a network, quite free of the need to commit to arbitrary assumptions on burst structure [32]. Future research work can be based on queueing models with batch renewal arrival processes in order to (i) characterise output processes of shared buffer, shared medium and space division ATM switches; (ii) obtain new simple performance approximations of good accuracy, and (iii) analyse general QNMs of ATM switches with correlated traffic.

Moreover, comparisons can be carried out on the effects of traffic propagation on QNMs and on the effects of the corresponding batch renewal arrival process. It is expected that these comparisons will lead to characterisation of the burst structure. Such characterisation would have significant practical relevance to ATM connection and admission control, flow and congestion control and routing as well as providing a sound theoretical basis for the performance modelling and evaluation of ATM networks. In this context, the effect of traffic control mechanisms on ATM switch performance can be investigated and suggestions for appropriate traffic control schemes can be made towards a fair bandwidth allocation and network protection against congestion due to the violation of the negotiated traffic parameters.

4.3 Asymptotic Approximation Technique

Correlation and burstiness are very important in characterisating ATM traffic and modelling ATM switch architectures. The solution space of a system of correlated processes comprises regions of low intensity and of high intensity, in which correlation has little effect and a region of intermediate intensity, in which correlation is significant. Kouvatsos et al [50] have recently defined locations of such regions. The work is based on an asymptotic approximation technique from Reliabiliy Theory [51] as applied to the performance analysis of a finite capacity, single server queue evolving within a randomly changing environment. All stochastic times in the system were considered to be distributed exponentially while the arrival and service rates are subject to random fluctuations. It was shown that, for a stable MMPP/MMPP/1/N queue the regions are clearly defined and the location of the boundaries are given as simple closed form expressions. The method also reveals that MPP/MMPP/1/N systems comprise families which are related by common features of the MMPP model parameters. The behaviour of all members of one family is practically identical in the regions of low intensity and of high intensity but may be markedly divergent in the region of intermediate intensity. The aforementioned asymptotic analysis in [50] can be extended to take into account simultaneous cell arrivals and departures in the discrete-time domain with more complex processes of batch events, corresponding to models of ATM traffic. Moreover, the ME and MRE performance methodology for QNMs (cf Section 4.1) can be advanced further to incorporate asymptotic analysis into building blocks (single switch models), especially for low intensity of ATM traffic (c.f. conventional LAN loadings of about 15%, where the effects of correlation are limited but not necessarily negligible). It is expected that the asymptotic technique can also be used to define performance bounds for discrete time queueing systems in the region of intermediate intensity, in which the effect of correlation appears to be significant.

References

1. Saha A. and Wagh M. Performance Analysis of Banyan Networks Based on Buffers of Various Sizes. In Proc. INFOCOM '90, pp.157-163
2. Theimer T., Rathgeb E. and Huber M. Performance Analysis of Buffered Banyan Networks. Symposium on Performance of Distributed and Parallel Systems. Kyoto, Japan, pp.57-72, 1988
3. Kim H. and Leon-Garcia A. Performance of Buffered Banyan Networks Under Nonuniform Traffic Patterns. IEEE Transactions on Communications; 38,2, pp.648-658, May 1990
4. Morris T.D. and Perros H.G. Performance Analysis of a Multi-Buffered Banyan ATM Switch Under Bursty Traffic. Research Report, Computer Science Department, North Carolina State University, July 1991
5. Devault M., Cochennec J.-Y. and Servel M. The Prelude ATD Experiment: Assessments and Future Prospects. IEEE JSAC 6(9), pp.1528-1537, December 1988
6. Boyer P., Lehnert M.R. and Kuehn P. J. Queueing in an ATM Basic Switch Element. Technical Report CNET-123-030-CD-CC, CNET, France 1988
7. Kuwahara H., Endo N., Ogino M. and Kozaki T. A Shared Buffer Memory Switch for an ATM Exchange. Int. Conf. on Communications, pp.441-445, Boston, MA, June 1989
8. Lee H., Kook K., Rim C.S., Jun K. and Lim S.-K. A Limited Shared Output Buffer Switch for ATM. Fourth Int. Conf. on Data Communication Systems and Their Performance, pp.163-179, Barcelona, 1990
9. Eckberg A.E. and Hou T.-C. Effects of Output Buffer Sharing on Buffer Requirements in an ATDM Packet Switch. INFOCOM '88, pp.459-466, March 1988
10. Petit G.H. and Desmet E.M. Performance Evaluation of Shared Buffer Multiserver Output Queue Switches used in ATM. 7th ITC Specialist Seminar, Paper 7.1, New Jersey, 1990
11. Harrison P.G. and Pinto A. de C. An Approximate Analysis of Asynchronous, Packet-Switched Buffered Banyan Networks with Blocking. Performance Evaluation 19(2-3), pp.223-258, 1994
12. Sriram K. and Whitt W. Characterizing Superposition Arrival Processes in Packet Multiplexers for Voice and Data. JSAC 4(6), pp.833-846, 1986
13. Heffes H. and Lucantoni D.M. A Markov Modulated Characterization of Packetized Voice and Data Traffic and Related Statistical Multiplexer Performance. JSAC 4(6), pp.856-868, 1986
14. Gussella R. Characterising the Variability of Arrival Processes with Indices of Dispersion. JSAC 9(2), pp.203-211, 1991
15. Elsayed K. On the Superposition of Discrete-time Markov Renewal Processes and Applications to Statistical Multiplexing of Bursty Traffic Sources. GLOBECOM'94
16. COST 224 Project Final Report. Performance Evaluation and Design of MultiService Networks. Roberts J.W. (e.d), Commission of the European Communities, 1992
17. Leland W.E. *et al.*, On the Self-Similar Nature of Ethernet Traffic. SIGCOMM 23(4), 1993

18. Fowler H.J. and Leland W.E. Local Area Network Traffic Characteristics, with Implications for Broadband Network Congestion Management. JSAC 9(7), pp.1139-1149, 1991

19. Perros H.G., Nilsson A.A. and Kuo H-C. Analysis of Traffic Measurement in the Vistanet Gigabit Networking Testbed. Proceedings of the High Performance Networking, pp.313-323, 1994

20. Jain R. and Routier S.A. Packet Trains - Measurements and a New Model for Computer Network Traffic. JSAC 4(6), 1986

21. Gagnaire M., Kofman D. and Korezlioglu H. An Analytic Description of the Packet Train Model for LAN Traffic Characterisation, Performance Modelling and Evaluation of ATM Networks, Kouvatsos, D.D. (ed), Chapman and Hall, Vol.1 pp3-13, 1995

22. Perros H.G. and Elsayed K.M. A Review of Call Admission Control Schemes. ATM Tutorial Handbook, Chapman and Hall, Kouvatsos D.D. (e.d), (to appear)

23. Haverkort B.R., Idzenga H.P. and Kim B.G. Performance Evaluation of ATM Cell Scheduling Policies using Stochastic Petri Nets, Performance Modelling and Evaluation of ATM Networks, Kouvatsos, D.D. (ed), Chapman and Hall, pp. 553-572, 1995

24. Balbo G. Combining Queueing Networks and Generalised Stochastic Petri Nets for the Solution of Complex Models of System Behaviour. IEEE Transactions on Computers 37(10), p.1251, 1988

25. Marsan M.A., Balbo G. and Chiola G. An Introduction to Generalised Stochastic Petri Nets. Microelectronics and Reliability 31(4), p.699, 1991

26. Harrison P.G. and Strulo B. Process Algebra for Discrete Event Simulation. in Proc. 2nd QMIPS Workshop, Gotz N., Herzog U. and Rettelbach M., (eds.), University of Erlangen, 1993

27. Yamashita H., Perros H.G. and Hong S.-W, Performance Modelling of a Shared Buffer ATM Switch Architecture, ITC-13, Teletraffic and Datatraffic n a Period of Change, ed. Jensen and Iverson, North-Holland, pp. 993-999, 1991

28. Hong S.-W., Perros H.G. and Yamashita H. A Discrete-Time Queueing Model of the Shared Buffer ATM Switch with Bursty Arrivals. Research Report, Computer Science Dept., North Carolina State University, 1992

29. Mitra D. Stochastic Fluid Models. Performance '87, Brussels, 1987

30. Kouvatsos D.D., Tabel-Aouel N.M. and Denazis S.G. ME Based Approximations for General Discrete-Time Queueing Models. Perf. Eval. 21, pp.81-109, 1994

31. Kouvatsos D.D. and Wilkinson J. A Product-Form Approximation for Discrete-Time Arbitrary Networks of ATM Switch Architectures. Performance Modelling and Evaluation of ATM Networks, Kouvatsos, D.D. (ed.), Chapman and Hall, London, Vol.1, pp.365-383, 1995

32. Kouvatsos D.D. and Fretwell R. Closed Form Performance Distribution of a Discrete-Time GI^G/D/1/N Queue with Correlated Traffic. IFIP Proceedings of 6th Conference on Enabling High Speed Networks, Istanbul, 23-26 October 1995 (to appear)

33. Gravey A. and Hebuterne G. Simultaneity in Discrete Time Single Server Queues with Bernoulli Inputs. Performance Evaluation 14, pp.123-131, 1992

34. Jaynes E.T. Information Theory and Statistical Mechanics. Phys. Rev. 106(4), pp.620-630, 1957

35. Kouvatsos D.D. Maximum Entropy Methods for General Queueing Networks. Modelling Techniques and Tools for Performance Analysis, Potier D. (ed.), North-Holland, pp.589-609, 1985

36. Kouvatsos D.D. Maximum Entropy and the G/G/1/N Queue. Acta Informatica 23, pp.545-565, 1986

37. Kouvatsos D.D. A Maximum Entropy Analysis of the G/G/1 Queue at Equilibrium. J. Op. Res. Soc. 39(2), pp.183-200, 1989

38. Kouvatsos D.D. and Xenios N.P. MEM for Arbitrary Queuing Networks with Multiple General Servers and Repetitive Service Blocking. Perf. Eval. 10, pp.169-195, 1989

39. Kouvatsos D.D. and Tabet-Aouel N.M. GGeo-type Approximations for General Discrete-Time Queueing Systems. IFIP Transactions C-15, Special Issue on Modelling and Performance Evaluation of ATM Technology, North-Holland, pp.469-483, 1993

40. Kouvatsos D.D. and Denazis S.G. A Universal Building Block for the Approximate Analysis of a Shared Buffer ATM Switch Architecture. Annals of Operations Research 44, Special Issue on Performance Analysis of High Speed Networks, pp.241-278, 1994

41. Kouvatsos D.D., Tabel-Aouel N.M. and Denazis S.G. Approximate Analysis of Discrete-Time Queueing Networks with or without Blocking. IFIP Transactions C-21, Special Issue on High Speed Networks and their Performance, North-Holland, pp.399-434, 1994

42. Jaynes E.T. Information Theory and Statistical Mechanics II. Phys. Rev. 108, pp.171-190, 1957

43. Shore J.E. and Johnson R.W. Axiomatic Derivation of the Principle of Maximum Entropy and the Principle of Minimum Cross-Entropy. IEEE Trans. Inf. Theory IT-26, pp.26-27, 1980

44. Williams A.C. and Bhandiwad R.A. A Generating Function Approach to Queueing Network Analysis of Multiprogrammed Computers. Networks 6, pp, 1-22, January 1976

45. Fowler H.J. and Leland W.E. Local Area Network Traffic Characteristics, with Implications of Broadband Network Congestion Management. IEEE JSAC 9(7), pp.1139-1149, September 1991

46. Andrade J. and Martinez-Pascua M.J. Use of the IDC to Characterise LAN Traffic. Proc. 2nd Workshop on Performance Modelling and Evaluation of ATM Networks, Kouvatsos D.D (ed.), pp.15/1-15/12, Bradford, July 1994

47. Kouvatsos D.D. and Fretwell R. Discrete Time Batch Renewal Processes with Application to ATM Performance. Proc. 10th UK Computer and Telecomms. Perf. Eng. Workshop, ed. Hillston J., Pooley R and King P, pp.187-192, September 1994

48. Kouvatsos D.D. and Fretwell R. Batch Renewal Process: Exact Model of Traffic Correlation. Proceedings of Scholl Dagstuhl Seminars on Architecture and Protocols for High Performance Networks, Danthine A. et al (eds.), Kluwer Academic Publishers, 1995 (to appear)

49. Czachorski, T., Foureneau J.M. and Pekergin, F. Diffusion Models to study Nonstationary Traffic and Cell Loss in ATM Networks, Performance

Modelling and Evaluation of ATM Networks, Kouvatsos, D.D. (ed.), Chapman and Hall, Vol. 1, pp. 347-364, 1995

50. Kouvatsos D.D., Fretwell, R. and Sztrik, J. Bounds on the Effects of Correlation in a Stable MMPP/MMPP/I/N Queue: An Asymptotic Approach, Performance Modelling and Evaluation of ATM Networks, Kouvatsos, D.D. (ed.), Chapman and Hall, Vol. 1, pp. 261-284, 1995

51. Anisimov, V.V. and Sztrik, J. Asymptotic Analysis of some Finite-Source Queueing Systems, Acta Cybernetica, Vol. 9, Issue 1, pp.27-38, 1989

ATM Networks

Evaluation of Source Policing Functions with Strict Enforcement Policies in ATM Networks

Edward Chan, K. K. Liu

Department of Computer Science
City University of Hong Kong
e-mail: csedchan@cityu.edu.hk

1. Introduction

In ATM networks, source policing functions are crucial for ensuring that user traffic does not exceed the negotiated quality of service. Many different source policing functions have been proposed, including the well known Leaky Bucket scheme [1]. However, recent studies have shown that there are a number of problems in using this scheme. One criticism is that it does not penalize violating traffic sufficiently to reduce its impact on non-violating traffic source [2]. The purpose of this paper is to evaluate modifications to the Leaky Bucket scheme which will enforce the policing function more strictly, and compare with some other proposed techniques such as the Peak Counter scheme [3]. Since analytical analysis is in most cases intractable, simulation is the main tool for this study.

2. Description of the Source Policing Functions

In this section a brief description of the key characteristics of the source policing functions to be evaluated in this paper will be presented.

2.1 Leaky Bucket

The Leaky Bucket (LB) is one of the most extensively studied source policing functions proposed for use in ATM networks. Its principle is quite simple: a counter is incremented by one when a cell arrives, and is decremented periodically at a suitable leak rate. A policing action (either discarding or tagging) is applied on arriving cells when the counter has reached its limit N (Fig. 1).

The performance of LB and its many variants has been studied, see for example [4,5,6] among others. The general characteristics of LB is shown in Fig. 2. This will be used primarily as a reference against which the schemes to be evaluated in this paper is measured. It can be seen quite clearly that the throughput of a violating source (i.e. a source where the nominal load offered by the load exceeds unity) levels off but does not drop. This may not be desirable in all cases, since

36

one can argue that a source that violates its service contract should be penalized, particularly in the presence of other, better behaved sources.

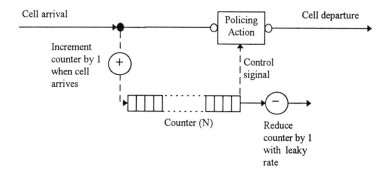

Figure 1 Mechanism of Leaky Bucket

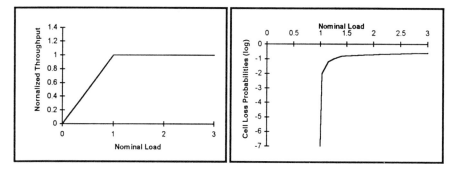

Figure 2 Nominal reference curve terms for throughput [2]

Figure 3(a) Nominal reference loss curve [2]

2.2 Generalized Leaky Bucket Scheme

The Generalized Leaky Bucket Scheme (GLB) proposed by Lague et. al. [2] is designed to penalize a violating source in a more aggressive manner. Lague argues that the typical reference cell loss curve, first postulated by Rathgeb [4] and shown in Fig 3(a), is too drastic for small deviations from the nominal mean cell rate and too lenient for major violations. He suggests a different reference loss curve and its corresponding throughput curve, shown in Fig, 3(b) and Fig. 3(c) respectively, as the goals for source policing functions.

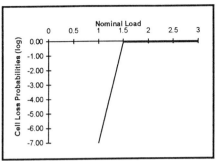
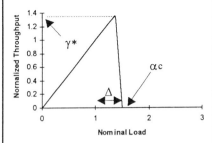

Figure 3(b) Compromise reference loss curve [2]

Figure 3(c) Compromise reference curve in terms of throughput [2]

The key modification of GLB (Fig. 4) is that the counter will be incremented even for cells which are discarded, up to a value K > N. The cell discard mechanism continues to be triggered when the counter value is N. The net effect is that the source policing scheme will penalize the source both by discarding violating cells and consuming the tokens: this ensures that the violating source will be severely penalized, resulting in throughput characteristics that closely match the reference curves in Fig. 3(c). What distinguishes GLB from most variants of LB is that it takes into consideration the information conveyed by the discarded cells, namely the system is overloaded, in its policing decision. It is worth noting that GLB is also superior to LB in several other ways, including a faster response time to overload conditions and a reduction of the maximum deviation from the nominal mean rate. However these factors are not the main focus of this study and will not be discussed further.

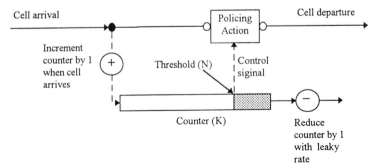

Figure 4 Mechanism of GLB

2.3 Peak Counter Scheme

The Peak Counter (PC) scheme shown in Fig. 5 is another variation of LB in which the decision to discard or tag is dependent on the duration that the traffic source has been operating above its predefined mean rate [3]. It can be implemented with two counters. The first is functionally the same as that in LB. The second one, called the Peak Counter, keeps track of how long the first counter is kept above its threshold. It will increment at a predefined rate (PC rate) if the first counter exceeds its threshold, and reduce at the same rate when it falls below the threshold. When the Peak Counter reaches its maximum value (PC limit) the arriving cells are discarded/tagged.

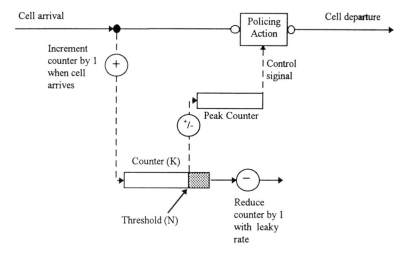

Figure 5 Mechanism of PC

The amount of time that the average counter is allowed to stay above its threshold before it starts the policing action (given by PC limit/PC rate) represents the dealy in initiating policing action. The evolution of the state of both PC counters is depicted in Fig. 6. PC is a more flexible scheme than LB since there are four control parameters: first counter limit, threshold, PC limit and PC rate. On the other hand, this leads to considerable complexity in tuning the system for maximal performance.

2.4 Leaky Bucket with Feedback

It has been pointed out by researchers [6] that policing functions should consider higher levels of abstraction in discarding traffic, since in many types of traffic (for example video frame sequences), it is meaningless to transmit a data unit which has a missing portion caused by a dropped cell. The Leaky Bucket with Feedback

Scheme (LBF) is based on this observation, and makes a simple assumption that a feedback signal is sent by the ATM multiplexer to the access node to drop the rest of the burst if one of its component cells is dropped (Fig. 7).

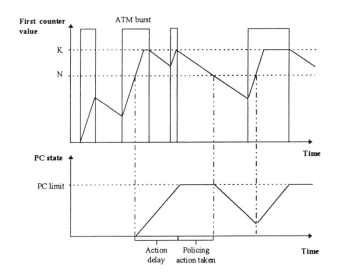

Figure 6 Evolution of PC's counters

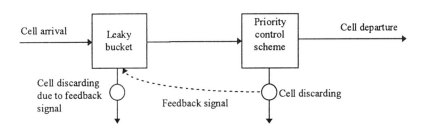

Figure 7 Mechanism of LBF

3. Simulation Model and Results

Since analytical results for the source policing functions in Section 2 are complex and in many cases intractable, we have decided to perform detailed simulation studies. The simulation is based on a modular graphcial simulator platform [7] built to study ATM networks. The platform is based on the proprietary graphical simulation package OPNET^{TM}, and contain modules which simulate the traffic source, the various ATM sublayers, the network access nodes, multiplexers and switches, all of which are clearly defined using State Transition Diagrams.

40

3.1 Simulation Model

Traffic source in the simulations are represented by a two-state Markov Modulated Poisson Process (MMPP) model. Peak bit rate is set to 30 Mbps, mean burst length is 100 cells and the state cell ratio is 0.4.

Two set of experiments are performed. In the first one, a single source is studied to determine the basic characteristics of the policing function. In the second set of experiments, multiple sources feed into an ATM multiplexer, as depicted in Fig. 8. The multiplexer uses a partial buffer sharing scheme for priority control. In this scheme, low priority (i.e. tagged) cells can enter the buffer only if the total buffer occupancy is less then a given threshold. On the other hand, high priority cells have access to the entire buffer. For our experiments, the buffer size is set to 10 and the threshold level to 5.

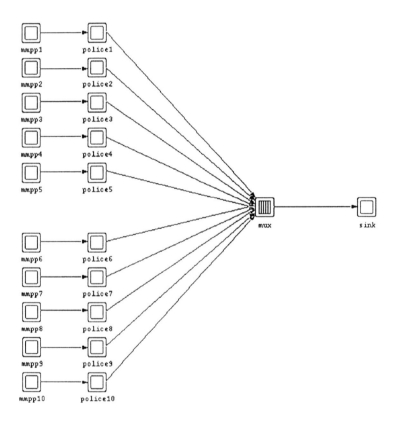

Figure 8 Configuration of multiple source evaluation

3.2 Performance of PC vs GLB

The performance of the GLB and PC schemes are first compared in the case of a single traffic source. Fig. 8(a) shows the performance of GLB with respect to different K values. In the GLB scheme, the threshold level N is set to 150 and leak rate 10 Mbps. As K increases, the policing function is more sensitive in that the source which exceeds the nominal mean rate by only a small amount will be penalized. It is also apparent that a very large K does not have a very significant effect.

In the case of PC, four parameters can be tuned to achieve optimal performance: counter limit K, threshold level N, PC rate and PC limit. Many different combinations of the parameters are possible, but a typical set of curves is shown in Fig. 8(b) with N = 150 and PC limit set to 10. It can be seen that PC has similar capabilities as GLB if the parameters are set appropriately. However, since GLB can achieve the same performance with a much simpler implementation, we conclude that GLB is in general superior to PC and hence will not evaulate PC in the next set of simulations.

3.3 Comparison of the Policies for Multiple Traffic Sources

The second set of experiments will evaluate the performance of LB, GLB and LBF schemes for multiple sources.

The results are shown in Fig. 9(a-i) Fig. 9(a) shows the throughput of the three schemes at the source: LBF has a lower throughput since a larger number of cells are dropped when a burst is discarded. The picture at the multiplexer is quite different. Detailed curves of cell loss probabilities are shown in Fig. 9(b-d). It is clear that both LBF and GLB has a much smaller loss in the multiplexer since the policing function has already penalized the violating source at the source level. This is desirable, since if excessive traffic is causing congestion in the network then it should be discarded as early as possible. By taking the cell discarding action as early as possible, the quality of service enjoyed by non-violating sources can be maintained since contention at the multiplexed link is reduced.

This effect is shown in the third set of experiments, in which half of the sources are well-behaved and generate traffic at a mean data rate of 15 Mbps. The rest are abusive sources which exceeds that mean cell rate. Fig. 9(e-g) shows the cell loss probabilities of both types of sources for the various schems. It can be seen that GLB and LBF can maintain the desired quality of services much better than LB.

Although both LBF and GLB appear to produce similar results, there is a major difference in the burst loss probability. Fig. 9(h) shows the burst loss characteristics in the multiplexer, and shows LBF's superiority clearly. However, this observation should be balanced by the fact that overall throughput of GLB is better, as shown in Fig. 9(i). Obviously, these two schemes can be applied to individual souces depending on whether burst or cell drop is more appropriate as dictated by the nature of the traffic to produce the best overall performance.

4. Conclusion and Areas for Further Study

The performance of a number of source policing functions with strict enforcement policies are evaluated and compared. The GLB scheme and PC scheme are both effective in shutting off a single abusive source; in the limiting case PC is very similar to the GLB scheme. However PC requires more complex implementation, making GLB the more desirable scheme of the two. When multiple sources are considered for LB, GLB and LBF, both GLB and LBF shows improvement, with the former performing better in handling cell loss sensitive traffic better and the latter giving superior results for burst loss sensistive traffic.

There has been some interesting results in the characterization of bursty traffic sources recently. It has been demonstrated that much data and video traffic exhibits behaviour, called Long Range Dependence (LRD [8]) that is very different from that of traditional traffic models. Instead of using standard Interrupted Poisson Process (IPP) and Markov-Modulated Poisson Process (MMPP) models, we have started to use trace driven source models in our simulation. Another possibility is to generate traffic exhibiting LRD characteristics using the algorithms described in [9]. However, simulation using these new traffic models is quite time consuming, and we are currently still studying the effect of using these models to investigate how they might affect the performance of ATM source policing functions which have previously been thought to be adequate.

References

[1]   Butto M, Cavallero E. Toniett A., "Effectiveness of the Leaky Bucket Policing Mechanim in ATM Networks", IEEE J. on Selected Areas in Comm., Vol 9, No. 3, April 1991, pp. 335-342.

[2]   Lague B et. al.. "On the Capability of the Policing Function to Release an ATM Connection", Telecommunication System 2(1994).

[3]   Monteiro J, "Input Rate Control for ATM Networks", Queueing, Performance and Control in ATM, pp. 117-122.

[4]   Rathgeb E P, "Modeling and Performance Comparison of Policing Mechanisms for ATM Networks", IEEE JSAC Vol. 9, No. 3, April 1991, pp. 325-334.

[5]   Chan E, Ng J, Liang T Y, Ho C C, "Effectiveness of Leaky Bucket as Source Policing Schemes for ATM Networks", Proc. IFIP 2nd Workshop on Performance Modelling and Evaluation of ATM Networks, Bradford, U.K., July 1994.

[6]   Liao K Q et. al., "Effectiveness of Leaky Bucket Policing Mechanism", Proc. ICC '92, pp. 1201-1205.

[7]   Ng J, Chan E, Yu P, Lam S K, "A Modular Simulator for Evaluating ATM-based Networks", Proc. 1993 European Simulation Symposium, Delft, The Netherlands, Oct 1993.

[8]   W. Leland et. al. "On the Self-Similar Nature of Ethernet Traffic", IEEE/ACM Trans. on Networking, Vol. 2 No. 1, Feb 1994, pp. 1-15.

[9]   Garrett M W, Willinger W., "Analysis, Modeling and Generation of Self-Similar VBR Video Traffic", Proc. 1994 ACM SigComm, London, U.K., Sep. 1994.

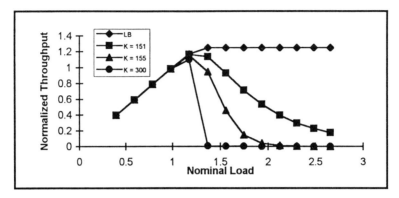

Figure 8(a) Performance of GLB with different K value

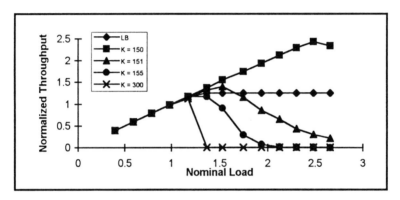

Figure 8(b) Performance of PC with different K and PC rate = 5

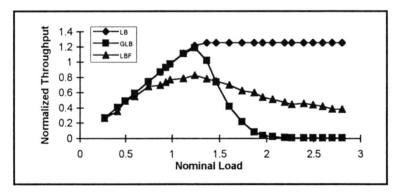

Figure 9(a) Normalized throughput of different policing functions

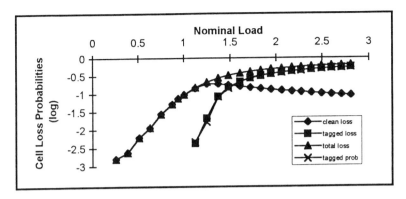

Figure 9(b) Cell loss probabilities of LB scheme

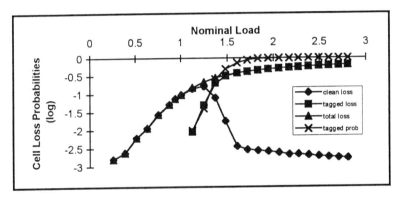

Figure 9(c) Cell loss probabilities of GLB scheme

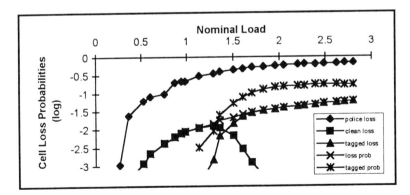

Figure 9(d) Cell loss probabilities of PC scheme

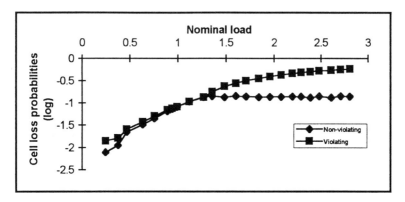

Figure 9(e) Fairness of LB scheme

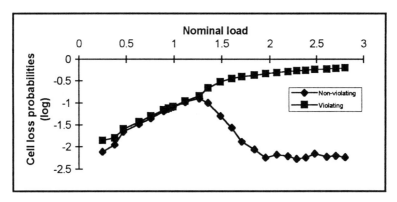

Figure 9(f) Fairness of GLB scheme

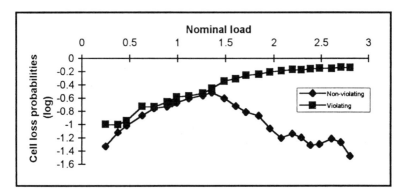

Figure 9(g) Fairness of LBF scheme

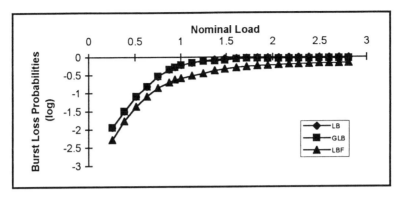

Figure 9(h) Burst loss probabilities in multiplexer

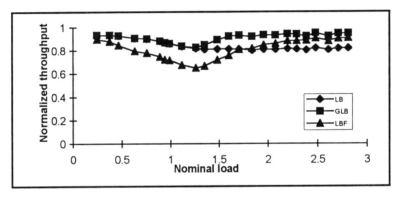

Figure 9(i) Normalized throughput of different policing functions

Performance Analysis of a Fault-tolerant ATM Switch in Single-fault Conditions

S. Segkhoonthod

M. C. Sinclair

Dept. of Electronic Systems Engineering, University of Essex,
Wivenhoe Park, Colchester, Essex C04 3SQ.

Tel: 01206-872477; Fax: 01206-872900; email: {segks,mcs}@essex.ac.uk

1   Introduction

One of the approaches for ATM switch design is to use a number of routing networks (multistage interconnection networks: MINs) in parallel so that multiple paths are created between inputs and outputs [1, 2]. This technique, called a replicated network, results in an improvement in both performance and reliability. However, it should be noticed that with this configuration, a great deal of complexity is needed in the Input Modules (IMs), which are used to distribute packets among routing networks, as they have to decide to which routing network packets should be sent.

In [3], we proposed a fault-tolerant ATM switch which adapts the idea of using several routing networks in parallel. The switch has a distribution network which is in charge of dispatching packets among the routing networks. As a result, the IMs are simply required to submit packets to the distribution network. Since the distribution network also carries out part of routing process, the routing networks themselves are truncated—resulting in a low overall complexity for the switch. In terms of fault tolerance, the distribution network is also in charge of re-routing packets in the case of faulty switching elements or links. Unlike some other fault-tolerant switches, where additional effort is required to avoid faults, in the proposed switch there is no modification in the routing algorithm when faults occur. In [3], we presented a probabilistic analysis of the proposed switch with re-routing deactivated, as well as simulation results for re-routing in the absence of faults. It was shown that the proposed switch gives both improved performance and reduced complexity when compared with replicated networks. In [4], we presented a performance analysis of the switch with re-routing in the absence of faults, based on the method of state diagrams, and compared the results with simulation. The results obtained showed reasonable agreement between the analysis and simulation. Further, the performance of the proposed switch with re-routing is significantly better than without. In this paper, we extend the work in [4] by taking into account the presence of a single fault in various parts of the switch.

2 Switch Architecture and Operation

An l-level $N \times N$ version of the proposed switch consists of an $N \times K$ distribution network (DN), K $N \times N$ routing networks (RNs) and N $K \times 1$ output modules (OMs) where $K = 2^l$. Shown in Fig. 1 is an $l = 2$, $K = 4$, 16×16 version of the proposed switch. Basically, the structure of the proposed switch is derived by interleaving the first few stages of K Banyan networks into a tree configuration by means of 2×4 switching elements (SEs). For example, in Fig. 1, there are 4 Banyan networks embedded with their stages 0 and 1 overlapped in the DN. The DN is constructed based on an l-level binary tree where each node is a plane of $N/2$ 2×4 SEs. In stage s, there will be 2^s such planes. The 2×4 SE used in the DN is not a typical 2×4 crossbar; it behaves like a 2×2 SE, but with two identical sets of two outputs, instead of only one set as in case of a 2×2 SE. Given an l-level DN, there will be $2^l = K$ RNs. Each output of the RNs is connected to one of the inputs of an OM. Here an OM can accept K packets at the same time.

The DN and RNs are both unbuffered. The switch operation is based on a request/acknowledgement protocol, consisting of two phases: path-establish and packet-transfer [5]. At the beginning of the first phase, an IM that has a packet generates a request, a form of short packet, and then submits it to the DN and waits for an acknowledgement (ACK). If the returned ACK is a positive one, the IM can transfer the packet to the appropriate OM during the second phase. If not, it waits for the next path-establish phase and starts the same procedure again. The DN is in charge of distributing, routing and re-routing. Basically, it both distributes an incoming request to one of the RNs according to some rule, and at the same time routes the request. After receiving a request from the DN, the RN then directs it to an appropriate OM. The routing algorithm used is the same as that of a Banyan network. The various options for the operation of the 2×4 SEs are given in [4]; here we consider the switch in re-routing mode with random selection. The 2×4 SEs employed in the DN have two sets of outputs, referred to as first and second choices. In each set, there are two outputs—upper and lower outputs like those of a 2×2 SE. In random selection, if a request is addressed to the upper output of a 2×4 SE, it is randomly granted one of the choices. If the request can not get through to the next stage via that choice, an attempt will be made to re-route it using the other choice—hence, re-routing mode.

3 Performance Evaluation

3.1 Assumptions and Fault Model

Before the analysis, a few definitions are required. A *stage cycle* is the time interval during which a request (or an acknowledgement) moves from one stage to an adjacent stage. A *network cycle* is the time interval for a packet received at an IM to pass through to an OM. In terms of the switch's operation described above, one network cycle then consists of the sum of the path-establish and packet-transfer phases. Further, we assume here that a packet may only arrive at an IM at the beginning of a network cycle. In which case, one network cycle consists of a finite number of stage cycles[1]. Another assumption, which applies to the situation when a 2×4 SE performs re-routing, is that it is assumed that re-routing cannot take place during the same cycle. This assumption, which was also adopted in [5], is needed to simplify the analysis, although in practice a request may be able to try an alternative path within one stage cycle. The fault model adopted is failure of an individual SE, leaving it totally unusable [6]. We will focus on a single SE fault in a given stage of the switch. Further, we will assume that the SEs in stage 0, the IMs and the OMs are all fault-free, as the failure of such components could lead to the failure of the whole switch. In addition, when a request encounters a faulty SE, we will assume that the faulty SE behaves as if it was always occupied; a request is therefore always blocked. In other words, a fault and blocking are treated identically. We analyse the proposed switch using the same assumptions as in [4]: independent, uniform traffic and infinite buffering at an OM. In addition, we assume that a fault occurs randomly. The performance measure used is the *probability of acceptance* (P_A) which is defined as the probability that an arbitrary request will be accepted [8, 9].

3.2 State Diagram Technique

In the method of state diagrams, two things are required [10]: (1) a finite number of states in a diagram and (2) transition probabilities between states. In the state diagram, the future state of a system depends only on its current state. Once the above two parts are completed, one can then obtain the probability of being in any desired state. To apply the method of state diagrams to the analysis of a re-routing version of the proposed switch, it is appropriate that the state diagram for the analysis be a diagram whose states are the "state" of a request in a particular stage (in a particular cycle) where the future state of the request (in the next cycle) only depends on its current state and the current state of the switch. So the task here is to first build a diagram with appropriate states, and then work out transitions probabilities among the states. Such a state diagram for an $l = 2$, $K = 4$, 16×16 version of the proposed switch is depicted in Fig. 2.

3.3 State Description

In Fig. 2, the states are identified by several components as follows:

- **Stage Number** (s) indicates the stage number where the request is.

- **Path Number** (p) indicates the path number a request is using. Recall that in the proposed switch, there exist 2^l paths (not disjoint) between every IM and OM pair. Here path number 1 is defined as the first path

---

[1]Hereafter, the term "cycle" refers to the stage cycle, unless otherwise stated.

that a request takes in attempting to go its required OM, path number 2 is the second one and so on.

- **Cycle Number** (c) indicates the cycle number that a request is in.

- **Blocking Number** (b) indicates the stage number where a request has been most recently blocked.

Consider an l-level $N \times N$ switch with $n = \log_2 N$ stages and $K = 2^l$. By means of the above components, the state of a request can be uniquely identified and classified into four groups:

Set-up: Being in a Set-up state means that a request has so far been successful in establishing a path up to the stage where it is currently, and it will be trying to establish a path to the next stage during the next cycle. In terms of the physical switch, this refers to when a request is allowed to occupy an output of an SE in a given stage. For example, state 1 in Fig. 2 may be interpreted as a request now being at one of the outputs of an SE in stage 0. Another example is state 7, which is a scenario when, after having been blocked previously, the request is able to be re-routed (*i.e.* set up). Set-up states can be further divided into three types as follows:

S_0 Request is at the IM and is being submitted to the first stage of the DN in cycle 1. This is state 0 in Fig. 2.

$S_{s,p,c}$ Request is in stage s, it is currently using path number p, it has not been blocked before, and it will be attempting to set up a path to stage $s + 1$ in cycle $c + 1$. These are states 1, 2, 3 and 4 in Fig. 2.

$S_{s,p,c}^b$ Request is in stage s, it is currently using path number p, it has been blocked most recently at stage b and it will be attempting to set up a path to stage $s + 1$ in cycle $c + 1$. Examples include states 7, 8 and 9.

Blocked: Being in a Blocked state has one of three meanings. The first refers to a situation when a request in stage s gets blocked when trying to establish a path to stage $s + 1$; a negative acknowledgement is received at stage s. From Fig. 2, states 6, 14 and 18 belong to this category. The second is when a request, after having been blocked previously, is blocked again as a result of the other output in a 2×4 SE being unavailable in an attempt to re-route the request. Examples include states 12 and 26. The third is when a request has already been blocked at some stage ahead of where it currently is and it is on the way back to some stage in the DN from where it may be re-routed. These include states 30 and 33. A Blocked state is defined as below:

$B_{s,p,c}^b$ Request is in stage s during cycle c, it was blocked at stage b when attempting to set up a path to stage $b + 1$ with path number p.

Hold: Being in a Hold state means that the request is at its required OM and a path is being held between its IM and OM for a number of cycles before the packet-transfer phase starts. Once it is in a Hold state, it will be in that state until the end of the path-establish phase. The Hold states are divided into two types:

$H_{p,c}$ Request has been successful in setting up a path from its IM to its requested OM by using path number p, and the path is being held between this source-destination. This is state 5 in Fig. 2.

$H_{p,c}^b$ Request is in its required OM after having been successful in setting up a path from its IM to the OM by using path number p, it has been blocked most recently at stage b and the path is being held between this source-destination. Examples include states 10, 24, 39 and 44.

Lost: When a request is in a Lost state, it is back at its originating IM after having failed on all paths. Once in a Lost state, it will be there until the last stage cycle of the path-establish phase. A Lost state is defined as below:

$L_{p,c}^b$ Request is at its originating IM after being unable to set up a path, the most recent path number used was p and it was most recently blocked at stage b. Examples include states 15 and 19.

3.4 Transition Probabilities Derivation

Denote the probability that a request will be in state A by $P[A]$ and the transition probability from state B to state A by $P[A|B]$ where A and B are any of the states defined in the previous section and A is not the same state as B. The transition probabilities of each state are given and described below:

3.4.1 Request is currently in a Set-up state

$$P[S_{0,1,1}|S_0] = 1 \quad \text{(At an IM)} \tag{1}$$

(1) Request has not been blocked before

With a fault in stage f, $1 \leq f \leq l - 1$:

$$\left. \begin{aligned} P[S_{s+1,1,c+1}|S_{s,1,c}] &= \frac{(2^{s+1})(\frac{N}{2})-1}{(2^{s+1})(\frac{N}{2})} \\ P[B_{s,1,c+1}^s|S_{s,1,c}] &= 1 - P[S_{s+1,1,c+1}|S_{s,1,c}] \end{aligned} \right\} (s = f - 1) \tag{2}$$

$$P[S_{s+1,1,c+1}|S_{s,1,c}] = 1 \quad (0 \leq s \leq l - 2, \, s \neq f - 1) \tag{3}$$

Otherwise:

$$P[S_{s+1,1,c+1}|S_{s,1,c}] = 1 \quad (0 \leq s \leq l - 2) \tag{4}$$

With a fault in stage f, $l \leq f \leq n - 1$:

$$\left. \begin{aligned} P[S_{s+1,1,c+1}|S_{s,1,c}] &= \left(\frac{\frac{KN}{2}-1}{\frac{KN}{2}}\right) \times \\ &\quad \left((1 - P[S_{s,1,c}]/K) + 0.75(P[S_{s,1,c}]/K)\right) \\ P[B_{s,1,c+1}^s|S_{s,1,c}] &= 1 - P[S_{s+1,1,c+1}|S_{s,1,c}] \end{aligned} \right\} (s = f - 1) \tag{5}$$

$$P[S_{s+1,1,c+1}|S_{s,1,c}] = (1 - P[S_{s,1,c}]/K) + 0.75(P[S_{s,1,c}]/K)$$

$$P[B^s_{s,1,c+1}|S_{s,1,c}] = 1 - P[S_{s+1,1,c+1}|S_{s,1,c}]$$

(6)

where $l - 1 \le s \le n - 2$ and $s \ne f\text{-}1$.

Otherwise:

$$P[S_{s+1,1,c+1}|S_{s,1,c}] = (1 - P[S_{s,1,c}]/K) + 0.75(P[S_{s,1,c}]/K)$$

$$P[B^s_{s,1,c+1}|S_{s,1,c}] = 1 - P[S_{s+1,1,c+1}|S_{s,1,c}]$$

(7)

where $l - 1 \le s \le n - 2$.

$$P[H_{1,c+1}|S_{n-1,1,c}] = 1 \quad (s = n - 1)$$

(8)

We will consider the fault-free case first. Eq. 1 states that whenever there is a request, the IM will always submit it to the first stage of the DN. $P[S_0]$ is also the input rate of the switch. Eq. 4 is derived from the fact that a request is always successful in establishing a path to a 2×4 SE in a stage of the DN because there are at most two requests coming to the inputs but there are four outputs available so there is always a free output. Eq. 7 comes from the fact that when a request is in stage s and wishes to set up a path via a 2×2 SE in stage $s + 1$, it can do so if firstly there is no other request that will be arriving at this particular SE (the first term), or secondly there is another request that will arrive at the SE, but it will want the other output of the SE that is not wanted by the considered request (the second term). The probability that there is another request arriving at the other input of the 2×2 SE under consideration is expressed as $P[S_{s,1,c}]/K$ since the random selection assumption is adopted. Eq. 8 means that a request in stage $n - 1$ can always set up a path to its addressed OM as a result of the multiple acceptance assumption. This is state 4 of Fig. 2. In the case where a fault is present in a given stage, the probability of a request getting through that stage now includes the probability that the request is routed to the faulty SE. If we let M be the number of SEs in a stage where a faulty SE is, then the probability that a request is not routed to the faulty SE is $\frac{M-1}{M}$ given that a fault occurs randomly. So Eq. 2 and 5 both follow the above argument. As an example, Fig. 3 shows a diagram of an $l=2$, $K=4$, 16×16 switch proposed switch with a fault in stage 1. Suppose a request is in stage 0, state 1; thus in the next cycle it is going to attempt to set up a path to stage 1. However, one of the SEs in stage 1 is unusable, and it may be the SE that this request requires. In which case, the request will be treated as blocked and will stay at the same stage, state 105. Otherwise, it can proceed to stage 1, state 2.

(2) Request has been blocked before

In order to derive transition probabilities for states that have been blocked before, the following terms are defined:

$$F1(s, c) = \sum_{i=s}^{n-1} \sum_{p=1}^{K} \sum_{b=0}^{n-2} \left(P[S^b_{i,p,c}] + P[S_{i,p,c}] + P[B^b_{i,p,c}] + P[H_{p,c}] + P[H^b_{p,c}] \right)$$

(9)

$$F2(s, c) = \sum_{p=1}^{K} \sum_{b=0}^{n-2} P[S_{s,p,c}^b] \tag{10}$$

Eq. 9 gives the probability that in cycle c, there is a request which may be in one of the stages starting from stage s and beyond (including the OM). In other words, it gives the probability that there are other requests ahead of the considered one. As in Fig. 2, suppose that we are considering state 7, then states that are ahead of it during the same cycle are states 4 and 18, $e.g.$ $F1(2, 4) = P[S_{3,1,4}] + P[B_{2,1,4}^2]$. Eq. 10 gives the probability that, in stage cycle c, requests in stage s will be setting up a path to stage $s + 1$ during the next cycle.

With a fault in stage f, $1 \leq f \leq l - 1$,
   a) a request has already encountered the fault:

$$P[S_{s+1,p,c+1}^b | S_{s,p,c}^b] = 1 \quad (s = f - 1) \tag{11}$$

   b) a request has not yet encountered the fault:

$$\left.\begin{aligned}
P[S_{s+1,1,c+1}^b | S_{s,1,c}^b] &= \frac{(2^{s+1})(\frac{N}{2})-1}{(2^{s+1})(\frac{N}{2})} \\
P[B_{s,1,c+1}^s | S_{s,1,c}] &= 1 - P[S_{s+1,1,c+1} | S_{s,1,c}]
\end{aligned}\right\} (s = f - 1) \tag{12}$$

$$P[S_{s+1,p,c+1}^b | S_{s,p,c}^b] = 1 \quad (0 \leq b \leq l - 2, s \neq f - 1) \tag{13}$$

Otherwise:

$$P[S_{s+1,p,c+1}^b | S_{s,p,c}^b] = 1 \quad (0 \leq s \leq l - 2) \tag{14}$$

Again, we consider the fault-free case first. Eq. 14 states that when a request is blocked, and then is in a Set-up state again in a stage in the DN and is attempting to set up a path to the next stage, it will always be successful in doing so. Considering a request which is now in a Set-up state in stage s, this means that the request is at the output of a 2×4 SE in that stage and is going to one of the inputs of a 2×4 SE in stage $s + 1$. Since there is at most one other request in the SE in stage $s + 1$ that can be in competition, the request always gets through stage $s + 1$. State 35 of Fig. 2 is an example of such a state. With a fault, however Eq. 12 follows the same argument as above, where the probability that a request is not routed to the faulty SE is $(2^{s+1} \frac{N}{2} - 1)/(2^{s+1} \frac{N}{2})$. Eq. 11 follows the same argument as Eq. 14 as a request has already encountered a fault. State 106 of Fig. 3 is an example of such a state.

With a fault in stage f, $l \leq f \leq n - 1$:

$$
\begin{aligned}
P[S^b_{s+1,p,c+1}|S^b_{s,p,c}] &= \\
\left(\tfrac{\frac{KN}{2}-1}{\frac{KN}{2}}\right) &\times \left\{0.75\big(P[S_{s,p,c}] + \mathrm{F2}(s,c)\big)/K\right. \\
&+\ 0.5\big(\mathrm{F1}(s+1,c)\big)/K \\
+\ \big(1 - P[S_{s,p,c}]/K &-\ \mathrm{F2}(s,c)/K - \mathrm{F1}(s+1,c)/K\big)\Big\} \\[4pt]
P[B^s_{s,p,c+1}|S^b_{s,p,c}] &=\ 1 - P[S^b_{s+1,p,c+1}|S^b_{s,p,c}]
\end{aligned}
\tag{15}
$$

where $s = f - 1$.

$$
\begin{aligned}
P[S^b_{s+1,p,c+1}|S^b_{s,p,c}] &=\ 0.75\big(P[S_{s,p,c}] + \mathrm{F2}(s,c)\big)/K \\
&+\ 0.5\big(\mathrm{F1}(s+1,c)\big)/K \\
+\ \big(1 - P[S_{s,p,c}]/K &-\ \mathrm{F2}(s,c)/K - \mathrm{F1}(s+1,c)/K\big) \\[4pt]
P[B^s_{s,p,c+1}|S^b_{s,p,c}] &=\ 1 - P[S^b_{s+1,p,c+1}|S^b_{s,p,c}]
\end{aligned}
\tag{16}
$$

where $l - 1 \leq s \leq n - 2$, $s \neq f - 1$.

Otherwise:

$$
\begin{aligned}
P[S^b_{s+1,p,c+1}|S^b_{s,p,c}] &=\ 0.75\big(P[S_{s,p,c}] + \mathrm{F2}(s,c)\big)/K \\
&+\ 0.5\big(\mathrm{F1}(s+1,c)\big)/K \\
+\ \big(1 - P[S_{s,p,c}]/K &-\ \mathrm{F2}(s,c)/K - \mathrm{F1}(s+1,c)/K\big) \\[4pt]
P[B^s_{s,p,c+1}|S^b_{s,p,c}] &=\ 1 - P[S^b_{s+1,p,c+1}|S^b_{s,p,c}]
\end{aligned}
\tag{17}
$$

where $l - 1 \leq s \leq n - 2$.

We explain the fault-free case first. Eq. 17 is the probability that the request is successful in setting up a path to a 2×2 SE in the next stage. In attempting to set up a path to the next stage, the request has to encounter one of the three situations depicted in Fig. 4. Case (c) will occur when there is another request in the same stage which is also going to set up a path in the same cycle. This is expressed as $(P[S_{s,p,c}] + \mathrm{F2}(s,c))/K$. The probability that case (b) will occur is the probability that there is another request in some stage ahead of the considered one during the same cycle, and is expressed as $\mathrm{F1}(s+1,c)/K$. For case (a), this is the probability that both cases (b) and (c) do not happen, which is expressed as $(1 - (P[S_{s,p,c}] + \mathrm{F2}(s,c) + \mathrm{F1}(s+1,c))/K)$. These probabilities are then weighted with the pass probabilities for each case as shown in Fig. 4. With regard to Fig. 2, an example transition from state 7 to state 8 has the probability of case (c) as $P[S^1_{1,2,4}]/K$, and the probability of case (b) as $\mathrm{F1}(2,4)/K = (P[S_{3,1,4}] + P[B^2_{2,1,4}])/K$. However, with a fault in a stage of the DN, Eq. 15 is derived using the probability that a request is not routed to the faulty SE being $(\frac{KN}{2} - 1)/(\frac{KN}{2})$.

$$P[H^b_{p,c+1}|S^b_{n-1,p,c}] = 1 \quad (s = n - 1) \tag{18}$$

Eq. 18 means that the request always gets through the OM when it is in the last stage of a RN. From Fig. 2, state 9 is an example of such a state.

3.4.2 Request is currently in a Blocked state

$$\left. \begin{array}{rcl} P[S^b_{0,p+1,c+1}|B^b_{0,p,c}] & = & 1 - 0.5(\mathrm{F1}(1, c)) \\ P[B^0_{0,p_{\mathrm{new}},c+1}|B^b_{0,p,c}] & = & 1 - P[S^b_{0,p+1,c+1}|B^b_{0,p,c}] \end{array} \right\} \tag{19}$$

where $s = 0$, $p \neq K$.

$$P[L^b_{p,c+1}|B^b_{0,p,c}] = 1 \quad (s = 0, p = K) \tag{20}$$

Eq. 19 expresses the probability of a request being in a position to be re-routed via an alternate path (one of the outputs of a 2×4 SE in stage 0). This is the probability that a particular output of the SE in stage 0 is not occupied. Recall that $\mathrm{F1}(1, c)$ is the probability that, in cycle c, there is another request at some stage ahead of stage 0. In addition, there is 50% chance that this request has made use of the particular output required. An example is the transition from states 11 to 35. Eq. 20 applies when a request has failed on all paths, so it has to return to its IM; examples include states 15 and 19.

$$\left. \begin{array}{rcl} P[S^b_{s,p_{\mathrm{new}},c+1}|B^b_{s,p,c}] & = & 1 - 0.5(\mathrm{F1}(s + 1, c)/(2^s)) \\ P[B^s_{s,p_{\mathrm{new}},c+1}|B^b_{s,p,c}] & = & 1 - P[S^b_{s,p_{\mathrm{new}},c+1}|B^b_{s,p,c}] \end{array} \right\} \tag{21}$$

where $1 \leq s \leq l - 1$ and $p \bmod 2^{(l-s)} \neq 0$.

$$P[B^b_{s-1,p,c+1}|B^b_{s,p,c}] = 1 \quad (1 \leq s \leq l - 1, p \bmod 2^{(l-s)} = 0) \tag{22}$$

Eq. 21 follows the same argument as Eq. 19 except that the probability that there is another request already occupying the required output is reduced by the factor of 2^s where s is the stage number. As an example, this refers to a transition from state 6 to state 7 in Fig. 2 in which the probability of another request already being ahead is the probability of its being in state 3 (*i.e.* a request in stage 2). Recall that there are 2^s planes in stage s ($l - 1 \leq s \leq 0$). As in Fig. 1, suppose that there is another request in stage 2, the 2×4 SE in stage 1 where this request has just successfully set up a path could have been either the one on the left plane or the other on the right. Thus, there is a 50% chance that this particular request has been occupying the same SE in which the considered request is going to be re-routed. Eq. 22 applies when a request has failed on all the paths that are available at a particular stage, thus it has to go back to a previous stage where it can try other paths. Examples include states 12 and 26.

$$P[B^b_{s-1,p,c+1}|B^b_{s,p,c}] = 1 \quad (l \leq s \leq n - 2) \tag{23}$$

When a request is in a Blocked state at a stage in the RN, it simply returns to the previous stage.

3.4.3 Request is currently in a Hold state

(1) Request arrived at an OM without being blocked

$$P[H_{p,c+1}|H_{p,c}] = 1 \tag{24}$$

(2) Request arrived at an OM having been blocked

$$P[H_{p,c+1}^{b}|H_{p,c}^{b}] = 1 \tag{25}$$

3.4.4 Request is currently in a Lost state

$$P[L_{p,c+1}^{b}|L_{p,c}^{b}] = 1 \tag{26}$$

3.5 Derivation of P_A of a Proposed Switch

P_A can be computed by adding all the probabilities that the request is in a Hold state.

$$P_A = \left(P[H_{1,m}] + \sum_{p=2}^{K} \sum_{b=0}^{n-2} P[H_{p,m}^{b}]\right)/P[S_0] \quad (c = m) \tag{27}$$

where m is the last cycle of the path-establish phase.

3.6 Simulation

Simulation was conducted in order to verify the analysis. The switch was simulated under the same assumptions as those adopted in the analytical model. Two independent runs were carried out, each consisting of five batches—altogether giving ten values to be averaged for each simulation point. Each batch consisted of 10,000 cycles. In each run, an additional 1,000 cycles were completed before data was collected in order to avoid transient affects. The 90% confidence intervals are very tight and therefore will not be shown. However, due to time limitations, only 16×16 and 32×32 switches were modelled.

4 Results and Discussion

Fig. 5 shows the P_A of a 16×16 with l=2 switch in re-routing mode fault-free and with a single fault present in various stages. Results for a 32×32 switch are depicted in Fig. 6. It is clear that the location of a fault determines the degradation of switch performance. It should be noticed that the performance of a 32×32 switch is less affected than a 16×16 switch. Fig. 7–9 compare the P_A of the switch with a single fault present as obtained from both analysis and simulation. They show reasonable agreement between the two. In both cases though, the analytical results gave a higher P_A than the simulation, which we presume to be due to failure of the independence assumptions. Curves of P_A of the proposed switch with a single fault in various stages as function of the number of stages are plotted in Fig. 10. Clearly, Fig. 10 indicates that the larger the switch size, the less affected the switch performance is.

5 Conclusions and Future Work

In this paper, we have presented an analysis of a fault tolerant ATM switch design based on the use of a number of routing networks in parallel when re-routing is performed and a single fault is present in various stages of the switch. The analysis is based on the state diagram method. The results obtained from the analysis were compared with the simulation results. It was shown that they were in reasonable agreement. These results also show that the degradation in performance of the switch depends on the location of a fault and that the performance is less affected with larger switches. There are a number of other aspects to be investigated in future. These include an analytical model for re-routing mode in the presence of multiple faults and reliability analysis of the switch.

Acknowledgements

The research described in this paper was undertaken by the first author (supervised by the second) as part of a Ph.D. project in the Dept. of Electronics Systems Engineering, University of Essex. The first author was initially supervised by Dr. E.A. Medova, and is sponsored by the Thai Government.

References

[1] Tobagi FA. Fast packet switch architectures for broadband integrated services networks. Proc. of the IEEE 1990; 78:133–167

[2] Venkatesan R, Mouftah HT. Performance analysis of multipath Banyan networks. Proc. IEEE International Conference on Communications 1992; 912–915

[3] Segkhoonthod S, Sinclair MC. Analysis of a fault-tolerant ATM switch based on a parallel architecture. Proc. UK Teletraffic Symposium, Old Windsor 1995; 22/1–22/9

[4] Segkhoonthod S, Sinclair MC. Re-routing analysis of a fault-tolerant ATM switch based on a parallel architecture. Third IFIP Workshop on Performance Modelling and Evaluation of ATM Networks, Ilkely, 1995; 73/1–73/12

[5] Padmanabhan K, Lawrie DH. Performance analysis of redundant-path networks for multiprocessor systems. ACM Trans. Computer Systems 1985; 3:117–144

[6] Kumar PV, Reibman LA. Failure dependent performance analysis of a fault-tolerant multistage interconnection network. IEEE Trans. Computers 1989; C-38: 1703–1713

[7] Wu CL, Lee ML. Performance analysis of multistage interconnection network configurations and operations. IEEE Trans. on Computers 1992; C-41:18–27

[8] Patel JH. Performance of processor-memory interconnections for multiprocessors. IEEE Trans. on Computers 1981; C-30:771–780

[9] Kumar M, Jump JR. Performance of unbuffered shuffle-exchange networks. IEEE Trans. Computers 1986; C-35:573–577

[10] Hamming RW. The Art of Probability for Scientists and Engineers. Addison-Wesley, 1992

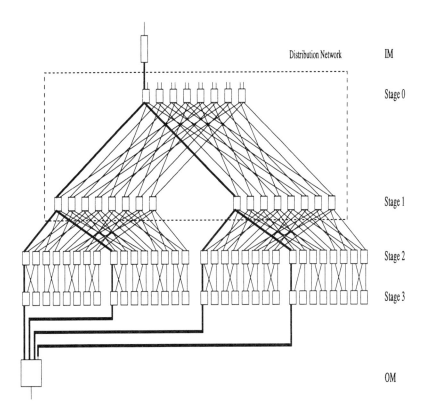

Figure 1: An $l = 2$, $K = 4$, 16×16 switch

60

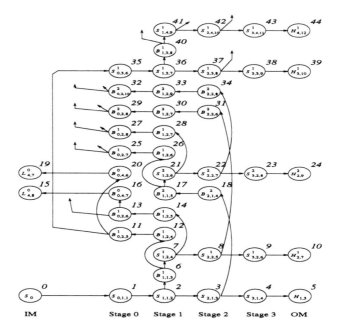

Figure 2: State diagram of an $l = 2$, $K = 4$, 16×16 switch with a fault in stage 2

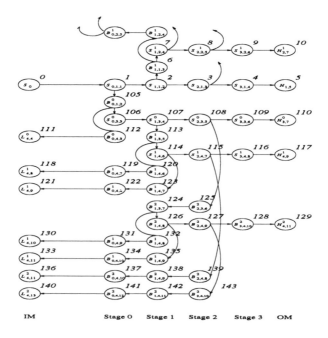

Figure 3: State diagram of an $l = 2$, $K = 4$, 16×16 switch with a fault in stage 1

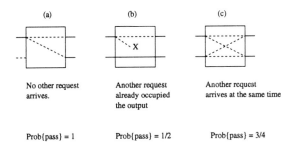

Figure 4: Probability of establishing a path in a 2×2 SE

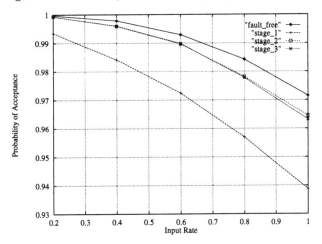

Figure 5: P_A of a 16×16 switch in re-routing mode with a fault in each stage (simulation results)

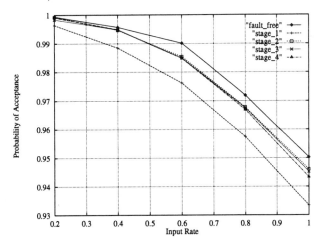

Figure 6: P_A of a 32×32 switch in re-routing mode with a fault in each stage (simulation results)

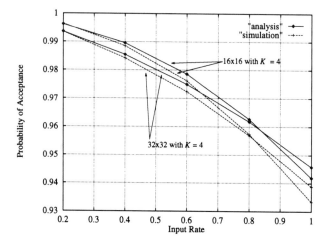

Figure 7: P_A in re-routing mode with a fault in stage 1 (simulation and analytical results)

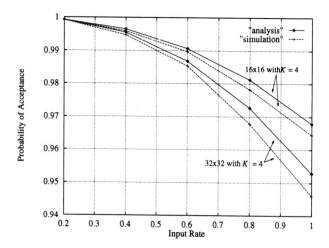

Figure 8: P_A in re-routing mode with a fault in stage 2 (simulation and analytical results)

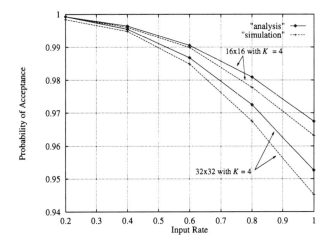

Figure 9: P_A in re-routing mode with a fault in stage 3 (simulation and analytical results)

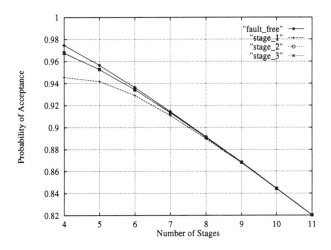

Figure 10: P_A of the proposed switch in re-routing with no faults and with a fault in each stage (analytical results)

Developing an Object Oriented Simulation Framework for ATM Network Performance Studies

Stephen Cusack and Rob Pooley
Department of Computer Science
The University of Edinburgh
email: *sdc@dcs.ed.ac.uk* & *rjp@dcs.ed.ac.uk*

September 11, 1995

Abstract

This paper outlines ongoing work in the performance evaluation of ATM networks through simulation. The use of analytical models to predict the performance of ATM networks is difficult due to the characteristics of the expected data traffic. Simulation is also problematic as the expected cell loss probability and call blocking rates in ATM networks are very low. Simulating at the individual cell level will require long and computationally intensive simulation runs to give statistically valid results. This work aims to investigate the feasibility of creating suitably efficient simulation tools which can be coupled with appropriate workload models to provide a greater insight into the performance of ATM networks.

1 Introduction

Asynchronous Transfer Mode (ATM) networks are destined to form the backbone for the provision of B-ISDN services [8, 10]. Before governments and public network operators can deliver this technology to the business and public sectors it will be vital to ensure that there is enough understanding of how to build these networks as efficiently as possible. Modelling such networks before their implementation will be one such means of attempting to gain insight into network design. It is important that the models produced are as accurate as possible. The use of inaccurate models could well lead to sub-optimal ATM network designs being adopted in the future due to 'worst case' design rules having been used.

Both analytical and simulation based methods are being actively used by ATM researchers in order to model both network topologies and the characteristics of expected workloads for such networks. Analytical models are difficult to solve due the complexity of the data which will typically be carried (i.e. autocorrelation and self-similarity in the traffic) and the simplifications necessary may detract from the accuracy of the results. Simulation, on the other hand, allows the models to be far more detailed but simulation is costly due to the design, implementation and validation cycle of the software and the long run times necessary to achieve statistically valid results. The long run times stem from the number of calls and cell transmissions which need to be simulated in order to observe valid cell loss ratios (CLR) and call blocking probabilities (CBP) in the network being modelled. In real ATM networks the CLR and CBP are likely to be of the order of 10^{-8} or lower.

The aim of this work is to design and implement an efficient, object oriented simulation tool (along with appropriate workload models) with which to model *total* ATM networks. Total network simulation, in this context, means that an entire network may be simulated at a level of abstraction typically seen by network management. Currently the **DEMOS** simulation package is being used to prototype the designs.

2 Background for the work

2.1 Overview

ATM technology has reached the stage where manufacturers are able to provide the equipment for ATM networks to be installed and used (eg. SuperJANET in the United Kingdom). Until fairly recently most simulation efforts have been focussed at the ATM switch level (eg. see [5]). This is understandable in the early days of such technology, but as this technology is now being gradually understood important issues such as the likely overall performance of networks are becoming more prominent. Issues related to the investigation of total network performance include:

- What are the possible effects of cell loss in an ATM network on the higher level protocols being used?

- What changes need to be made to workload models such that the addition of LAN and MAN networks connected to an ATM network via a bridge can be simulated?

- What are the dangerous operational states for an ATM network (which could possibly cause network failure)? Can these states be recognised and possibly avoided?

Clearly these issues can only be investigated with the simulation of total ATM networks. In other words the level of abstraction of the models has to be raised from the switch level to a level which incorporates the entire network.

2.2 Existing Simulation Methodologies

Work on providing total network simulation tools has mainly focussed on simulating ATM networks at the individual cell level (eg. *RC* [3] and *TeleSim* [12]). This is the most detailed form of ATM network simulation possible and should yield the most accurate results. However, the fundamental problem is that of having enough computational power to run the appropriate experiments in reasonable times. The TeleSim solution is to have two versions of the software: one highly optimised sequential version for model creation and validation and a parallel version for performing larger experiments (see [6] and [12] for more details). TeleSim utilises discrete event simulation and the techniques of *Time Warp* [7] to provide the parallel version.

3 The Design and Implementation Methodology in this work

3.1 Overview

The aim of this work is to provide an efficient simulation tool for total ATM network simulation. Rather than trying to implement a cell level simulator, it was decided to try and abstract above this level. In this work the most basic form of data transmission is the *Unit*. At its simplest the Unit is a descriptor of a time period covering a group of related cells in transmission. One way to visualise a Unit is as a typical 'burst' of cells. Necessarily, a Unit's characteristics will change on its passage through the simulation and these changes are described in more detail in section 4.2. One of the hopes of using the Unit approach is that the efficiency will increase over cell level simulation such that large numbers of cell transmissions may be simulated in reasonable time.

An object oriented design methodology is being used in the design. Network elements such as data sources, data sinks, switches and multiplexors are being designed as objects which can be arbitrarily connected to model networks. The advantages of using objects include: the reuse of existing code for modelling objects which have similar behaviours and the scalability made possible by keeping object interfaces consistent. The actions of the objects can be validated in small scale experiments before large scale experiments which utilise the features are performed. The other advantage of keeping object interfaces consistent is that it allows for the straightforward addition of new features and the modification of existing objects. This allows, for example, for a library of different traffic profiles to be created (eg. one for MPEG video, one for LAN traffic, etc.) which can be appropriately 'plugged' into the simulation model.

It was decided early on to use an existing simulation package rather than produce a bespoke simulator to aid rapid prototyping of the ideas. The requirements of such a package were as follows:

- Process based using discrete events.

- Object oriented design.

- Efficiency.

- Simplicity to aid prototyping and modification.

The DEMOS (Discrete Event Modelling on SIMULA) simulation package satisfies most of these criteria and is being used to test the ideas. SIMULA [2] itself contains sufficient simulation primitives to create simulation models but leaves many of the implementation issues with the user. DEMOS [1] provides a standardised approach by providing for the creation of models composed of *entities*. Entities may compete for specified resources, produce and consume items from buffers, interrupt and restart each other, wait for specified events to occur and hold for specified periods of simulation time. Section 3.2 describes some of the features of DEMOS.

3.2   Features of the DEMOS simulation system

DEMOS is a process based discrete event simulator for uniprocessor systems. This means that a simulation run follows a time ordered event list of actions. Entities which exhibit dynamic behaviour are inserted in the event list at the appropriate time of their next action. The DEMOS *Hold* primitive is used to deschedule the current entity and insert it into the event list at some specified time interval from the current simulation time.

As well as providing a simplified structure for modelling systems with entities, DEMOS also provides several useful features for manipulating the actions of the entities. Amongst the most useful DEMOS objects are the following:

- **Bin** A Bin object represents an unbounded buffer. Although a Bin object cannot hold object information, it is effectively an integer counter for storing the *number* of objects in a buffer. Primitives exist to allow entities to place or remove items (increment or decrement the counter) in the Bin.

- **WaitQ** A WaitQ provides entities with a controllable master/slave relationship. Entities can be placed in a WaitQ where they wait as slaves. Other entities, acting as masters, may coopt (reawaken) slaves from a WaitQ.

- **CondQ** An entity using a CondQ may use the *WaitUntil* primitive to wait for a specified logical condition to become true before proceeding. Entities which can alter the condition issue *Signal* commands to any waiting entities which prompts them to reevaluate the condition.

- **Interrupt** If one entity issues an interrupt to another, this forces the interrupted entity to wake from a *Hold* condition and check a flag set by the interrupt.

DEMOS also provides several integer and real random number generators and extensive reporting facilities. In addition the authors have added a simple graphics class which allows timing diagrams to be produced in a simulation. An example of the use of DEMOS to model WAN protocols may be seen in [11].

4   Current Design and Implementation status

4.1   Overview

This is a work in progress although the ideas presented here are an evolution of the earlier work presented in [4]. Much work still remains before the design aims are met, but the authors are confident that the remaining design issues and problems can be solved.

4.2 The basic Unit approach

As previously mentioned, a Unit represents a number of individual cells in transmission in the simulation. The motivation for this abstraction above the individual cell level was to try and improve the efficiency of the simulator. However, care needs to be taken with this approach if cell loss is to be accurately recorded; a trivial problem at the individual cell level. The first simulator [4] was originally designed with the Unit approach but was also designed to effectively degenerate to a cell level simulator around times of *critical* activity. Critical activity is defined to be when cells have to be dropped due to buffer overflow somewhere in the system. During non-critical activity cells are not lost. It was hoped that this approach would ensure high simulator efficiency during non-critical activity (as only Unit markers are passed around the system) and lower efficiency during critical activity (more entity interaction required) so that cell loss was accurately modelled. A fundamental test of this approach was to model a producer-buffer-consumer system as shown in figure 1.

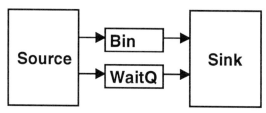

Cell loss determined by complex interactions between objects modelling the buffer as a separate entity.

Figure 1: The first simulation of a simple producer-buffer-consumer system

In this simulation the Source produces Units describing random numbers of cells with random delays between subsequent Units. The transmission time of the cells from the Source to the buffer is represented by a *Hold* in the Source. A new Unit is passed into the WaitQ to await the attention of the Sink. The Bin is used to model the buffer and is incremented by the Source to represent the number of cells in the Unit. The action of the Sink is to coopt Units from the WaitQ, decrement the Bin by the number of cells in the Unit and to *Hold* for a time representing the transmission of the cells from the buffer to the Sink. Interactions between the objects are few during non-critical activity, but the number increases when critical activity occurs (when the cells in a Unit will fill or overload the buffer). In this case, the Source uses interrupts to force the Sink to update the state of the buffer such that the number of cells reported to be in the buffer is accurate at that time. This is where the model can effectively degenerate into a cell level simulator with the Source interrupting the sink at every cell transmission time if the buffer is still full when it tries to send. The source keeps a tally of the number of cells in a unit it cannot pass to the buffer and this total is passed to the Sink as a 'lost cell' count.

The use of a Bin and WaitQ as constituents to model a buffer was deemed to be too expensive during critical activity. Other constraints (described in the next sections) forced by the concerns of multiplexing Units together and using the Sink entity in various call admission schemes suggested that precalculation of the 'lost cell' count would be a better alternative. Examination of the inter-actions involving the entities during critical activity led to the development of an algorithm to mimic this behaviour which could be placed in either the Source or the Sink. It was decided to place this algorithm in the Source as the Sink may be involved in complex interactions with other entities. The basic idea is now that the Bin is dispensed with and the Source models the entire status of the queue (see figure 2). When the Source produces a Unit it executes the cell loss algorithm to determine exactly how many of the cells in the Unit will be lost during transmission. When the Unit is placed in the WaitQ it contains not only the original cell count but also the number of lost cells for that Unit. When the Sink coopts the Unit it does not have to interact further with the Source as all the information it needs to pass the Unit on is readily available within the Unit. The Source does not have to concern itself with the activities of the Sink (apart from checking to see if the Sink is in a receiving state) as it is fully aware of the state of the queue and can just pass Unit information to the WaitQ for receipt by the Sink.

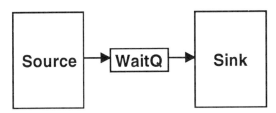

Cell loss per Unit precalculated by Source as buffer
modelled within Source by an algorithm.

Figure 2: The current simulation of a simple producer-buffer-consumer system

Other strategies were attempted before the cell loss precalculation method was implemented. These included the passing of Unit start and finish 'flags' designed to start and stop cell flows. The idea here was that the Source and Sink are aware of their production and consumption rates respectively and can use these markers to calculate the number of cells transmitted. This approach was abandoned due to problems with the marking of Units with small numbers of cells, the reliance of a non-blocking Sink to process the flags and the difficulty of modelling cell delay variation.

Cell loss precalculation is currently being used as the simple basis for the next stages of development. This method forces entities to pass Units between them through WaitQs. It is clear that realistic workload models for describing *constant bit rate* (CBR) and *variable bit rate* (VBR) sources will have to be tailored to fit the Unit approach. CBR services are particularly straightforward to model with this approach (eg. Units describing large numbers of cells typic-ally covering the duration of the call), whereas the impact of the Unit approach on VBR models needs to be investigated further. Questions to be answered will include just how many cells can be described by a Unit before you alter the characteristics of the traffic generator.

4.3 Multiplexing Units together

4.3.1 Overview

The issue of multiplexing poses quite a challenge for the Unit approach attempted in this work. Once again this problem is fairly trivial in a cell level simulation as the multiplexor can simply arbitrate between active sources on a cell by cell basis (eg. round robin scheduling). Multiplexing also raises the problems of dealing with *cell delay variation* (CDV also known as 'jitter'). CDV occurs when the inter-arrival times between related cells (eg. cells forming part of a higher level data packet) are altered due to the cell stream having been multiplexed with other streams. CDV is of great importance when considerations such as *Quality of Service* (QoS) for real-time applications are taken into account. Cell inter-arrival times can also be altered when a cell stream is forced to wait in a buffer somewhere in the transmission path (for example at the output buffer of a switch).

4.3.2 Technique under investigation

The approach currently being investigated in this work is to use *multiple* Units. A multiple Unit is very similar to a Unit object but instead of holding a description of one sequence of cells (a normal Unit) it holds an *array* of descriptions. To illustrate, let's suppose that three sources are to be multiplexed onto one transmission line. In this case the line may carry one, two or three coincident cell streams depending on the transmission states of the sources. With the multiple Unit approach, the multiplexor will produce multiple Units each of which describes a time period with one, two or three cells streams described. Each Unit descriptor entry in a multiple Unit describes a portion of the original Unit, i.e. the number of cells from the original Unit which would be transmitted in the time period described by the multiple Unit. Figure 3 shows an example of how multiple Units are produced for three multiplexed sources. It is clear that a multiple Unit starts or finishes when an incoming Unit starts or finishes transmission. Recalling that the Source entity simply passes one Unit descriptor to the Sink, it will be the job of the Sink to pass an appropriate 'finish' marker to the multiplexor to provide this signal.

This approach should allow the modelling of cell delay variation to be possible. Each Unit carries variables denoting the total Unit transmission time and the cell arrival rate in that Unit. When a multiple Unit is created the cell arrival rates for each of the constituent Units will be known. Utilising this information, together with the maximum transmission rate of the multiplexed line, will allow the individual cell arrival rates for each constituent Unit of the multiple Unit to be adjusted such that the CDV is modelled for the period of transmission time covered by the multiple Unit.

Figure 4 shows the DEMOS objects required to build a multiplexor to merge the Unit streams from two independent sources. As multiple Units begin and end on the receipt of Unit start and finish signals it is the job of the Sink to provide the Unit finish signal to enable the multiplexor to create the multiple Units. Section 4.3.3 details the problems which still need to be overcome with this approach.

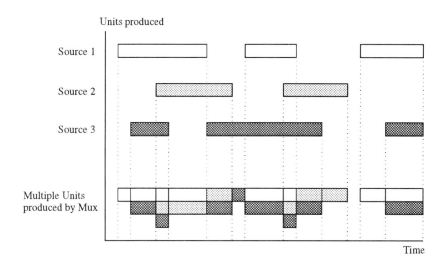

Figure 3: Timing diagram showing the transmission times of three sources and the appropriate multiple Units to be produced by the multiplexor.

4.3.3 Problems to be solved

The most obvious problem is that of deciding what fraction of the cells of a Unit ought to be carried in the appropriate multiple Unit. All of the information required for each Unit is known when the Unit (start) descriptor is passed to the multiplexor, namely: the number of cells, the number of lost cells and the period of transmission. However, the multiplexor is unaware of when, and from which source, the next Unit start descriptor will be passed. The introduction of the notion of *complete* and *incomplete* multiple Units is being investigated as a possible solution. A complete multiple (or individual) Unit contains all of the information required in the start of Unit descriptor. For example, the Units passed from a Source to a Sink are complete as the counts of numbers of cells and lost cells contained in the Unit are accurate. An incomplete multiple (or individual) Unit contains only some of this information in the start of Unit descriptor. Final counts of the number of cells carried and the numbers of lost cells in the incomplete Unit are carried by the corresponding finish marker. A promising interim measure seems to be to set the cell count in an incomplete Unit to be the total number of cells left to transmit from that Unit (if some cells have already been sent in a previous multiple Unit). Other entities along the transmission path use this value of the cell count in their calculations but are also aware that the actual value may change. When the finish marker is received for a multiple Unit, any calculations which have been based on incomplete data can be appropriately 'rolled back' to reflect the true value. As no 'real' data is being carried, only cell counts, it is hoped that this approach will still yield accurate cell loss counts. It will be important to ensure that phantom extra cell counts are not introduced by this technique. CDV calculations should still be valid as they will only rely on the cell arrival rate of the constituent Units.

Another problem to be overcome is that of how other entities in the simulator deal with multiple Units. Complete multiple Units will be fairly straight-

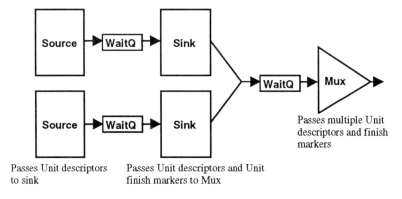

Figure 4: Multiplexing two independent sources

forward to deal with as all of the information required will be available. Incomplete multiple Units will cause more problems especially in buffer interactions (i.e. Source-buffer-Sink) where the Source precalculates cell loss. Work is progressing on the provision of state roll back in this situation as a solution. A feature of the approach which makes this feasible is the fact that multiple Units are always described by a pair of markers; a start marker and a finish marker. A finish marker will always be received before the start marker of the next multiple Unit thus allowing for a roll back to occur without causing interference to the next multiple Unit.

An effect which will have to be investigated is that of Unit fragmentation. Each Unit en route from its source to its destination will be multiplexed into multiple Units which in turn may be multiplexed with other Units and multiple Units and so on. It is possible that Units will be fragmented such that the efficiency advantage of having Units in the first place may be diminished. One possible solution may be to allow entities to try and rejoin fragments of original Units during processing. This could be achieved by allowing the entity to scan its input WaitQ for start/finish Unit pairs to try and build maximally sized (multiple) Units for transmission. The use of WaitQ scanning may also allow for entities to mark incomplete multiple Units as complete if the finish marker for that Unit is also available in the WaitQ.

4.4 Abstraction from real networks

4.4.1 Overview

The Unit approach has been chosen as the basic means of passing cell information around the simulator. The next step, after fine-tuning the Unit approach, is to provide entities to model real network components. The previous work [4] provided simple entities allowing for very simple networks to be created. However, the new flexibility of the multiple Unit approach requires a new set of entities to be created.

A real ATM network will comprise of some ATM infrastructure to which clients will connect their equipment. Client equipment will typically comprise of elements such as workstations and real-time applications (such as video-conferencing and video on demand) either directly connected to the network or connected to the network via a LAN or MAN which is itself connected via a bridge or router. An introduction to the proposed *user-network interface* (UNI) may be seen in [9].

Modelling a typical ATM network will require at least the following components:

- **ATM network client entities**

- **Shared line multiplexors and demultiplexors**

- **ATM network switches**

The following sections describe the proposed implementations of these entity types in the simulation model.

4.4.2 ATM network clients

As this work aims to examine the effects of adding client LAN and MAN networks to an ATM network, it is important that the generic network client used can be configurable. In some senses we shall be interested in individual applications running on one workstation (for example) as well as the aggregate effect of the traffic from a LAN or MAN. The approach first attempted in [4] is being continued in this work, that is to have a library of 'plug in' traffic characterisation modules. Depending on the module used it is hoped that the network client will be able to model the range of workloads from individual applications to LAN and MAN networks.

In addition to sending Units onto the network, the network client will also have to be able to receive Units from the network and have some means of reporting information such as cell loss, call connect failure and the number of higher level data packets which have to be retransmitted. Rather than try to produce one complex entity to provide these features, a composite entity will be produced instead using the building blocks available. Figure 5 shows the current state of the composite ATM network client. The client is built up of three distinct components. The first is the transmission Source-buffer-Sink component which has the job of transmitting Unit information onto the network. This part of the client will also be responsible for implementing such features as *call admission control* (CAC) and *traffic shaping*. As the Sink entity in the transmission component may be allowed to block (in combination with a controlling external entity such as a switch or multiplexor) this will provide the opportunity for CAC negotiation. Traffic shaping (where the cell stream from a transmitting source is buffered and fed onto the network according to some time-based shaping function) may also be implemented by altering the cell consumption rate (from the virtual buffer) of the Sink entity. This value may be set during the CAC phase of a call. Any buffer size required is implemented by simply setting a value in the Source entity.

The second component of the client is the Unit receiving component. Once again this component consists of a Source-buffer-Sink combination. The job of this component is to receive incoming Units from the network and buffer them if appropriate.

The third component of the client is the control and reporting module. The job of this component is to instruct the transmission component to call the

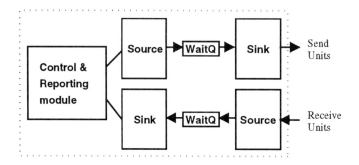

Figure 5: A composite network client entity

network (and start transmission when the call is accepted) and to monitor the incoming Units from the network and produce statistics. Further actions such as retransmission of failed data packets may also be undertaken if such messages are passed back via the receiving component (this will require extra data fields in the Unit objects in order to specify a type of message). By a suitable choice of control and reporting module it is intended that the client entity will be flexible enough to model various traffic profiles within the constraints of the Unit approach.

4.4.3 Shared line multiplexors and demultiplexors

Section 4.3 describes the technique under investigation to multiplex independent Unit streams together to produce multiple Units. Figure 4 effectively shows how two client entity transmission components would be multiplexed together. Of course, more than two clients may be multiplexed in this way. By appropriately setting the buffer sizes in the client entity transmission component (or by the provision of separate buffers between the client and the multiplexor) the input queues of the multiplexor may be modelled.

The action of the demultiplexor entity is to receive incoming Units or multiple Units and send the appropriate sections of each Unit to the correct output WaitQs of the demultiplexor. The demultiplexor and the switch entities have the job of breaking up multiple Units (if necessary) depending on the destination addresses of each Unit description contained within the multiple Unit.

4.4.4 ATM network switches

In the previous work [4] the job of switching was performed by the appropriate connection of a multiplexor and a demultiplexor. While this adequately models the function of a simple switch it provides little flexibility for modelling different switch architectures. Work is in progress to design a structure to allow for various architectures to be modelled. Figure 6 shows one test composite entity for a perfect non-blocking 2 × 2 switch.

Each I/O port in the switch is modelled using a pair of multiplexors. One multiplexor handles incoming multiple or individual Units (from possibly more than one client) while the other produces multiple Units from the Units switched to that output port. The output multiplexor is part of the switch

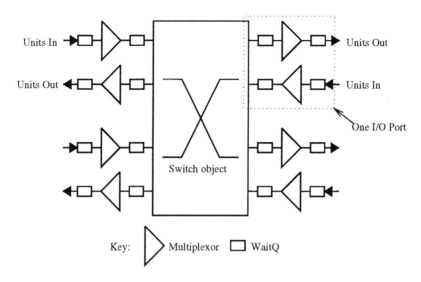

Figure 6: A composite perfect non-blocking 2×2 switch

whereas the function of the input multiplexor is unnecessary if a network client (or the output from a port of another switch) is directly connected. The reason for the use of a multiplexor at each output port is to allow the switch object to pass Unit start and finish markers without having to worry about creating multiple Units when more than one cell stream is switched to a particular output port.

The switch object itself uses a form of call admission to allow inputs to pass Units into the switch. The call admission currently uses a CondQ object to inform the input multiplexors when they can and cannot proceed based on a maximum switching capacity in the switch (the algorithm simply allows the port with the longest wait time to proceed). Figure 7 shows how the input multiplexors of each port are connected to the CondQ. Switch buffering is not directly supported in the basic switch design. Output buffering, for example, can be implemented by establishing queues in the input components of directly connected network clients or by adding Source-buffer-Sink components to each output port.

Providing true call admission control schemes in the simulator will require the switch objects to engage in some form of negotiation before calls are accepted. Such negotiation could occur through the use of specially flagged Units passing on separate links between switches (to model *signalling virtual channels* (SVC)).

By utilising a separate switch object it is hoped that various models of real switch designs can be plugged into the simulator. This could allow various switch designs and strategies to be evaluated in otherwise identical testbeds.

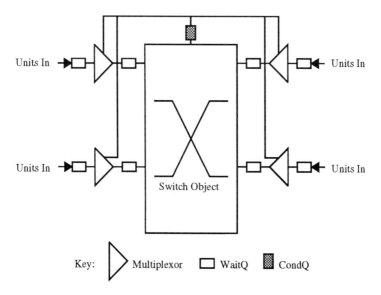

Figure 7: Using a CondQ to provide simple call admission

5 Future Work

Clearly the ideas presented here are in their early stages of development. Much work and refinement will be required such that a simulator powerful enough to perform the experiments envisioned can be produced, specifically:

Units A critical test of the Unit approach will be to try and describe workload models in terms of Units. CBR sources are very easy to model with Units, whereas some experimentation will have to be performed to assess the impact of Units on other varieties of traffic

Multiple Units The use of multiple Units will have to be validated in the system to ensure that cell counts, especially lost cell counts, are accurate. The major problem will be to ensure that the mechanisms for state rollback when dealing with incomplete multiple Units are robust enough not to introduce inaccuracies.

Cell Delay Variation The switch and multiplexor entities will have the job of adjusting cell inter-arrival times to enable CDV to be quantified. The only problem with this approach will be in how Unit finish markers are dealt with. If the cell inter-arrival time is altered in a Unit then the transmission time for the Unit is increased. This consideration may force input buffering to be critical in multiplexors. Switches will be enabled with some buffering and hence shouldn't suffer from the same problem.

Entity construction Presently most of the requisite entities exist in basic experimental forms. This will remain necessary while the Unit transmission approaches are investigated further. Once the designs are finalised, updated versions of the entities (with fully consistent interfaces) will have to be produced.

Efficiency The whole reason for the approaches outlined here is to try and improve the efficiency of the simulator when compared to a cell level simulator. Comparative studies will have to be performed in order to confirm if the efficiency really does increase and to try and quantify the effect.

Optimal number of objects One of the keys to improving the efficiency will be through the use of an optimal number of basic building objects. Adding greater functionality (optionally selected through parameters passed to the object) to the standard objects could allow the number of objects necessary to describe an ad hoc network to be reduced. For example, the multiplexor (when used on an output port) could be given the virtual buffer algorithm for use when connected to a switch thus providing implicit output buffering.

Looking further into the future, several other questions also arise. For example, it is likely that the ideas implemented in the DEMOS system may have to be replicated in a bespoke simulation environment to ensure that efficiency and portability issues are fully addressed. Implementation in an object-oriented language, such as C++, would open up the number of hardware platforms available for the system, possibly including very high performance machines.

It could be possible that the Unit approach will still be too detailed to provide the efficiency required. It may be necessary to replace some of the more complex entities with simpler entities which use a statistical distribution (based on real network studies) to determine variables such as CBP and CLR within the simulation. Such techniques may, however, lower the confidence in the results obtained.

Other practical considerations, which are irrelevant during this early stage of the work, include issues such as translating network descriptions into simulation code and the visualisation of experimental results.

6 Conclusions

The work presented in this paper represents the first few steps towards trying to produce an efficient simulation framework for total ATM network simulation. At this early stage it is impossible to say if the approaches and ideas outlined are sufficient, and indeed realisable, for achieving this goal. Much work still needs to be performed to produce an environment in a state serious enough to assess the validity of the work. As problems and drawbacks arise it may be necessary to completely change some of the ideas. However, the authors are confident that a fruitful set of ideas are being investigated which should help provide better insight into the design and performance of ATM networks.

References

[1] G.M. Birtwistle "Discrete Event Modelling on SIMULA", Macmillan Press, 1979.

[2] G.M. Birtwistle, O-J. Dahl, B. Myhrhaug and K. Nygaard "SIMULA BEGIN", Studentletteratur, Lund, Sweden, 1973.

[3] R. Lo Cigno and M. Munafò "RC - A Flexible Language for the Specification of ATM Networks Simulation Experiments", Paper 65, Third IFIP Workshop on the Performance and Evaluation of ATM Networks, Ilkley (U.K.), July 1995.

[4] Stephen Cusack and Rob Pooley "Simulation Experiments with Protocol Interactions in ATM Networks", Paper 30, Third IFIP workshop on the Performance and Evaluation of ATM Networks, Ilkley (U.K.), July 1995.

[5] Joan García-Hara, Rocío Marín-Sillué and José Luis Melús-Moreno "ATMSWSIM An Efficient, Portable and Expandable ATM SWitch SIMulator Tool", Lecture Notes in Computer Science 794, Austria, May 1994

[6] P. Gburzynski, T. Ono-Tesfaye and S. Ramaswarmy "Modelling ATM networks in a parallel simulation environment: a case study", SCSC'95, Ottawa, Ontario, Canada, July 24-26, 1995.

[7] F. Gomes, S. Franks, B. Unger, Z. Xiao, (Univ. of Calgary); J. Cleary, (Waikato, NZ); A. Covington (JADE) "SimKit: A High Performance Logical Process Simulation Class Library in C++". Submitted to WSC 95.

[8] Rainer Händel, Manfred N. Huber and Stefan Schröder "ATM Networks - Concepts, Protocols, Applications", 2nd Edition, Addison-Wesley, 1994.

[9] ITU-T: Recommendation I.413. "B-ISDN User-Network Interface". Rev. 1, Geneva, 1993.

[10] Craig Partridge "Gigabit Networking", Addison-Wesley, 1994.

[11] R.J. Pooley and G.M. Birtwistle "Process Based Modelling of Communications Protocols" in S. Schoemaker Ed *Computer Networks and Simulation III*, North Holland, 1986, pp 81-101

[12] Brian Unger (Project Director), "ATM-TN System Design: Simulation tools for Telecommunications Design and Analysis", WurcNet Inc. For details refer to the URL http://bungee.cpsc.ucalgary.ca/Publications/ATM_TN.ps

Multimedia Systems

Workload Modelling for Cost-Effective Resource Management in ATM Networks

Mourad Kara and *Peter M. Dew*
School of Computer Studies, The University of LEEDS
Leeds, LS2 9JT, UK
{mourad, dew}@scs.ac.leeds.uk

Abstract

In this paper, we discuss the design and implementation of a workload model for the performance evaluation of a cost-effective resource manager for distributed multimedia applications over ATM networks. Potentially, these applications could undertake some local multiplexing of their traffic requirements, if the pricing structure for the integrated services network makes it more cost-effective. To explore this issue, this paper presents the implementation of an *application level* traffic generator (workload model). This latter distinguishes itself from other traffic generators because it aims to capture the behaviour of these applications in its traffic generation.

1. Introduction

In ATM networks, a suitable resource management mechanism is required to allocate network resources such as bandwidth, buffering, and end-to-end delay at set-up time. A resource reservation mechanism will allocate network resources at the end systems and at all intermediate switches between source and destination. A policing scheme is also required to monitor the user behaviour at the network edge[Par94]. The network has to exercise flexible resource allocation and congestion control to exploit the potential increase in network efficiency resulting from the use of statistical multiplexing. In addition, a pricing structure is required to reflect the quality of Service (QoS) provided to the user. An important element is the fact that users should be aware of the pricing structure to enable them to cost their services. In this paper, we discuss the resources required for distributed multimedia services, and associated pricing policies[PKF92]. Potentially, an application could undertake some local multiplexing of its traffic requirements, if the pricing structure for the integrated services network makes it more cost-effective. The conditions for which some local multiplexing, *at the application level,* is desirable are discussed.

Next section presents QoS requirements for distributed multimedia applications, the algorithms and protocols needed to provide performance guarantees to these traffic requirements. Pricing policies and structures matching the QoS requested at the admission control are discussed in Section 3. After that, a discussion of the cost effectiveness of internal multiplexing is outlined in Section 4. In section 5, we present the objectives, design and implementation of the workload model. We conclude with the current implementation of the cost-effective resource manager and future work.

2. Resources Allocations and Admission Control

Future high-speed networks are expected to use the ATM technology in which the information is transmitted using short fixed-size cells consisting of 48 bytes of payload and 5 bytes of header. The fixed size of the cells reduces the variance of delay making the networks suitable for integrating traffic consisting of voice, video and data Providing the desired QoS for these various traffic types is much more complex than the data networks of today. Proper traffic management helps ensure efficient and fair operation of networks in spite of constantly varying demand. This is particularly important for data traffic which has very little predictability, and therefore, cannot reserve resources in advance as in the case of voice telecommunications networks.

ATM networks provide virtual circuits that offer consistent performance in the presence of stochastically varying loads on the network. In principle, this objective can be achieved by requiring that users specify traffic characteristics when a virtual circuit is established, so that the network can select a route that is compatible with the specified traffic and allocates resources as needed. While this introduces the possibility that a particular virtuai circuit will be blocked or delayed, it allows established circuits to receive consistent performance as long as they remain active. In multimedia networks, traffic performance requirements for real time traffic demand a congestion control strategy that can provide guaranteed transmission, bounded end-to-end delay and small and controllable delay-jitter. In order to efficiently use network resources, loss-free transmission is guaranteed only to the class of traffic with zero-loss requirement. For the remainder real time traffic, loss performance is provided on an as needed basis. Small delay-jitter and bounded delay performance however, are retained by for all real-time traffic by performing a special form of statistical multiplexing.

Integrated service packet networks offer guarantees for the services provided. It is essential to develop an understanding of the guarantees provided. For example, Clark, Shenker and Zhang [CSZ92] describe an Integrated Service Packet Network which can support three levels of service of commitment: (i) *Guaranteed service* for real-time applications, (ii) *Predicted service* which uses the measured performance of delays and is targeted towards adaptive and continuous media applications which can compensate for momentary loss of QoS; and (iii) *Best effort datagram service* where no QoS guarantees are provided. Also, a unified traffic queuing scheduling mechanism is developed which is based on a combination of weighted fair queueing and static priority algorithms.

In order to provide real-time services, applications need to declare their traffic characteristics, and performance requirements at the time of channel establishment. The network model is viewed as consisting of an interconnection of nodes. Nodes are connected to each other using links. Nodes can be switches or hosts. In each case, the resources used by the channels in each node are bandwidth, buffer space, CPU cycles. The traffic model and performance parameters are [FV90]:

- X_{min} is the minimum packet interarrival time on the channel.
- X_{ave} is the minimum value of the average packet interarrival time over an interval of duration I.
- S_{max} is the maximum packet size in bytes.
- T_{max} is the maximum service time in the node for the packets.
- D is the source to destination delay bound for the packets.
- Z is the minimum probability that the delay of a packet is smaller than the given bound D.
- W_{min} is the minimum packet loss probability due to buffer overflow.

The last three parameters define the QoS offered by the network. X_{ave} is the average interarrival time during the channel's busiest interval duration I. X_{min} and X_{ave} together with S_{max} corresponds to requesting that the network provide a certain peak bandwidth and a certain long-term average bandwidth respectively. If the channel request is accepted, i.e. if the desired bandwidth are allocated to the channel, the application is expected to satisfy the interarrival bounds by the offered load parameters, whereas the delay bounds are to be guaranteed by the network. Various types of guarantees can be defined, corresponding to different types of delay bounds. For example:

1. *Deterministic*: The bound D is an absolute one. This is necessary for hard real time applications.

2. *Statistical:* The bound is expressed in statistical terms: For instance, the probability that the delay of a packet is smaller than the given bound D must be greater than a value Z.

3. Pricing Policies   in Multimedia Networks

The pricing structure in current data networks is complex, and publicly available information regarding the basis of these prices are often not available. For example, the current pricing structures, in the TCP/IP Internet, regional networks are supported by governmental funding, and local LAN costs are supported by participating institutions. Therefore, users pay only for their LAN costs and the membership fees of the regional network (which are usage insensitive). This system operates reasonably well, since the majority of the costs to end users are relatively small. Pricing in the telephone network has been better researched, but the telephone offers *only a single QoS*, and the costs of providing the service are available. So it is fairly simple to set a price such as the costs (plus a margin for profit) are recovered. While the basic pricing structure in telephone networks is simple, the present regulatory structures and the monopoly situation has meant that most of the research in pricing for telephone networks has been to study the effect of regulation rather than the effect of pricing on network performance [CDS+91].

With the ever increasing desire for multimedia applications on computer networks, the need for real-time communications services (i.e. services with guaranteed performance) is evident. In conjunction with the fulfilment of this need, efforts are being   made   to define pricing schemes suitable for use in integrated services network. While pricing policies exist for non-realtime networks, (datagram networks) as well as for conventional public utilities networks (telephone and power system networks), they cannot fulfil the requirements for a pricing policy for an integrated-services network. For example, some of the requirements for a network pricing policy are:

- Allow clients to predetermine the charges for the services they need.
- Discourage client actions that will decrease the efficiency of the network.
- Allow the policy mechanism to be implemented with minimal overhead.
- Allow a simple relationship between the performance characteristics of the network and the return derived from the network.

The   pricing policy may be expressed as $Total\_Charge_i$ = $Reservation\_Charge_i$, where the $Reservation\_Charge_i$ is the charge applied for the reservation of channel i. All of the resources reserved for the channel available for immediate usage by that channel. Then, this equation can be written as [PKF92];

$$total\_Charge_i = TOS_i * t_i * tod_i$$

where

- TOS_i is the service required by the client for channel I.
- t_i *is the lifetime of the channel.*
- tod_i is the period factor (time of day) associated with the channel i.

TOS_i is selected from a set of performance parameters of the service provided. There are n different TOS and TOSi is the service from 1..n requested by channel i. Each of the n TOS has a fixed cost per unit time associated with it, and these prices are made available to applications. Tod_i is the factor associated with peak periods. This factor is used to motivate applications with flexible needs to shift their workloads to periods where the tod factor is less. The tod_i for each period are available to the client. Ti is the duration of the channel. (TOSi * ti) gives the cost for reserving the resources for the service type given by TOS_i for time t_i. With this pricing policy, applications with higher performance service requests get charged at higher rate. The service provider can scale the value of TOS_i to recover the costs and achieve a reasonable profit.

TOS_i and tod_i charges are fixed and available to the applications. These latter know the pricing policy, and usually the lifetime of each channel (more or less), and can therefore define the costs that their channels will incur., as well as check the charges applied to those channels by the network. The TOS value assigned to each channel type is based on the resources reserved by the channel and the channel's effect on the admission of other channels into the network. The resources reserved directly related to the applications' requirements, as can be seen in the equations below. If all nodes are identical, it is sufficient that the resources be determined for a single node for a complete analysis of the network:

$$\text{Bandwidth reserved} = \sum (1 / X_{min}).S_{max}$$

$$\text{Buffer space reserved} = \sum Buffers().S_{max}$$

$$\text{CPU time reserved} = \sum (1 / X_{min}).T_{max}$$

$$\text{Delay resource reserved} = \sum (1 / D).Z.K(D,Z)$$

where:

- N_p is the number of nodes traversed by the channel.
- Buffers() is the buffer space function that calculates the number of buffers required by the channel at each node based on the values of client parameters.
- K(D.Z) is a heuristically determined delay function dependent of the parameters D and Z.

The bandwidth reserved in a node is determined by multiplying the maximum packet sending rate, by the maximum packet size. The total bandwidth reserved is the sum across the nodes in the path. If the nodes are identical, this sum is achieved by simply multiplying the bandwidth reserved in a node and by the total number of nodes in the path. The total amount of buffer space is determined by summing the result of buffers() function across all nodes multiplied by S_{max} . The TOS value is computed as follows:

TOS = α. Bandwidth reserved + ß. Buffers reserved + γ. CPU time reserved + δ. Delay

The values of the coefficients α, ß, δ, and γ are determined by the network designers to reflect the relative scarcity of each of the resources. For instance, if the buffer space is the bottleneck, ß must be changed accordingly, i.e. larger than the other constants in order to encourage applications to use the resource efficiently.

The resources used (bandwidth, buffer space, and CPU time) are simply defined and are related to the quality of service requirements. However, the delay resource cannot be directly associated to the quality. The delay resource can represented as a qualification of the loss of potential for the network to accept other channels due to the admission of the channel being created. The admission of a channel into the network may reduce the network's ability to accept other channels due to the increased difficulty the node scheduler encounters in satisfying all the local deadlines. Therefore, the new channel consumes some of the delay resource of the network. The delay resource is related to the delay (D) and the probability of no lateness (Z) requirements of the channel. The delay resource is made inversely proportional to the delay parameter, since the shorter the delay required by the channel, the greater the restrictions placed on the scheduler, and directly proportional to the probability parameter since the greater this probability, the greater the restrictions placed on the scheduler.

4. Formulation of the Problem

When a collaborative working applications undertake its own resource allocations, it uses the cost-effective resource manager. This latter gets as input all the QoS parameters of the traffic requested. Let us assume that for a particular application, it requires n different services, defined as $(X_{min, i}, X_{ave,i} , S_{max, i} , T_{max,i} , D_i , Z_i)$

where the parameter i varies from 1 to n. The program can evaluate the resources needed to combine these traffics together, by calculating the effective bandwidths, buffer space, CPU and delay. One method for undertaking this is to use Ferrari and Verma's guarantees tests (statistical and deterministic), the delay bound test, and the destination host test [FV90]. Let us assume that the resulting channel has an index k $(k \neq 1..n)$. The program needs to compare the following:

$$\sum_{i=1}^{n} Total\_ch \arg e_i \geq Total\_ch \arg e_k$$

and this is equivalent to :

$$\sum_{i=1}^{n} TOS_i \geq TOS_k$$

as the tod_i and the t_i parameters are the same for i and k. (Recall that $total\_Charge_i = TOS_i * t_i * tod_i$.). If this inequation is satisfied, then it is cost-effective to undertake a local multiplexing at the application level. Therefore, if we look at the TOS definition, this implies that the local multiplexing can use statistical multiplexing to reduce the amount of resources requested while maintaining the level of guarantees promised, then it is a viable option.

5. Workload Modelling

An important factor in the performance evaluation process of the cost-effective resource manager is the input workload selected. This latter is made up of traffic flows generated by sources of different media types (audio, video, objects, text, etc.) with different characteristics determining their behaviour and QoS requirements. Choosing the right workload for the performance evaluation of algorithm s and protocols is crucial as different workloads will lead to selecting different algorithms and protocols (as well as their internal parameters). Most performance experiments tend to select traffic sources without considering profiles generated by applications, hence obtaining results from system-level sources. While the best algorithms and protocols would have been selected for the particular workload selected, it remains to be seen how these will perform under the workload generated by real applications.

The *application level* traffic generator distinguishes itself from other traffic generators because it aims to capture the behaviour of these applications in its traffic generation. Two key design issues addressed in the traffic generator are *calibration* and *validation*. The calibration process determines the profile and patterns of the different media types required by an application. Validation is the process of comparing the traffic generated by scenarios from the traffic generator with those of real applications.

5.1 Objectives

The design objectives of the workload model are:

- *Application-oriented traffic generation.* The aim of this architecture is to provide a realistic set of scenarios to algorithms and protocols designer from which they can test their new designs.

- *Calibration.* The traffic pattern generated by applications needs to be characterised as a set of services with temporal relationships. This objective is to ensure that the architecture features a calibration process is in place to extract the essential features of an application.

- *Validation.* This is the process of verifying that the traffic generated by the workload software reflects the real application it is emulating. A level of confidence as well as an interval of confidence are usually provided with the validation process. The level of confidence placed upon this tool as a reliable instrument to drive experiments rests in the validation objective.

- *Standard compliant (ITU-T service classes definition).* B-ISDN will support services with both constant and variable bit rates, data voice, (sound), still and moving picture transmission and, in particular, multimedia applications which may combine data, voice, and video service components. The ITU-T recommendations (I.211) divides ATM-based services and applications into four major areas: messaging services, retrieval services, conversational services, and distribution services [ITU93].

- *To provide an X/Open Transport Interface XTI for algorithms and protocols.* This is to ensure that the interface between the architecture and the algorithms developers is (1) independent of any specific transport provider, (2) conforms to the OSI RM, (3) is widely implemented across different platforms.

5.2 Design

The overall architecture of the workload model is depicted in Figure 1 below. The workload model operates by specifying :

1. The configuration of the network (i.e. the participating nodes),

2. A profile of the scenarios, and
3. A set of libraries of the algorithms and protocols to be tested.

A profile file contains the specification of the experiments, and acts as a binding for the scenario, configuration, and algorithms and protocols required, as well as experimental settings for repetitive experiments. The *template level* provides a generic model for a class of applications (e.g. visual surveillance, distributed virtual engineering). The *basic services level* is made up of media types and subtypes (e.g. video; colour full scan) , and is aimed to be ITU-T standard compliant. A scenario is an instance of an application, and is translated into a profile of traffic services with timing sequences. The Quality of Service (QoS) negotiation and traffic generation component acts upon the services level to deliver the flow. The calibration and validation phases are undertaken at the three different levels. The *scenario level* defines the profile of an application. The Sessions handler is a configuration database that holds all current operating sessions (or experiments).

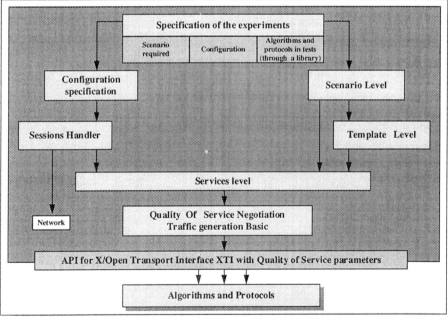

Figure 1 : Traffic characterisation and generation architecture

5.3 Implementation

We have undertaken an initial implementation of the traffic generation architecture on the Leeds ATM LAN testbed at the University of Leeds (see Figure 2). The current setting has several Silicon Graphics R4000 and Sun Sparc2 workstations with first and second generation ATM Fore interface cards (GIA-100, GIA-200 and SBA200) , linked by two ATM switches (Fore ASX 100 and Synoptics Lattis-Cell 1024).   The ATM interface cards all operate the same API (Application Programming Interface) from Fore systems. The ATM LAN contains a mixture of TAXI 4B5B 100 Mbits/sec and also SONET OC3 (155 Mbits/sec).

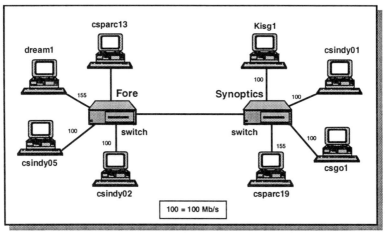

Figure 2 : Leeds University ATM Testbed.

The Session handler is based on a client/server model with a centralised database. It uses the interprocess communications facilities available on most Unix platforms (such as SunOS and IRIX) that support the University of Berkeley Software Distribution 4.3BSD. The socket abstraction allows networked applications to be developed independently of the underlying networks and protocols. The implementation of traffic shaping available are token bucket with leaky bucket rate control, and simple leaky bucket. They are both implemented with the Fore API using the ATM adaptation layer (AAL4). Some of the targeted applications for profiling, and calibration are visual surveillance, distance learning, and distributed virtual engineering [KDH+94].

A typical distributed multimedia application is made up of different basic level services. The   mapping function   is used to translate basic level services' parameters into network system attributes. Three parameters are used to define theses services: (1) the service class (retrieving, distributing, conversational or distributing services [ITU93]), (2) the service type (Video, Audio, ...), and (3) the quality required (Colour Full Scan, CD Quality, ...). A default traffic class (CBR, VBR, UBR or ABR) is associated with each media type in an internal database. CBR , VBR , UBR and ABR stand for Constant Bit Rate, Variable Bit Rate, Undefined Bit Rate, and Available Bit Rate, and they are defined by the ATM forum as the four classes of service. A default traffic class is provided but this parameter can be changed if a more suitable traffic is required by the application. The mapping function returns four parameters : the QoS parameters, the ATM Adaptation layer, the traffic shaping scheme which promises the network that the flow it sends into the network will conform to the description agreed, and a statistical function which generate datagrams .

The current implementation of the session handler is based on multiple *TCP daemons,* (each residing on a separate node of the network). These daemons await clients' connections to generate the traffic of a service. Four stages are used for the implementation of the basic services level (see figure 3 below). The mapping function described above provides systems parameters that will be used by the others stages. QoS and AAL parameters are used during the ATM connection to negotiate traffic. Traffic shaping parameter is used by initialisation stage during which resources (buffer and new processes) are allocated. An independent process reads the buffer and shape the traffic before sending it into the network. The last stage is the traffic generating operation which used the statistical function to generate datagrams and send them to the buffer.

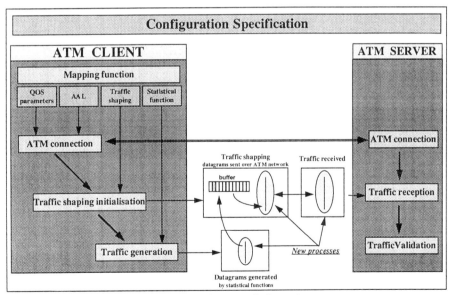

Figure 3: Traffic generator implementation.

6. Implementation of the Cost-Effective Resource Manager

An implementation of the cost-effective resource manager is currently under way. This implementation already runs the bandwidth allocation on the Leeds ATM LAN, using simple bandwidths allocation algorithms developed by Turner [Tur92]. The aim of prototyping was to undertake statistical multiplexing over an ATM network, as the current setting does not permit switched virtual circuits (SVC) connections over ATM. A multimedia application generating multiple traffic streams (through multithreading) with different QoS has been tested with two different options. In the first option, each flow has a one to one mapping with an ATM AAL5 virtual circuit. In the second option, the application requests only one large virtual circuit, and undertake its internal resource management.

7. Conclusions and Further Work

In this paper, we have proposed a design to provide cost-effective services for collaborative working applications. The design is based on the idea that the application should undertake some local multiplexing of its traffic requirements, if the pricing structure for the integrated services network makes it more cost-effective. we have also discussed the importance of an application level traffic generator. Performance experiments for evaluating cost-effective resource managers for distributed multimedia applications are complex to plan and undertake. We believe that in the majority of cases, to attain a certain level of confidence in the results obtained in evaluating these resource algorithms, it is essential to undertake repetitive and long experiments. This can only be possible through the use of traffic generator at the first stage. However, we argue that this generator have to be calibrated to reflect the applications they will be supporting.

We are currently completing the implementation of this traffic generator over the Unix platform and using the Leeds ATM LAN. There are several tasks to be undertaken in the near future following the first implementation. The first is to characterise some of the applications (e.g. visual surveillance, distributed virtual engineering) in terms of QoS, and then encapsulate these into a suitable representation. From this work, should emerge a methodology for undertaking the mapping from the scenario level to the template level (figure 1). Another important future task is the implementation of this traffic generator over an ATM MAN (between Leeds and Wakefield initially), using the IBM Lakes collaborative architecture. This work is part of a joint industrial-academic project which aims to deliver an ATM metropolitan area network (MAN) testbed with QoS performance guarantees, security and billing features. The aims are to provide realistic scenarios running on an ATM MAN, and demonstrating the potential of QoS support through the implementation of various resource management algorithms and flow and congestion protocols.

Acknowledgements

The authors would like to thank Jim Jackson for the installation and maintenance of the Leeds ATM Network, and Arnaud Tardy (from INSA Lyon - France) for undertaking the first implementation of the workload model.

References

[CDS+91] R Cocchi, D Estrin, S Shenker, L Zhang, A Study of Priority Pricing in Multiclass networks, Proceedings of the SIGCOMM'91. September 1991.

[Eck92] A E Eckerg, B-ISDN/ATM Traffic and Congestion Control, IEEE Network September 1992.

[FV90] D Ferrari, D Verna, A Scheme for Real-Time Channel Establishment in WANs, IEEE JSAC 18(3) April 1990

[HHS94] R Handel, M N Huber, S Schroder, ATM Networks, 2nd Edition, Addison Wesley 1994.

[Par94] C Partridge, Gigabit Networks, Addison Wesley 1994

[PKF92] C Parris, S Keshav, and D Ferrari, A Framework for the Study of Pricing in Integrated Networks, Technical Report TR-92-016, International Computer Science Institute, Berkeley March 92.

[Tur92] J S Turner, Managing Bandwidth in ATM Networks with Bursty Traffic, IEEE Network, September 1992

[ZK91] H Zhang, S Keshav, A Comparison of Rate-Based Service Disciplines, Proceedings ACM SIGCOMM 1991.

SUPPORTING QUALITY OF SERVICE GUARANTEES ACROSS HETEROGENEOUS LANS

Frank Ball and David Hutchison
Computing Department
Lancaster University
Engineering Building
Lancaster LA1 4YR
E.mail: <frank, dh>@comp.lancs.ac.uk

1. INTRODUCTION

Developments in multimedia workstations and high speed networks, and advances in digital audio/video compression technology, make possible new classes of distributed applications, eg. distance learning, video-conferencing and remote multimedia database access.The communications needs of these applications will lead to a diversity of traffic types with various Quality of Service (QoS) requirements such that new classes of communication services will be required to support this new traffic. Digitised audio and video traffic are generally classified as continuous media. Continuous media place a high demand on both the network and the workstation, requiring not only high throughput, but also timely delivery. Therefore communications performance and predictability are important issues, particularly if these new applications are to be provided with a guaranteed QoS.

Existing communication architectures, and in particular OSI, are deficient in support for continuous media services. This has led to the development of a Quality of Service Architecture (QoS-A) which provides a framework for QoS specification and resource control, over all architectural layers, from the application platform to the network [1]. QoS is seen as a full end-to-end issue requiring support within both the end system host and the network. QoS-A allows for different levels of commitment to the QoS parameters, giving broadly three types of service commitment:-

1) Deterministic service: offering a guaranteed QoS, typically for hard real time communications.

2) Probabilistic service: may suffer from QoS degradation due to the statistical nature of the network.

3) Best effort service: no guaranteed level of service.

In order to provide a guaranteed end-to-end QoS to an individual connection, resource allocation will be required at every node in the network, along the path of that connection. Resource reservation protocols have been proposed for use in the internet environment [2], and end-to-end methods of bounding delay in packet switched networks have been developed [3] [4]. However, these cases are based on the assumption that either the network comprises homogeneous nodes, or that packets remain unchanged as they migrate through the network. For a network comprising heterogeneous nodes the implications of different resource allocation mechanisms and the effects of fragmentation/packetisation will need to be considered.

Within the context of the QoS-A an Enhanced Network Layer Architecture (ENLA) has been proposed [5], to support guaranteed services to continuous media across multi-hop heterogeneous networks. This architecture proposes an number of new mechanisms which aid the resources reservation protocol in matching the QoS requirements of the source to the performance characteristics of individual sub-nets.

This paper presents the ENLA and outlines its associated resource reservation model. Two of the main mechanisms of the ENLA, namely QoS and Temporal Mapping, are described in detail. Work in progress toward the development of a performance model for use by the QoS and temporal mapping mechanisms in bounding loss/delay is described, and some provisional results are reported.

2. ENHANCED NETWORK LAYER ARCHITECTURE (ENLA)

The ENLA comprises three horizontal layers, which correspond to the three sublayers of the OSI network layer [6] and three vertical planes which correspond to the planes of the QoS-A. Figure 1 illustrates the ENLA and its associated mechanisms, many of which are common to existing resource reservation and network level protocols. Although modification may be needed to a number of these existing mechanisms, in this paper we will limit discussion to resource reservation and the two new mechanism which are introduced by the ENLA, viz. QoS Mapping and Temporal Mapping.

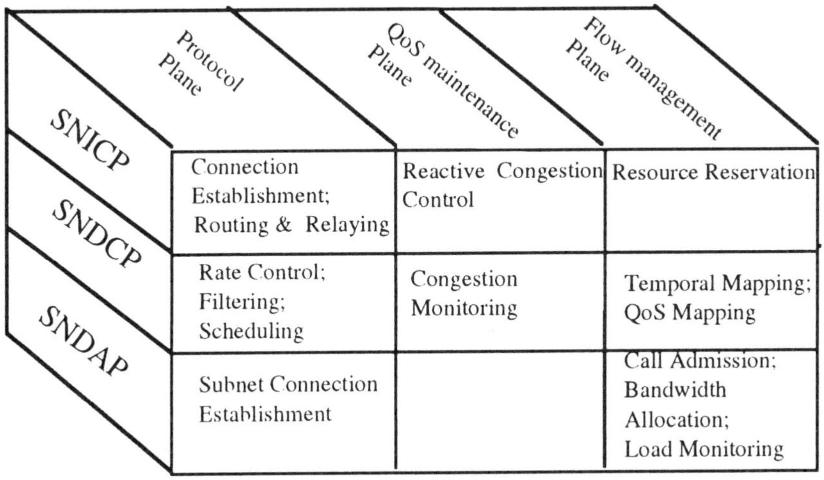

Fig 1. Enhanced Network Layer Architecture (ENLA)

Within the context of the ENLA, the role of QoS Mapping is to match the QoS requirements of the connection to the services which are offered by the underlying subnetwork. The QoS Mapping function would choose the most appropriate service offered by the subnetwork in order to meet the QoS requirements of the connection.

Traffic flows through a network have been shown to comprise a multi-layer temporal structure [7]. This structure may change as it passes through the network, a transformation in structure may occur as traffic is passed between protocol layers, or between different subnets in a heterogeneous network. The capacity offered by a particular subnetwork may also possess temporal properties, ie the subnetwork may give a variable response time to the transmission of data units. The traffic flow and the subnetwork may also operate at a different temporal granularity, and therefore mapping between the two may be required. In this case fragmentation of data units may be necessary in order to match the temporal granularity of the incoming traffic to that of the subnetwork. This mapping may affect the traffic structure to be presented to the next node along the path of the connection.

Temporal Mapping will match the temporal structure of the traffic onto the temporal characteristics of the subnetwork. Generally this mapping will need to be carried out before a call admission test can be applied. The Temporal Mapping process will also report any resulting change in the temporal structure of the traffic to Resource Reservation. This report will then be used by Resource Reservation to update the traffic characteristics contained in the flow specification.

Although the functions of temporal and QoS mapping are logically separate, it may be more convenient to combine them into a single processes as shown below in the resource reservation model of the ENLA

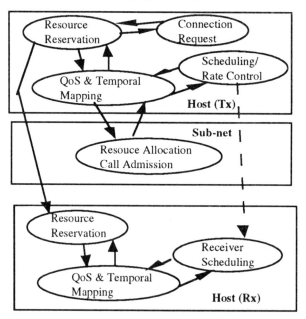

Fig.2 The ENLA Resource Reservation Model

The resource reservation protocol will be sub-net independent therefore existing resource reservation protocols may be used with minor modification. All the other components of the model are both sub-net and implementation dependent.and as such will have to be taylored to meet the needs of each individual node.

A resource reservation entity is associated with each transmitting and receiving host At the transmitter the QoS and temporal mapping function will map the requirements of the source onto the characteristics of sub-net and request the appropriate resources from both the sub-net.and the end system scheduling process. At the receiver the QoS and temporal mapping function will map the predicted temporal structure of the connection's traffic onto the receiver scheduling process. Again the appropriate resources will be requested to ensure that the QoS requirements of the connection are met.

Resource reservation at a gateway or bridge can be modelled as the combination of a receiver and a transmitter as shown below.

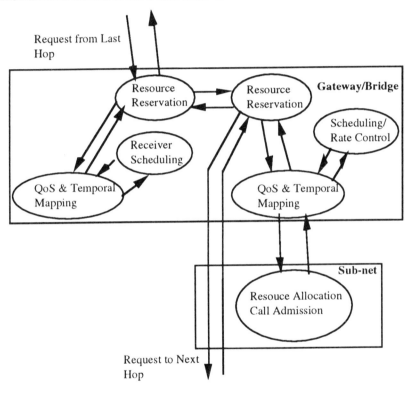

Fig.3 Gateway/Bridge Resource Reservation Model

3. TEMPORAL AND QOS MAPPING

Filipiak [7] identifies three layers in the temporal structure of traffic flows: the call layer, the burst layer and the packet layer. Each layer is characterised by its time scale, which is defined by the mean interarrival time of the layer entities. The flow of entities can be described by two random variables, the interarrival time of the entity (I) and the entity size (L). For the purpose of resource reservation and temporal mapping, only the burst layer and packet layer are of interest.

Certain types of networks provide a transmission channel which also exhibits temporal characteristics. Packets which are awaiting transmission will experience an access delay before being transmitted across the network. This access delay, which is the response time of the network, is mainly influenced by the time between successive transmission opportunities (T^0) at the access point to the network.

At each transmission opportunity a station may transmit a given amount of data (A_x), the maximum size of which is determined by the rules of the particular network access protocol. The temporal characteristics of a number of network types are given below.

Network	T^0	A_x
FDDI-II (Isoch) DBDQ PA	Fixed (125µs)	Fixed (1,2, ..,N octets)
FDDI (synchronous)	Bounded (Min, Ave, Max)	Pre-allocate max. Multiple-frames.
Token Ring	Bounded (Min, Max)	One frame of up to maximum length
ATM	Arbitrary, Pre-defined	Maximum number of cells
CSMA/CD	Non- Deterministic	One frame of up to maximum length

Table 1 Network Temporal Characteristics

In some cases, eg FDDI, A_x may be determined by negotiation, whilst in others, eg Token Ring, A_x is influenced by the size of the frame. The transmission time (τ) of A_x will generally be much less than T^0 and is determined by both A_x and the transmission rate of the network.

Consider packets from a number of sources that arrive at the access point to a subnetwork and which join a queue before being transmitted. The combined average packet interarrival time must be greater than T^o_{ave} of the network, otherwise the size of the transmission queue would continue to grow without limit resulting in eventually buffer overflow for a finite buffer. However, in most cases the average interarrival time of an individual connection ($I_{p.ave}$) will be much greater than T^o_{ave}, and in this case two options are possible:-

1) Transmit a full packet at the first transmission opportunity, then ignore all subsequent transmission opportunities until the next packet becomes available.

2) Fragment the packet into a number of smaller units and transmit these units over a number of transmission opportunities.

If the first option is chosen it may result in an inefficient utilisation of bandwidth. However, if the second option is chosen care must be taken to ensure that the QoS requirements of the connection are met. It will be the role of the QoS and temporal mapping function to select the appropriate strategy that ensures the QoS requirements of the connection are respected, whilst at the same time the efficiency with respect to bandwidth allocation is maximised.

3.1 A Fragmentation Queueing Model

In order to select the most appropriate mapping strategy, the QoS and temporal mapping function may consider the following queuing model.

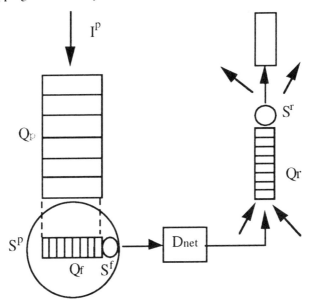

Fig. 4 Fragmentation Queueing Model

Figure 4 shows a general model for the fragmentation, transmission and reassembly of packets across a network. Specific applications of this model may omit certain sections, eg. re-assembly, as required. Also in certain cases it may be necessary to add implementation specific processing delays.

The incoming packets join the packet queue (Q_p) with an inter-arrival time of I^p. When a packet at the head of Q_p goes into service it is instantaneously formed into n equal size fragments which are immediately loaded into the transmission queue (Q_f). Q_f is then served until empty with a service time of S^f. Between the time of the initial loading of Q_f and the time that Q_f becomes empty, no further fragments are allowed to join the Q_f. This model can be considered as a two dimensional queuing model in which the time taken to empty Qf becomes the service time of Qp (S^p).

For networks such as FDDI S^f represents T^0 which will generally be a random variable. Alternatively for an ATM network S^f represents the inter-cell transmission time over a particular VC and will generally be a multiple of the minimum inter-cell transmission time of the ATM link and therefore constant. Where S^f is a random variable, subscripts are used to denote a particular instance of S^f.

eg. $S^f_{j,i}$ = the service time as experienced by the ith fragment of packet j

The timing relationships of the system are given below and illustrated in figure 5.

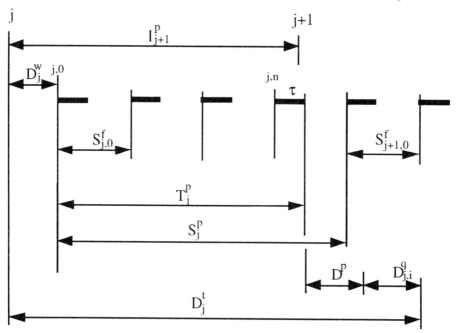

Fig. 5 Fragmentation Timing Diagram

j = the arrival time of the jth packet and I^p_{j+1} = the inter-arrival time between the jth and the j+1th packet.

D^w_j = The waiting time experienced by packet j. Even when the jth packet arrives to find an empty queue with no packets in service, there may still be a waiting time due to the slotted nature of the system. In this case D^w_j will be bounded by S^f_{max}.

S_j^p = the service time of the jth packet and is given by :-

$$\sum_{i=0}^{n} S_{j,i}^f \qquad \text{alternatively, if } S^f \text{ is a constant } \forall\, j,i, \; S^f = nS^f$$

T_j^p = the time taken to transmit the jth packet and is given by :-

$$\sum_{i=0}^{n-1} S_{j,i}^f + \tau \qquad \text{alternately, if } S^f \text{ is a constant } \forall\, j,i, \; (n-1)S^f + \tau$$

D_j^t = the transit delay experiences by the jth packet and is given by :-

$$D_j^w + T_j^p + D^p + D_{j,n}^q$$

Where D^p = A fixed delay due to the latency in the network

and $\qquad D_{j,n}^q$ = A variable delay due to queueing delays within the network.

For shared medium networks, eg FDDI, there will be no internal network queuing delay, however, a delay may still be experienced at the receiving queue.

3.2 Traffic Characterisation Parameters

For ATM networks the most popular model for source characterisation is the two state ON/OFF model. During the ON state cells are generated at a fixed rate, which represents the peak rate of the source, and during the OFF state no cells are generated. Extensions of this model have been used to characterise variable bit rate (VBR) traffic. These models have multiple ON states each with its own cell generation rate. Based on the ON/OFF paradigm, ITU (formerly the CCITT) have recommended the following parameters for source characterisation [8]:-

R_p: The peak arrival rate of the cells

R_a : The average arrival rate of cells

β : The burstiness, defined as the ratio between the peak and average cell rate.

t_{on} : The duration of the active state.

An alternative but equivalent set of parameters is as follows :-

R_p: The peak arrival rate of the cells.

m : The mean number of cells in a burst.

T_i : The average interarrival time between two consecutive bursts.

Traffic characterisation at the transport level, or for packet switched networks in general, requires a different set of parameters from those above. The QoS-A [9], suggest the following transport level source characterisation parameters :-

Frame_Size : The average frame size.

Frame_Rate : The average frame rate. (token generation rate)

Peak_Rate   : The maximum transmission rate.

Burst   : The maximum number of frame that may be transmitted at the peak rate.

Damaskos and Gevras [10] propose a similar set of parameters for multimedia transport services as follows:-

SDU Size   : The maximum service data unit (SDU) size.

SDU Interval   : The interval between two consecutive SDUs.

SDU Burst   : The maximum SDU which may be sent before their expected time

QoS and temporal mapping requires a more general set of parameters which can characterise traffic for both packet and cell based networks. We propose a set of parameters which are based on the multi-layer temporal structure of traffic as identified by Filipiak [7]. This set is as follows :-

$$\text{Traff.Char.Desc} = \{I_C, L_C, I_B, L_B, I_P, L_P\}$$

This traffic descriptor gives the inter-arrival time (I) and length/duration (L) of layer entities for the Call, Burst and packet level respectively. For QoS and temporal mapping purposes only the burst and packet level are of interest. However the call level parameters may be useful to other functions, eg. advanced resource reservation, and therefore are include for completeness.

Generally, each of the parameters in the traffic descriptor will represent a random variable. A random variable may be described by a pdf/pmf together with certain parameters of this pdf/pmf. eg mean, variance. Any parameter from the traffic descriptor set which is a random variable will be further represented by the following set of parameters :-

$$\text{RVar.Desc} = \{ \text{pdf/pmf, Max, Mean, Min, } \sigma^2, -,-,- \}$$

The definition of this set is not complete at this stage and remains an open ended issue; other parameters (denoted -) may be added as required; eg. higher moments of the pdf, autocorrelation function etc.

In most cases it is unlikely that a network user will be able to provide full statistical information about its traffic source, and where this is the cases, instances of the above sets may contain a number of null parameters. Therefore, the QoS and temporal mapping process should be designed to still be able to operate when provided with only a minimal sub-set of the proposed parameters. The process should also be capable of taking advantage of cases where more complete information is provided.

These proposed new parameter sets do not represent a departure from the previously proposed sets, but rather a more suitable generalisation for QoS and temporal mapping purposes. Previously proposed parameters can be easily mapped onto these new sets.

3.3 The General Operation of Temporal Mapping

When choosing the appropriate mapping strategy at the transmitter, the main parameters of concern for the QoS and temporal mapping function will be delay and loss at the packet transmission queue Q_p. Similarly the QoS and temporal mapping process at the receiver will be concerned with the same factors at the receiving queue Q_r. If the predicted loss and delay values are within the required limits, then in both cases, the QoS and temporal mapping function will report these values to the resource allocation process for use in the final end-to-end QoS evaluation. In this paper we will only discuss the operations of QoS and temporal mapping at the transmission end.

The QoS and temporal mapping process at the transmitter will also consider the internal delays within the sub-net. This information will be used in the prediction of the change in the traffic structure and will also be reported to the resources reservation process for the calculation of end-to-end delay.

At the transmitter, for Q_p to be stable the utilisation ρ must be less than unity. This requirement is given by the following :-

$$\rho = \frac{S_{ave}^{p}}{I_{ave}^{p}} < 1,$$

where $S_{ave}^{p} = n S_{ave}^{f}, n \in N.$

I_{ave}^{p} is determined by the characteristics of incoming traffic, therefore the QoS and temporal mapping process will need to chose a combination of n and S_{ave}^{f} which complies with the above requirement. In order to maximise the efficiency in bandwidth allocation, the QoS and temporal mapping process should choose a combination of n and S_{ave}^{f} which gives the highest value of ρ, that still meets the QoS requirements of the connection. Therefore, once an initial combination of n and S_{ave}^{f} has been chosen, further tests will be required to ensure that these QoS requirements will be met .

There are two main cases to consider when choosing the combination of n and S_{ave}^f :-

1) ATM-like subnetworks which use a fixed length packet.

2) FDDI-like subnetworks which allow a variable length packet, and where the response time is determined by the subnetwork.

With fixed length packet type subnets, n will be determined by the length of the incoming packets and the fixed packet size used by the subnetwork. In this case it is the value for S_{ave}^f which must then be chosen . Conversely, for subnetworks with given T^0 and variable length packets, it is n which must be chosen so as to meet the required criteria. In this paper we will focus on QoS and temporal mapping with respect to this second type of subnetwork. Mapping onto ATM like subnetworks will be the subject of future work.

Another point to consider for the type of networks in question is the influence of variable length packets. Since n is chosen to optimally match the temporal granularity of I^p with that of T^o, packet length will have no effects on the choice. Each packet will be divided into n approximately equal sized fragments, the size of which may vary from packet to packet. Bandwidth allocation (A_x) will be based on the maximum possible size of fragment given by :-

$$f_{max} = \left\lceil \frac{I_{max}^p}{n} \right\rceil$$

The QoS and temporal mapping process will initially choose the maximum value of n for which the following constraint holds:-

$$n < \frac{S_{ave}^p}{I_{ave}^p}$$

This is given by the following :-

$$\text{If } I_{ave}^p \text{ MOD } S_{ave}^p = 0 \quad n = \frac{I_{ave}^p}{S_{ave}^f} -1 \quad \text{else } n= \left\lfloor \frac{I_{ave}^p}{S_{ave}^f} \right\rfloor$$

Once the initial value for n has been derived, estimates for delay and loss in the transmission queue will be obtained and if necessary the value of n will be decreased. The sequence of operations of the QoS and temporal mapping process will be as follows:-

1) Given I^p_{ave} and S^f_{ave} choose the largest n which meets the constraint $\rho < 1$.

2) Estimate the delay and loss at Q_p for this n.

3) If the QoS requirements of the call cannot be met, then decrement n and repeat 2.

4) Request necessary resources from the sub-net and within the host, terminate and report if these are not available.

5) Predict the temporal structure of the traffic arriving at the receiving queue.

6) Report predicted delays, loss and change in traffic temporal structure to the resource reservation process.

3.4 Estimating Delay/Loss at the Transmission Queue

Having chosen an initial value for n we then need to ensure that the delay/loss at the transmission queue complies with the QoS requirements of the connection. An estimate for delay/loss we be based on the queuing model described in 3.1, information provided about he arrival characteristics of the traffic and knowledge of the temporal properties of the individual sub-network.

Although both delay and loss are of importance, in this paper we will consider only delay, and assume buffer sizes are sufficiently large to prevent loss in all cases. The mean delay through the system may be of interest, however the most important parameter of concern is the q-percentile of delay which gives a bound for the maximum delay with a given probability.

If full statistical information of the arriving traffic can be provided and full statistical information is available for the timings of the network, then this information may be used directly in the estimation of delay. However, in many cases it is more likely that only a limited amount of information will be available and worst case assumption may need to be made. Given the maximum, mean, and minimum values of a random variable, the pmf of a distribution for that random variable can be formed, which for those three values, will give a worst case variance [11]. This pmf, say A(t), is given by the following :-

$$A(t) = \begin{cases} \dfrac{mean-max}{max-min} & t = max \\\\ 1 - \dfrac{mean-max}{max-min} & t = min \\\\ 0 & \text{otherwise} \end{cases}$$

For networks where the maximum, mean and minimum values of T^o are known, eg FDDI, a pmf for S^f ($S^f(t)$) can be formed as above. A pmf for S^p ($S^p(t)$) can then be formed by the n fold convolution of $S^f(t)$. Similarly we can form the pmf for the packet arrival process $I^P(t)$, assuming of course that maximum, mean, and minimum values of I^P are provided, and that no further information on the distribution of I^P has been given. At this point variation in packet lengths is ignored, for reasons stated in section 3.3.

The resulting queueing model may be M/G/1, D/G/1, G/G/1 etc, whatever the case we need to estimate the q-percentile delay. We are currently investigating techniques which will allow us to make the appropriate delay estimates, the emphasis being on practical solutions which can be readily implemented.

A simulation study is also in progress, to compare the delay estimates obtained by the fragmentation queueing model described above, with those of a more detailed model of the fragmention process which is being simulated. Some provisional results are reported in the next section.

4. THE SIMULATION MODEL

A simulation model has been built to test the temporal mapping queueing model described in this paper. The simulation models the arrival of transport layer frames which are then fragmented into MAC layer data unit and placed in the transmission queue of a shared medium type network. T^o is generated randomly according to some distribution, current only three distributions are supported, exponential, uniform and the type described by A(t) in section 3.4. The latter represents a situation where T^o_{max}, T^o_{ave}, and T^o_{min} are known but no other information on the behaviour of the network is available.

The next stage in the development of the simulation model will be to generate T^o according to the known characteristics of actual networks, the first of which will be FDDI. The current simulation model will allow us to validate our delay estimates based on the assumed distributions, and the next development will allow us to compare the delay estimates with those of the actual network.

An initial experiment to validate the fragmentation queuing model has been carried out for an M/G/1 with the following parameters :-

Packet arrival process poisson with $1/\lambda = 40$ ms

$T^o_{max} = 20$ms

$T^o_{ave'} = 10$ms

$T^o_{min} = 3$ms

$n = 1, 2, 3$

Population 25000 packets (approximately 15 mins of video)

The expected delay E[T] for an M/G/1 queue was calculated for each value of n, and compared with the corresponding result from the simulation run. The two sets of results were sufficiently close for optimism; however further testing is still required, and the accuracy with q-percentile delays has not yet been tested.

5. CONCLUSION AND FUTURE WORK

In this paper we have presented our proposal for an Enhanced Network Layer Architecture capable of supporting guaranteed services to continuous media traffic across multi-hop heterogeneous networks. We have presented this architecture within the framework of a more general Quality of Service Architecture, the QoS-A, developed at Lancaster University.

Two of the main mechanisms of the this architecture, viz QoS and temporal mapping, have been described in detail and a supporting performance model has been presented. Some provisional results have been reported.

The next stage of this work will be to complete the investigation into delay estimation techniques, following which we will consider temporal mapping from the perspective of the receiver. The application of QoS and temporal mapping to ATM-like networks will be investigated. Finally, in a simulation study, QoS and temporal mapping will be applied end-to-end across a number of heterogeneous interconnected networks.

ACKNOWLEDGEMENTS

The work presented in this paper has been partly carried out within the Quality of Service Architecture (QoS-A) project, which is funded as part of the UK EPSRC Specially Promoted Programme in Integrated Multiservice Communication Networks (GR/H77194) in co-operation with GDC (formerly Netcomm Ltd). We also gratefully acknowledge funding from BT labs as part of their University Research Initiative to investigate QoS management of multiservice networks.

References

1. A. Campbell, G. Coulson and D. Hutchison. "A Quality of Service Architecture", **Computer Communication Review**, 24, 2 (April 1994) 6-27

2. D. J. Mitzel, D. Estrin, S. Shenker and L. Zhang. "An Archectural Comparison of ST-11 and RSVP", **INFOCOM 94**, Vol 2, 716-725 (June 1994) Toronto, Canada

3. D. Ferrari and D. C. Verma. "A Scheme for Real-Time Channel Establishment in Wide-Area Networks.", **IEEE Journal on Selected Areas in Communications**, 8, 3 (April 1990) 368-379

4. J. Kurose. "On Computing Per-session Performance Bounds in High Speed Multi-hop Computer Networks.", **Performance Evaluation Review**, 20, 1 (June 1992) 128-139.

5. F. Ball and D. Hutchison, "An Architecture For Supporting Guaranteed Services in Heterogeneous Networks", **EFOC & N 95,** 124-127 (27-30 June 1995) Brighton, UK.

6. F. Halsall. "Data Communications, Computer Networks and Open Systems", Third Edition, Addison Wesley, 1992, ISBN 0-201-56506-4

7. J. Filipiak, "Structure of Traffic Flow in Multiservice Networks", **INFOCOM 88,** 425-429 (March 1988) New Orleans, USA

8. S. Damaskos and A. Gavras, "A simplified QoS Model for Multimedia Protocols over ATM", **High performance Networking V,** Grenoble, France

9. A. Campbell, G. Coulson and D. Hutchison, "A Multimedia Enhanced Transport Service in a Quality of Service Architecture", **NOSSDAV 93,** Lancaster, UK.

10. H. Saito and K. Shiormoto. "Dynamic Call Admission Control in ATM networks", **IEEE Journal on Selected Areas in Communications,** 9, 7, 982-989.

11. G. D. Stamoulis, M. E. Anagostou and A. D. Georgantas. "Traffic Source Models for ATM: A Survey", **Computer Communications,** 17, 6, 428-438.

Analytical Models

MRE Analysis of Open Exponential Queueing Networks with Server Vacations and Blocking[*]

DEMETRES D. KOUVATSOS and CHARALABOS A. SKIANIS
Computer Systems Modelling Research Group,
University of Bradford,
Bradford BD7 1DP, West Yorkshire,
England

Abstract

Queueing network models (QNMs) are widely recognised as powerful evaluation tools for representing computer and communication systems and predicting their performance. This paper focuses on arbitrary open exponential QNMs with server vacations and repetitive-service blocking with random destination (RS-RD). A finite capacity M/M/1/N queue with exponential server vacation periods - solved via the principle of minimum relative entropy (MRE) and classical queueing theory - plays the role of an efficient building block in the solution process. A cost-effective algorithm for the analysis of such QNMs is step-wise described and numerical results against simulation are included to demonstrate the credibility of the MRE approach.

1 Introduction

Queueing Network Models (QNMs) are widely recognised as powerful evaluation tools for representing computer and communication systems and predicting their performance. Exact closed-form solutions for the queue length distribution are not generally attainable. As a consequence, numerical techniques and analytic approximations are used to study open queueing networks under different blocking mechanisms (c.f. [1]). However, exact numerical methods become computationally prohibited as the number of servers and the buffer capacity increases.

Classical queueing theory provides a conventional but limited framework for formulating and solving QNMs. Alternative methods, however, analogous to those applied in the field of Statistical Mechanics, have been proposed in the literature for the analysis of more complex queueing systems (c.f. [2]).

In this context, the information theoretic principle of Minimum Relative Entropy (MRE) can be used as a uniquely correct, self-consistent method of inference

---

[*] Supported by the Engineering and Physical Sciences Research Council (EPSRC), UK, under grants GR/H/18609 and GR/K/67809

for characterising the form of a true but unknown probability distribution based on information expressed in the form of mean values and a prior estimate of the unknown distribution (c.f. [3]). The principle has already been applied towards the analysis of single queues with vacations, where customers have to wait for service not only when the server is busy but also when the server goes away from the service facility for a randomly distributed period of time (c.f. [4, 5]). Moreover, the principle has been applied towards the analysis of arbitrary open and closed QNMs without server vacations (c.f. [5]).

In this paper the MRE principle is used towards the analysis of an open queueing network with arbitrary configuration and repetitive service blocking of random destination (RS-RD). An exact MRE solution for the state probability of an M/M/1/N queue with exponentially distributed vacation periods under an exhaustive service scheme is used as an effective building block for the solution of the entire network.

The principle of MRE is introduced in Section 2. In Section 3 the MRE methodology is applied to QNMs with finite capacity. In particular the MRE joint queue length distribution (qld) for the network is derived in Section 3.1 and the first moments of the effective flow are determined. The MRE algorithm for open QNMs with blocking is described in Section 3.2, while in Section 3.3 numerical results are introduced to illustrate the credibility and relative accuracy of the algorithm against simulation. Finally, conclusions and comments on future work follow in Section 4.

2 The Principle of Minimum Relative Entropy (MRE)

Let x be the state of one system with a set D of feasible states. Let D* be the set of all the probability density functions (pdf) f on D such that $f(x) \geq 0 \ \forall x \in D$ and

$$\int_D f(x)dx = 1 \qquad (2.1)$$

Suppose that this system is described by a true but unknown density function $f^* \in D^*$ and that $g \in D^*$ is a prior density that is a current estimate of the f^* such that $g(x) > 0 \ \forall x \in D$.

In addition new information for the system places a number of constraints on f^*, in the form of expectations defined on a set of k suitable functions $\{a_i(x), i=1,...,k\}$ with known values $\{<a_i>, i=1,...,k\}$, namely

$$\int_D a_i(x)f^*(x) = \langle a_i \rangle \ , i = 1, \ ..., k \qquad (2.2)$$

where k is less than the number of the possible states.

Since the above set of constraints (2.1) and (2.2) denoted by $I=(f^* \in \Phi)$, do not determine the form of $f^*(x)$ completely, they are satisfied by a density set $\Phi \subseteq D^*$.

The principle of MRE states that of all pdfs that satisfy constraints I, the suitable one is the posterior pdf f∈ Φ that minimises the relative entropy function, H(f,g) in the set Φ

$$H(f,g) = \min_{f' \in \Phi} \{ \int_D f'(x)\log\{\frac{f'(x)}{g(x)}\}dx \} \qquad (2.3)$$

By applying the Lagrange's method of undetermined multipliers, the form of the posterior pdf f is

$$f(x) = g(x)\exp\{-\beta_0 - \sum_{i=1}^{k} \beta_i a_i(x)\} \qquad (2.4)$$

where β_0 and $\{\beta_i\}$, i=1,2, ..., k are the Lagrangian multipliers whose values are determined by the constraints (2.1) and (2.2).

From (2.1) the normalising constant is given by

$$\exp\{\beta_0\} = \int_D g(x)\exp\{-\sum_{i=1}^{k} \beta_i a_i(x)\}dx \qquad (2.5)$$

If the integral in (2.5) is solved analytically, closed form expressions could be derived for $\{\beta_i\}$, i=1,2, ..., k, in terms of the mean values $<a_i>$.

The MRE principle has been characterised in [3], as a uniquely correct method of inference by means of four consistency axioms. They showed that if an other functional is used in an inference procedure satisfying the four axioms, then the resulting posterior density is identical to the proposed by the MRE principle given by (2.4).

When D is a discrete state space and in the presence of a uniform prior density (when D is finite), the principle of ME can be used as a special case of relative entropy minimisation (c.f. [3]).

3 MRE for QNMs with Finite Capacity

This section considers the applicability of the principle of MRE for the analysis of arbitrary exponential open single class QNMs, at equilibrium, with M finite capacity single server queues with vacation periods and repetitive service blocking of random destination, (RS-RD). This type of blocking occurs when a job upon service completion at a queue i attempts to join a destination queue, j, whose capacity is full. As a result, the job is rejected by queue j and immediately receives another service at queue i. This process is repeated until the job completes service at queue i and the destination queue j is not full. Under the RS-RD blocking mechanism, each time the job completes service at queue i, a downstream queue is selected independently of the previously chosen destination queue, j.

The MRE approximation algorithm is based on the decomposition of a general QNM into individual finite capacity queues with server vacation periods and revised interarrival and service time distributions, under RS-RD blocking. Each queue in turn plays the role of a building block and is solved separately via relative entropy minimisation, subject to mean value constraints and prior distribution.

3.1 MRE Open QNMs with RS Blocking

Consider an arbitrary open queueing network under RS-RD blocking mechanism of M FCFS single server queues with vacation periods and exponential external interarrival time, service time and vacation time distributions. At any given time the state of the network is described by a vector $\underline{n} = (n_1, n_2, ..., n_M)$, where n_i denotes the number of jobs at queue i, i=1,2,...,M, such that $0 \leq n_i \leq N_i$, where N_i is the buffer capacity for the queue i. Let S, be the set of all feasible states $\underline{n}$, of the network, and $p(\underline{n})$ be the equilibrium probability that the queueing network is in state $\underline{n}$. In addition, let $p_i(n_i)$ be the equilibrium marginal probability that queue i is at state n_i. Suppose all that is known for the state probabilities, $p(\underline{n})$, $\underline{n} \in S$, is

(i) The Normalisation Constant

$$\sum_{\underline{n} \in S} p(\underline{n}) = 1 \tag{3.1.1}$$

(ii) The Full Buffer State Constraints, $\Phi_i \equiv p_i(n_i), 0 < \Phi_i < 1$,

$$\sum_{\underline{n} \in S} f_i(\underline{n}) p(\underline{n}) = \Phi_i \tag{3.1.2}$$

where

$$f_i(\underline{n}) = \begin{cases} 0 & , \quad n_i < N_i \\ 1 & , \quad n_i = N_i \end{cases} \tag{3.1.3}$$

Moreover, $q(\underline{n})$ is a pdf used as a prior estimate for the true but yet unknown pdf of our system, $p(\underline{n})$ and $q_i(n_i)$ the corresponding marginal probabilities of the prior estimate. In our case we consider as prior the queue length distribution of the corresponding queueing network with infinite capacity, which is proven to provide as with satisfactory results in previous cases (c.f. [4, 5]).

The form of the MRE joint state probability, $p(\underline{n})$, can be completely specified by minimising the cross entropy, H(p,q), written as

$$H(p,q) = \sum_{\underline{n} \in S} p(\underline{n}) \log \frac{p(\underline{n})}{q(\underline{n})} \qquad (3.1.4)$$

subject to the M+1 constraints (3.1.1), (3.1.2), and is given by,

$$p(\underline{n}) = \frac{1}{Z} \prod_{i=1}^{M} q_i(n_i) y_i^{f_i(\underline{n})} \qquad (3.1.5)$$

where the normalising constant, Z, is given by

$$Z = \sum_{\underline{n} \in S} \prod_{i=1}^{M} q_i(n_i) y_i^{f_i(\underline{n})} \qquad (3.1.6)$$

and y_i is the Lagrangian coefficient that correspond to constraint (3.1.2).

Since the auxiliary function $f_i(\underline{n})$, $\forall i=1,2,...,M$, and the prior estimate $q(\underline{n})$, depend only on the variable, n_i, it is possible to interchange the sum and the product in eq. (3.1.6), defining also Z_i, $i=1,2,...,M$, namely,

$$Z = \prod_{i=1}^{M} \left\{ \sum_{n_i=0}^{N_i} q_i(n_i) y_i^{f_i(n_i)} \right\} = \prod_{i=1}^{M} Z_i \qquad (3.1.7)$$

which implies that $p(\underline{n})$ is the product of the marginal probabilities $p_i(n_i)$ since eq.(3.1.5) can be written as

$$p(\underline{n}) = \prod_{i=1}^{M} \frac{1}{Z_i} q_i(n_i) y_i^{f_i(\underline{n})} = \prod_{i=1}^{M} p_i(n_i) \qquad (3.1.8)$$

So

$$p_i(n_i) = \frac{1}{Z_i} q_i(n_i) y_i^{f_i(\underline{n})} \qquad (3.1.9)$$

Note that eq. (3.1.9) is the MRE solution for an $M/G/1/N_i$ queue with vacation periods, subject to constraints of normalisation and full buffer state probability and, in addition, a prior distribution determined by the exact queue length solution for the corresponding queue with infinite capacity (c.f. [4, 5]). This MRE solution is of a closed form and becomes exact in the case of exponential service and vacation times. Thus, the MRE solution offers a cost-effective building block - as opposed to the recursive solution obtained via exact queueing theory (c.f. [6]) - for the analysis of the entire network.

Thus, the entropy minimisation suggests - via the product form solution (3.1.8) - a decomposition of the original open network into individual $M/M/1/N_i$ queues with exponentially distributed vacation periods with revised interarrival time and service time distributions.

The evaluation of the Lagrangian coefficient y_i, $i=1,2,...,M$, of the MRE solution requires the iterative application of the following steps:

(i)   Consider each queue i within the network under exponentially distributed service times and vacation periods

(ii) Determine the effective service time at each queue (ie. total service time received by each customer on average)

(iii) Revise the overall arrival process at each queue by merging all contributing arrival streams

(iv) Solve each queue i, $i=1,...,M$ as an isolated $M/M/1/N_i$ queue with exponentially distributed vacation periods, using as overall arrival and effective service processes obtained in steps (ii) and (iii) and obtain output statistics.

Such a queue serving as a building block for our network can be seen in Figure 1 and provides an indication of how steps (i-iv) can be carried out.

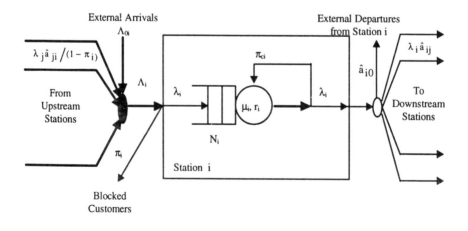

Figure 1. The Queueing Station i within the network

Assuming that all flow processes (i.e. merge, split, departure) are renewal each queue i, i=1,2,...,M, can be seen as a M/M/1/N_i queue with (a) exponentially distributed vacation periods, with mean rate r_i (b) exponential overall interarrival process (including "rejected" jobs), formed by the merging of departing streams for queue i, generated by queues $\{j\}, \forall j \in A_i$, where A_i is the set of upstream queues of queue i (which may include external arrival), and (c) an effective service time distribution reflecting the total time during which a server of queue i is occupied by a particular job.

The rate of the effective service time depends on (a) the type of RS blocking mechanism enforced, and (b) all blocking probabilities $\pi_j \equiv p_j(N_j), \forall j \in D(i)$ (i.e. the probabilities that a completer from queue i is blocked by queue j, where D(i) is the set of all downstream queues of queue i,i=1,2,...,M). It can be seen that the rate, $\hat{\mu}_i$, of the effective service time is given by [8]

$$\hat{\mu}_i = \mu_i(1 - \pi_{ci}) \tag{3.1.10}$$

for all i=1,2,...,M where π_{ci} is the blocking probability that a completer from queue i is blocked under RS-RD blocking mechanism, i.e. $\pi_{ci} = \sum_{j \in D(i)} a_{ij} \pi_j$, with $\{a_{ij}\}$, i=1,2,...,M being the associated transition probabilities.

The effective arrival stream of jobs at the queue can be seen as the result of a two-way splitting of the overall merging stream with parameter π_i (i.e. the probability that an external arriver or a completer from an upstream queue will be blocked by queue i). By denoting as λ_i the effective interarrival rate, the overall interarrival process is given by

$$\Lambda_i = \frac{\lambda_i}{1 - \pi_i} \tag{3.1.11}$$

Moreover $\{\lambda_i\}$, i=1,2,...,M must satisfy the effective job flow rate equations, i.e.

$$\lambda_i = \lambda_{0i} + \sum_{j=1}^{M} \hat{a}_{ji} \lambda_j \tag{3.1.12}$$

where under RS-RD blocking $\hat{a}_{ij}$ is the effective transition probability given by [9]

$$\hat{a}_{ij} = \begin{cases} \dfrac{a_{ij}(1 - \pi_j)}{(1 - \pi_{ci})} &, \quad j \in D(i) - \{0\} \\ \dfrac{a_{i0}}{(1 - \pi_{ci})} &, \quad j = 0 \end{cases} , i=1,...,M. \tag{3.1.13}$$

3.2 The Minimum Relative Entropy Algorithm for Open QNMs-B

In this section the MRE algorithm is derived for arbitrary open exponential QNMs with single server queues, vacation periods, exponential interarrival and service times, under RS-RD blocking.

The procedure is based on the previously developed Maximum Entropy algorithm (c.f. [8-10]) for analysing QNMs without vacation periods and is presented below in a stepwise manner.

{ <u>MRE Algorithm</u> }
BEGIN

Inputs $M, \Lambda_{0i}, \mu_i, r_i, N_i, a_{ij}, \quad \forall i = 1,2,..,M, j = 0,1,..,M$

Step 1: Initialise $\{\pi_i\}, i = 1,2,..,M$, any value in $(0,1)$

Step 2: $\{\pi_i\}, i = 1,2,..,M$ convergence

Step 2.1: Calculate effective routing probabilities

$$\hat{a}_{ij} = \begin{cases} \dfrac{a_{ij}(1-\pi_j)}{(1-\pi_{ci})} & , \quad j \in D(i) - \{0\} \\ \dfrac{a_{i0}}{(1-\pi_{ci})} & , \quad j = 0 \end{cases} \quad , i=1,2,..,M$$

Step 2.2: Solve effective job flow balance equations

$$\lambda_{0i} = \Lambda_{0i}(1-\pi_i)$$

$$\lambda_i = \lambda_{0i} + \sum_{j=1}^{M} \hat{a}_{ji}\lambda_j \quad \forall i = 1,2,..,M$$

Step 2.3: Calculate effective service parameter

$$\pi_{ci} = \sum_{j \in D(i)} a_{ij}\pi_j$$

$$\hat{\mu}_i = \mu_i(1-\pi_{ci})$$

Step 2.4: Calculate overall actual parameters

$$\Lambda_i = \frac{\lambda_i}{1-\pi_i}$$

Step 2.5: Obtain new values for $\{\pi_i\}, i = 1,2,..,M$

$$\pi_i \leftarrow p_i(N_i)$$

Step 2.6: Return to Step 2.1 until convergence of π_i

Step 3: Apply MRE to evaluate each queue i, i=1,2,..,M as

$M(\Lambda_i)/M(\hat{\mu}_i)/1/N_i$, queue with exponential vacation periods.

END.

3.3 Numerical Tests

The credibility of the proposed MRE Algorithm in Section 3.2 is demonstrated against simulation via typical numerical experiments, presented in Tables 1-2 involving an open exponential QNM with five exponential queueing stations and server vacations. The following statistics for each queue are considered: (i) throughput, $\{\lambda_i\}$, (ii) mean queue length, $\{<n>_i\}$, (iii) probability of an empty queue, $\{p_i(0)\}$, and (iv) full buffer state probability, $\{p_i(N_i)\}$.

Relative comparisons are carried out between MRE solutions and simulation, (SIM) in Tables 1-2 and Figures 2-3. The simulation in each case is obtained via QNAP-2, given at 95% confidence intervals (c.f. [11]).

For validation purposes two kinds of error indicators are used:

(a) The tolerance (TOL) for mean queue lengths, $\{<n>_i\}$, is defined as the absolute ratio:

$$TOL(<n_i>) = \left| \frac{SIM(<n_i>) - MRE(<n_i>)}{\sum_{j=1}^{M} SIM(<n_j>)} \right|$$

whereas for the state probabilities, $\{p_i(0)\}, \{p_i(N_i)\}$, as the absolute difference:

$$TOL(p_i(k)) = \left| SIM(p_i(k)) - MRE(p_i(k)) \right| , k=0, N_i$$

(b) the relative difference (REL-DIFF) for the throughputs, $\{\lambda_i\}$, is defined as the absolute ratio:

$$REL - DIFF(\lambda_i) = \left| \frac{SIM(\lambda_i) - MRE(\lambda_i)}{SIM(\lambda_i)} \right|$$

Figures 2-3 indicate the tolerances and the relative differences in the form of bar charts for each queue, Q_i. It can be seen that the MRE solution is very much comparable in accuracy to that of simulation for all the statistics under consideration.

4 Conclusion and Future Work

In this paper analytic approximations are produced for arbitrary open exponential queueing networks at equilibrium with RS-RD blocking. Each queue of the network is considered as an M/M/1/N queue with finite capacity, N, exhaustive service scheme and exponential server vacation periods. The accuracy of the MRE algorithm is very much comparable to that of simulation for typical performance measures such as throughput, mean queue length and the state probabilities.

Future work will involve the use of MRE methodology for the analysis of queueing networks with vacations and generalised exponential (GE) type of interarrival, service and vacation times. In cases of such networks with bursty traffics, it is essential to determine the first two moments for the departing, merging and splitting streams at each queueing station within the network. Applications will also be carried out towards the analysis of polling systems and Asynchronous Transfer Mode (ATM) networks with shared medium switch architectures.

References

1. Perros, H.G., Approximation algorithms for open queueing networks with blocking, Stochastic Analysis of Computer and Communication Systems, ed. H.Takagi (North-Holland, 1990), pp.451-494
2. Pinsky, E., Yemini, Y., The canonical approximation in performance analysis, Computer Networking and Performance Evaluation, eds. T. Hasegawa et al. (North-Holland, 1986) pp.125-137
3. Shore, J.E., Johnson, R.W., Properties of cross entropy minimisation, IEEE Trans. Inf. Theory IT-27(5), pp.472-482
4. Kouvatsos, D.D., Skianis, C.A., Minimum Relative Entropy Analysis of single server Queues at Equilibrium, Technical Report CS-22-94 Comp. Systems Modelling Res. Group, Univ. of Bradford, 1994
5. Kouvatsos, D.D., Skianis, C.A., Minimum Relative Entropy Analysis and Queues with Vacation Time, 10th UKPEW '94, Sept. 94, pp.67-78
6. Lee, T.T. M/G/1/N queue with vacation time and exhaustive service discipline, Operations Research, Vol.32, No.4(July-August 1984), pp.774-784
7. Kouvatsos, D.D., Entropy maximisation and queueing network models, Annals of Operations Research 48(1994), pp.63-126
8. Kouvatsos, D.D., Xenios, N.P., MEM for arbitrary queueing networks with multiple general servers and repetitive-service blocking, Perf. Eval. 10(1989), pp.106-195
9. Kouvatsos, D.D., Denazis S.G., Entropy maximised queueing networks with blocking and multiple job classes, Perf. Eval. 17 (1993), pp.189-205
10. Kouvatsos, D.D., Denazis S.G., Comments on and tuning to: MEM for arbitrary queueing networks with repetitive-service blocking, Technical Report CS-18-91, University of Bradford (May 1991)
11. Veran, M., Potier, D., QNAP-2: A Portable Environment for Queueing Network Modelling, Proc. Int. Conf. on Modelling Techniques and Tools for Performance Analysis, INRIA, 1984

Table 1: Inputs of QNM

Queue i	1	2	3	4	5
N_i	2	2	2	2	2
λ_i	0.5	0.5	0.5	0.5	0.5
r_i	2.0	2.0	2.0	2.0	2.0
μ_i	1.0	1.0	1.0	1.0	1.0

a_{ij}	1	2	3	4	5
1	0.0	0.5	0.0	0.0	0.0
2	0.0	0.0	0.2	0.2	0.0
3	0.2	0.2	0.0	0.0	0.2
4	0.0	0.0	0.0	0.0	0.5
5	0.0	0.0	0.0	0.5	0.0

Table 2: Performance Statistics of QNM

QUEUE	STAT.	SIM	MREM
1	λ	0.441	0.432
	$<n>$	0.903	0.940
	$p(0)$	0.376	0.354
	$p(N)$	0.279	0.294
2	λ	0.546	0.533
	$<n>$	1.106	1.137
	$p(0)$	0.275	0.258
	$p(N)$	0.381	0.395
3	λ	0.446	0.431
	$<n>$	0.921	0.974
	$p(0)$	0.368	0.337
	$p(N)$	0.289	0.311
4	λ	0.545	0.530
	$<n>$	1.211	1.237
	$p(0)$	0.227	0.214
	$p(N)$	0.439	0.451
5	λ	0.542	0.527
	$<n>$	1.203	1.231
	$p(0)$	0.231	0.217
	$p(N)$	0.434	0.447

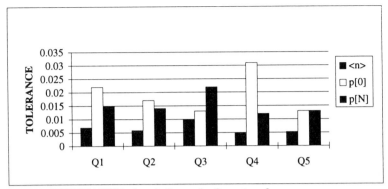

Figure 2: TOLERANCE of each Queue in the network

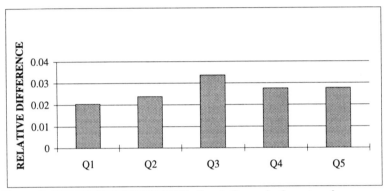

Figure 3: RELATIVE DIFFERENCE of each Queue in the network

IPP Overflow Traffic on a Two-stage Queueing Network

Fred Hung-Ming Chen[1], Zsehong Tsai[2]*, John Mellor[3], Philip Mars[4]
School of Engineering, University of Durham, South Road,
Durham DH1 3LE, UK.
[1]H.M.Chen@durham.ac.uk [2]ztsai@cc.ntu.edu.tw
[3]J.E.Mellor@durham.ac.uk [4]Philip.Mars@durham.ac.uk

Abstract

In a two-stage queueing network the state space is very large such that the exact analytic solution is very hard to obtain. The concept of the IPP overflow stream is a good method to decompose the two-cascade-stage network into two independent queueing systems. The transition matrix will reduce from the size with all the states in the state space of the system to the size depending on the size of the trunk group and the number of the operators in the service centres. In order to present the characteristic of the call stream passing the trunk group, the approximate analytic method revises the concept of the overflow stream to match it. Using the call stream and the IPP overflow call stream from the other service centre, the blocking probability of the second stage can be obtained. Then, the blocking probability of the first stage is also calculated after reducing the mutual effect between the two stages. The overall blocking probability of this network can be calculated very easily.

1 Introduction

Under the architecture of intelligent network (IN) [1–3], a private network consists of two private branch exchanges (PBXs) connected to public network via trunk groups, called $TG0$ and $TG1$. There are two service centres with some operators to serve special request calls, called type 2 calls. In addition, regular calls, called type 21 calls, compete with the special request calls for the limited number of links in each trunk group. Between two PBXs an internal network is setup to deal with the internal calls. In order to provide more convenient service the special request calls can use the resource of the internal network; i.e., if one special request call enters the network from one trunk group and can not find an operator whose phone is connected to the PBX of this trunk group, the PBX can alternate this call to remote service centre when there are

---

*Prof. Tsai is with the Department of Electrical Engineering, National Taiwan University, Taipei, Taiwan.

some idle operators. This route is called alternative route. If the incoming type 2 call is served by the local operators of the local service centre, it takes the direct route. Fig.1 is the network model described above. A routing mechanism can decide which route will be taken when a type 2 call arrives. This routing mechanism can be the service control point (SCP) in IN.

Probabilistic routing algorithm has been studied to analysis the performance of this network [4]. In this algorithm a probabilistic parameter is used to randomly determine the route of the incoming type 2 calls. However, how to determine this parameter is an important issue. The false position method has been used to get an optimal parameter that makes the blocking probability of the type 2 calls smallest. However, when the size of trunk groups and the number of the operators increase, the problem becomes intractable under the false position method. In this paper we use interrupted Poisson process (IPP) to decompose the traffic of the special request class in this network.

In alternative routing algorithms the call stream can not enter a certain backbone due to the limited resource and try another one to setup the calls. The call stream with this behavior is called overflow stream. The overflow call stream does not always exit. When the resource in a certain backbone or trunk group is occupied, the overflow stream generates, called ON state, otherwise the call stream does not exit, called OFF state.. This kind of call stream can be described as a Interrupted Poisson process (IPP) if the rate of the ON and OFF states are independent random variables with exponential distribution. The characteristic of the IPP can be described by three parameters. In addition to the sojourn time of the ON, OFF state, the arrival rate of this call stream during the ON state must be an independent Poisson process.

The concept of the IPP first appearing in Kuczura's paper [5], the author used some stochastic quantities, such as peakness, to approximately describe the characteristics of the call stream. Then, several articles discussed the behavior and approximate solutions of the IPP [6–10]. Not until 1986 had the combination of the IPP and other call stream not been discussed[11]. Therefore, applying the IPP to approximately solve the optimization problem of this network is a possible method.

In probabilistic routing algorithm, to obtain the optimization probabilistic parameter, all the steady state probabilities of the network system must be calculated. However, in the approximate analytic method, instead of solving the steady state probability of each state, we separately calculate the blocking probability of the type 2 calls on each trunk group, and then obtain the total blocking probability of type 2 calls by the probabilistic parameter. In this analytic algorithm the call stream passing via each trunk group is with the IPP characteristics and the parameters can be obtained. Then, the call stream that the calls pass the trunk group but can not be served by the local service centre and try to find an idle operator in the remote service center is an IPP overflow call stream. From the two IPP call streams the blocking probability of the type 2 calls caused by no available operator, called destination block, can be obtained. In addition, the blocking probability of type 2 calls caused by the internal network, called internal block, and no available link on the trunk group, called trunk group block, can be calculated by the parameters of the IPP call streams. After getting the blocking probabilities of the trunk group block, the internal block, and the destination block, the overall blocking probability of the type 2 call stream can be obtained.

The following two sections describe how to calculate the parameters of the IPP call streams passing via the trunk group and the overflow call streams. Then, the numerical results of two typical network configuration will be examined.

2 The IPP Parameters

On each trunk group there are two different kinds of calls arriving, and their arrival processes are independent Poisson processes. The arrival rate of the type 1 calls is $\lambda_{1i}, i \in \{0, 1\}$. Assume that the probability of the type 2 calls passing TGi and served by the operators in the service centres is $(1 - \eta_i)$. Therefore, the arrival rate of the type 2 calls to the network subsystem is $\lambda'_{2i} = (1 - \eta_i)\lambda_{2i}$, where λ_{2i} is the arrival rate of the type 2 calls allocated to localtion i, i.e. $\lambda_{2i} = p_i\lambda_2$, and P_i is the probabilistic parameter. The holding times of the two types of calls are independent variables with exponential distribution. The average holding time of the type 1 calls on TGi is $1/\mu_{1i}$ and the service rate of the type 2 calls is μ_2. Using these two Poisson process call streams, we can obtain the IPP parameters $(\rho_i, \alpha_i, \beta_i)$ of type 2 calls passing TGi, where ρ_i is the arrival rate of type 2 calls on the PBX of location i when the TGi is in the ON state, α_i is the rate that the TGi is in the ON state for type 2 calls, i.e. there are type 2 calls passing via TGi and arriving the PBX connected with this trunk group, and β_i is the rate that the TGi is in the OFF state for type 2 calls.

To be convenient, let x_t be the number of the type 1 calls on TGi at time t and y_t be the number of the type 2 calls on TGi at time t. Due to all incoming calls will be accepted if there is any idle link, let $W_t = (x_t, y_t)$. Thus, the continuous time Markov chain of this trunk group can be represented as $W = \{W_t, t \geq 0\}$. All the states of the continuous-time Markov chain communicate with each other, i.e. it is an irreducible Markov chain, therefore, there exists a product form to obtain the steady state probabilities of the states. However, in this case to calculate the IPP parameters of the call streams which pass the trunk group by using Heffes' time-constant approximation method, so the state transition matrix will be obtained from the steady state equations. Let P_{mn} be the steady state probability of the state (x_t, y_t), where $n = x_t, m = y_t$, and place it on the $\frac{(m+n+1)(m+n)}{2}$th row. The steady state equations list as follow:

$$0 = -P_{0,0}(\lambda_{1i} + \lambda'_{2i}) + P_{0,1}\mu_{1i} + P_{1,0}\mu_2,$$

$$0 = -P_{0,n}(\lambda_{1i} + \lambda'_{2i} + n\mu_{1i}) + P_{0,n-1}\lambda_{1i} + $$
$$P_{1,n}\mu_2 + (n+1)P_{0,n+1}\mu_{1i}, 0 < n < N_i,$$

$$0 = -P_{m,0}(\lambda_{1i} + \lambda'_{2i} + m\mu_2) + P_{m-1,0}\lambda'_{2i} + $$
$$P_{m,1}\mu_{1i} + (m+1)P_{m+1,0}\mu_2, 0 < m < N_i,$$

$$0 = -P_{m,n}(\lambda_{1i} + \lambda'_{2i} + n\mu_{1i} + m\mu_2) + P_{m,n-1}\lambda_{i1} + P_{m-1,n}\lambda'_{2i} + $$
$$(n+1)P_{m,n+1}\mu_{1i} + (m+1)P_{m+1,n}\mu_2,$$
$$0 < m + n < N_i, o < m, n < N_i,$$

$$0 = -P_{m,n}(n\mu_{i1} + m\mu_2) + P_{m-1,n}\lambda'_{2i} + $$

$$P_{m,n-1}\lambda_{i1}, m+n = N_i, 0 < m, n < N_i,$$
$$0 = -N_i P_{0,N_i}\mu_{1i} + P_{0,N_i-1}\lambda_{1i},$$
$$0 = -N_i P_{N_i,0}\mu_2 + P_{N_i-1,0}\lambda'_{2i}.$$

From above equations we can obtain

$$\Pi' Q' = 0, \tag{1}$$

where

$$\Pi' = [\Pi_0, \Pi_1, ..., \Pi_{L-1}],$$
$$Q' = [q'_{r(m,n),s}],$$

and

$$L = \frac{(N_i+2)(N_i+1)}{2},$$
$$\Pi_{r(m,n)} \equiv \text{the steady-state probability of } TGi$$
$$\text{with } m \text{ type 1 calls and } n \text{ type 2 calls.}$$

In above equations the symbol $r(m,n) = \frac{(m+n+1)(m+n)}{2}$, so

$$q'_{r(m,n),s} = \begin{cases} n\mu_{1i}\bar{\delta}_{n,0} & s = r(m,n-1), \\ m\mu_2\bar{\delta}_{m,0} & s = r(m-1,n), \\ -[n\mu_{1i}\bar{\delta}_{n,0} + m\mu_2\bar{\delta}_{m,0} \\ \quad +\lambda_{1i}\bar{\delta}_{n,N_i} + \lambda_2\bar{\delta}_{m,N_i}] & s = r(m,n), \\ \lambda_{1i}\bar{\delta}_{n,N_i} & s = r(m,n+1), \\ \lambda_2\bar{\delta}_{m,N_i} & s = r(m+1,n), \end{cases}$$

From the steady state equations and normalization condition

$$\sum_{i=0}^{L-1} \Pi_i = 1, \tag{2}$$

we can obtain the steady state probability Π' and the blocking probability of the type 2 calls on the TGi, t_i. By using the blocking probability the first two moments of the instantaneous arrival random variable are obtained [12]:

$$\tilde{\alpha}_1 = (1-t_i)\lambda'_{2i}, \tag{3}$$
$$\tilde{\alpha}_2 = (1-t_i)(\lambda'_{2i})^2. \tag{4}$$

Then, using Heffes' formulas,

$$\tau_c = [\Pi'\overline{\Lambda}(e\Pi' - Q')^{-1}\overline{\Lambda}e - m_1^2]v^{-1}, \tag{5}$$

the time constant τ_c is obtained, where

$$\overline{\Lambda} = \begin{bmatrix} 0 \\ \lambda'_{2i}I \end{bmatrix},$$

in which 0 is a $\frac{N_i(N_i+1)}{2} \times \frac{N_i(N_i+1)}{2}$ matrix and I is a $(N_i + 1) \times (N_i + 1)$ identity matrix,

$$e = \begin{bmatrix} 1 \\ 1 \\ \vdots \\ 1 \end{bmatrix},$$

$$m_1 = \tilde{\alpha}_1,$$
$$v = \tilde{\alpha}_2 - (\tilde{\alpha}_1)^2.$$

By the following two equations,

$$\alpha_i = \tau_c^{-1} t_i, \tag{6}$$
$$\beta_i = \tau_c^{-1}(1 - t_i), \tag{7}$$

the ON and OFF rates of the type 2 call stream passing via the TGi are obtained, and the arrival rate is still λ'_{2i}, that is

$$\rho_i = \lambda'_{2i}, \tag{8}$$

$(\rho_i, \alpha_i, \beta_i)$ is the IPP parameters of the type 2 call stream that passes the TGi and arrives the service centre on location i.

3 The IPP Overflow Parameters

When the IPP call stream passing via TGi arrives the service centre, there is another call stream with IPP characteristic which overflows from the remote service centre to compete with the main IPP call stream. When all the operators in one service centre are all busy, the arriving IPP call stream from the TGi will form an IPP overflow stream to the other service centre. At this moment the IPP overflow stream will be blocked. The two IPP overflow streams may be blocked by the internal network.

The main IPP call stream passing the TGi is called $IPP1(i)$, with the IPP parameters $(\rho_i, \alpha_i, \beta_i), i \in \{0,1\}$. The overflow stream with IPP characteristic from the other service center is with the IPP parameters $(\tau_{\bar{i}}, \gamma_{\bar{i}}, \omega_{\bar{i}})$, where

$$\bar{i} = \begin{cases} 1 & \text{if } i = 0, \\ 0 & \text{if } i = 1, \end{cases}$$

and let $\tau_i = \rho_i(1 - b_i)$ in which b_i is the blocking probability of the internal network from location i to location $\bar{i}$. τ_i is the overflow rate of type 2 call which pass TGi but cannot be served by the local service centre, $1/\gamma_i$ is the average ON time of the overflow IPP stream, and $1/\omega_i$ is the average OFF time of the overflow IPP stream. The overflow stream is also called $IPP2(i), i \in \{0,1\}$.

Let $(s; n_1, n_2)$ present that s calls are served by the operators in the service centre on location i, and $IPP1(i)$ and $IPP2(i)$ are in the system state with the combination of n_1 and n_2 states, where $0 \leq s \leq S_i$ and

$$n_1 = \begin{cases} 0 & \text{if } IPP1(i) \text{ is OFF,} \\ 1 & \text{if } IPP1(i) \text{ is ON,} \end{cases} \quad n_2 = \begin{cases} 0 & \text{if } IPP2(\bar{i}) \text{ is OFF,} \\ 1 & \text{if } IPP2(\bar{i}) \text{ is ON,} \end{cases}$$

Due to the two type 2 call streams can be represented by the independent IPP and their holding time are independent random variables with exponential distribution, we know the system is a continuous-time Markov chain with the system state $(s; n_1, n_2)$. Because the system is a continuous-time Markov chain, the steady state probability $\tilde{\Pi}$ matches the steady state equations and normalization condition. Fig.2 is the state transition relationship of IPP1 and IPP2 between the ON and OFF states when type 2 calls are served by the operators on location i. Therefore, different number of the type 2 calls served by the operator has itself ON and OFF state transition matrix $\overline{Q}$,

$$\overline{Q} = \begin{bmatrix} -\gamma_{\overline{i}} - \alpha_i & \gamma_{\overline{i}} & \alpha_i & 0 \\ \omega_{\overline{i}} & -\omega_{\overline{i}} - \alpha_i & 0 & \alpha_i \\ \beta_i & 0 & -\beta_i - \gamma_{\overline{i}} & \gamma_{\overline{i}} \\ 0 & \beta_i & \omega_{\overline{i}} & -\beta_i - \omega_{\overline{i}} \end{bmatrix}.$$

Every state has its own steady state probability vector $\tilde{\Pi}_{is} = [\tilde{\Pi}_{is0}\tilde{\Pi}_{is1}\tilde{\Pi}_{is2}\tilde{\Pi}_{is3}]$, where s is the number of the type 2 calls served by the operators, 0 represents both of the $IPP1(i)$ and $IPP2(i)$ are in ON state, 1 means $IPP1(i)$ is in the ON state and $IPP2(i)$ is in the OFF state, 2 is the opposite condition of 1, and 3 is in the situation that both of them are in OFF state. Thus, $\tilde{\Pi}_{isj}$ represents the steady state probability of the service centre on location i when s type 2 calls are served by the operators and $IPP1(i)$, $IPP2(i)$ are in the j state. From the relationship among different number of type 2 calls served by the operators, the state transition matrix of the whole system is

$$Q = \begin{bmatrix} \overline{Q} - \Lambda & \Lambda & & \\ \mu_2 I & \overline{Q} - \Lambda - \mu_2 I & \Lambda & \\ & & \cdots & \\ & & S_i\mu_2 I & \overline{Q} - S_i\mu_2 I \end{bmatrix}. \tag{9}$$

where

$$\Lambda = \begin{bmatrix} \rho_i + \tau_i & 0 & 0 & 0 \\ 0 & \rho_i & 0 & 0 \\ 0 & 0 & \tau_i & 0 \\ 0 & 0 & 0 & 0 \end{bmatrix},$$

and I is a 4×4 identity matrix. From the state transition matrix and normalization condition, the steady state probability vector of the system is

$$\tilde{\Pi}_i = [\tilde{\Pi}_{i0}\tilde{\Pi}_{i1} \cdots \tilde{\Pi}_{iS_i}],$$

in which $\tilde{\Pi}_{is} = [\tilde{\Pi}_{is0}\tilde{\Pi}_{is1}\tilde{\Pi}_{is2}\tilde{\Pi}_{is3}]$, $s = 0, \cdots, S_i$. By using the probability vector, the first two moments of the instantaneous arrival rate of the overflow IPP stream are [12]

$$\tilde{\alpha_1} = \lambda'_{2i}\Pi_{iA} \cdot e, \tag{10}$$

$$\tilde{\alpha_2} = (\lambda'_{2i})^2\Pi_{iA} \cdot e, \tag{11}$$

where Π_{iA} is the probability when the main IPP call stream is in ON state but all the operators are busy on the location i, i.e.

$$\Pi_{iA} = [0 \cdots 0\Pi_{iS_i0}\Pi_{iS_i1}00].$$

Therefore, the mean and variance of the instantaneous arrival rate of the overflow stream are $\tilde{\alpha}_1$ and $\tilde{\alpha}_2 - \tilde{\alpha}_1^2$. Let

$$\overline{\Lambda} = \begin{bmatrix} 0 & \\ & \Lambda' \end{bmatrix},$$

where 0 is a $4S_i \times 4S_i$ zero matrix,

$$\Lambda' = \begin{bmatrix} \rho_i & 0 & 0 & 0 \\ 0 & \rho_i & 0 & 0 \\ 0 & 0 & 0 & 0 \\ 0 & 0 & 0 & 0 \end{bmatrix},$$

and

$$\begin{aligned} m_1 &= \tilde{\alpha}_1, \\ v &= \tilde{\alpha}_2 - \tilde{\alpha}_1^2. \end{aligned}$$

Using Eq.5 again, substitute $\overline{\Lambda}, m$, and v into this equation and replace Π' by $\tilde{\Pi}$, Q' by Q. The time constant τ_c of the IPP overflow stream is obtained. Then, using the following three equation, the IPP parameters of the type 2 call stream passing the TGi and overflowing to the service center on location i can be obtained

$$\gamma_i = \tau_c^{-1}(1 - \tilde{\Pi}_{iS_i} \cdot e), \tag{12}$$

$$\omega_i = \tau_c^{-1}\tilde{\Pi}_{iS_i} \cdot e, \tag{13}$$

$$\tau_i = (1 - b_i)\lambda'_{2i}, \tag{14}$$

From the above calculation procedure, the IPP parameters of the overflow streams can be obtained. Then, iteratively feed the two IPP parameters into the procedure to get the new IPP parameters until they are converge. The blocking probability of the type 2 call stream caused by no available operators can be calculated by using the obtained IPP parameters.

4   Approximate Analytic Algorithm

In approximate solution we calculate the blocking probabilities of the type 2 call stream on each trunk group respectively, then combine the two blocking probabilities to get the total blocking probability of type 2 call stream. To represent conveniently, the whole algorithm is divided into three procedures. The first procedure is the algorithm to calculate the blocking probability of the network. In this algorithm it needs some parameters and values that get from the second and third procedures. The second one is used to get the IPP parameters of the type 2 call stream passing via the trunk group and the blocking probabilities of the type 2 calls caused by the trunk groups. The third procedure is the one calculating the parameters of the overflow IPP call stream and the blocking probabilities caused by the internal and destination blocks.

The blocking probability of type 2 call stream can be divided into four major parts: the blocking probabilities of the internal block with different directions, the blocking probability of the calls blocked by the trunk group, P_t, and the blocking probability caused by the destination block, P_d. In the approximate analytic solution the system is separated into several independent subsystems to get its blocking probability of the type 2 calls. The ON and OFF parameters of the IPP parameters imply the blocking probabilities of the type 2 call stream in every subsystem. So, the four blocking probabilities can be represented by the IPP parameters. Let P_{ai} be the blocking probability that the type 2 calls choose the alternative route entering from TGi but blocked by the internal network,

$$P_{ai} = \frac{\alpha_i}{\alpha_i + \beta_i} \frac{\gamma_i}{\gamma_i + \omega i} b_i p_i. \tag{15}$$

And define P_{ti} as the blocking probability of type 2 calls caused by TGi, and P_{di} as the blocking probability caused by the destination block. Their formulas are listed as follow:

$$
\begin{aligned}
P_{ti} &= \frac{\beta_i}{\alpha_i + \beta i} p_i, \\
P_{di} &= \frac{\alpha_i}{\alpha_i + \beta_i} \frac{\gamma_i}{\gamma_i + \omega i} (1 - b_i) p_i D_i,
\end{aligned}
$$

where $D_i = \tilde{\Pi}_{iS_i 0} + \tilde{\Pi}_{iS_i 2}$. Then, let

$$
\begin{aligned}
P_t &= P_{t0} + P_{t1}, \\
P_d &= P_{d0} + P_{d1}.
\end{aligned}
$$

From above equations the blocking probability of type 2 call stream is

$$B_2 = P_{a0} + P_{a1} + P_t + P_d.$$

The approximate analytic algorithm is

Procedure 1 Calculate the blocking probability of the type 2 call stream. Let η_i be the probability that type 2 calls pass TGi but cannot be served by the operators.

> **Step 1** Set $\eta_0 = 0$ and use Procedure 2 to get the IPP parameters $(\rho_0, \alpha_0, \beta_0)$ of the type 2 calls passing $TG0$.

> **Step 2** Set $\eta_1 = 0$ and use Procedure 2 to get the IPP parameters $(\rho_1, \alpha_1, \beta_1)$ of the type 2 calls passing $TG1$.

> **Step 3** By using $(\rho_0, \alpha_0, \beta_0)$, $(\rho_1, \alpha_1, \beta_1)$, and Procedure 3, calculate the blocking probability of type 2 calls, B_2, and η_0, η_1.

> **Step 4** Substitute η_0 and η_1 into Procedure 2 to obtain the main IPP parameters $(\rho_i, \alpha_i, \beta_i)$ of the type 2 calls on each trunk group, where $i \in \{0, 1\}$.

> **Step 5** Substitute the IPP parameters obtained from Step 4 into Procedure 3 to obtain the new values of B_2, η_0, and η_1.

> **Step 6** Repeat Step 4 and 5 until the difference of the two newest B_2 is less than a pre-setting value, T_h.

Procedure 2 Calculate the IPP parameters of the type 2 calls passing the TGi and the blocking probability of type 2 calls caused by the TGi, P_{ti}.

Step 1 Use the arrival rate of type 1 and 2 calls on the TGi, λ'_{1i} and λ'_{2i}, and their average holding time, $1/\mu_{1i}$ and $1/\mu_{2i}$ to obtain the state transition matrix $\overline{Q}$.

Step 2 Use the steady state equations Eq.1 and the normalization condition Eq.2 to get the steady state probability.

Step 3 Obtain the blocking probability of the type 2 calls on TGi from the steady state probabilities, then substitute the value into Eq.3 and Eq.4 to get the mean and variance of the instantaneous arrival rate of the type 2 calls passing the TGi.

Step 4 Use Eq.5 to obtain the time constant τ_c of the type 2 calls passing the TGi.

Step 5 Substitute the time constant into Eq.6 and Eq.7 to get the ON and OFF rate of the IPP call stream, α_i and β_i, then obtain the instantaneous arrival rate from Eq.8 and the blocking probability of the type 2 calls passing TGi, P_{ti}.

Procedure 3 Use two IPP call streams to obtain the overflow IPP parameters of the type 2 calls passing TGi, the blocking probability of type 2 calls caused by the internal network, P_a, and the blocking probability of type 2 calls caused by the destination block, P_d.

Step 1 Set $\tau_c = 0$, i.e. the overflow stream $IPP2(1)$ does not exit at first, then get the state transition matrix Q by Eq.3 and Eq.9.

Step 2 Get $\tilde{\Pi}_0$ from the state transition matrix and normalization condition.

Step 3 Use Eq.10 and Eq.11 to get the mean and variance of the instantaneous arrival rate, m_1 and v.

Step 4 Apply Eq.5 to get the time constant τ_c of the overflow stream from this location.

Step 5 Substitute the time constant τ_c into Eq.12 and Eq.13 to get the ON and OFF parameters of the overflow stream $IPP2(0)$, γ_0 and ω_0.

Step 6 Substitute the parameters of the $IPP1(1)$ and $IPP2(0)$ into Eq.3 and Eq.9 to get the state transition matrix, the follow the same procedure as Step2 to Step 5 to obtain the parameters of the $IPP2(1)$, $(\tau_1, \gamma_1, \omega_1)$, and the blocking probability of the type 2 calls caused by no idle operators, P_{d1}.

Step 7 Use the $IPP1(0)$ and $IPP2(1)$ as the arriving call streams and follow the same procedure as Step 6 to obtain the parameters of the $IPP2(0)$, $(\tau_0, \gamma_0, \omega_0)$, and P_{d0}.

Step 8 Repeat Step 6 and Step 8 until the differences of the two newest values of the P_{d0} and P_{d1} are smaller than the two pre-setting values, T_{h0} and T_{h1}, respectively.

Step 9 Apply Eq.15 to get the blocking probability of the type 2 calls caused by the internal network, P_{ai}, and let $\eta_i = P_{ai} + P_{di}, i \in \{0, 1\}$.

Fig.3, Fig.4, and Fig.5 are the flow charts of the three procedures. Applying these three procedures and the false position method, the optimization probabilistic parameter can be obtained.

5 Numerical Result

From above algorithm each time the dimension of the transition matrix is related to the service centres and is independent of the size of trunk groups for the call stream entering the service centres. And the dimension of the transition matrix of the Poisson stream entering the trunk group is related to the size of trunk group and is independent of the number of operators in the service centres. Thus, we can easily deal with a large network.

At first a symmetrical network is examined. In this network there are 4 links on each trunk group, $N_0 = N_1 = 4$, and 2 clients in each service center, $S_0 = S_1 = 2$. To obtain a better utilization of type 2 calls is the goal. Thus, false position method is applied to get near optimal solution. The arrival rate of type 1 calls on $TG0$ is a variable. The arrival rate on the other trunk group is 4 calls per minutes entering $TG1$. Every minute there are 8 type 2 calls entering the network. The holding time of all calls is average 1 minute. Thus, $\lambda_{11} = 4$ calls/min, $\lambda_2 = 8$ calls/min, and $1/\mu_{10} = 1/\mu_{11} = 1/\mu_2 = 1$ min/call. The blocking probabilities of internal network are 0.1, i.e. every ten calls enter the internal network there is one call blocked. Fig.6 is the comparison of the exact solution of the type 2 call blocking probability obtained by the false position method and the optimal solution of the approximate analytic method.

From the diagram the error between the exact and approximate solutions decreases when the the arrival rate of the type 1 call stream on the $TG0$ increases. From previous investigation [4] the blocking probability of the destination block is very large at first then gradually decreases. When the blocking probability of the destination block is a larger ratio to the other blocking situations, this approximate analytic method has a larger error with the exact solution. It is because that the blocking probability of the trunk group block is over-estimated. In the Procedure 2 the trunk group blocking probability is obtained by the competition of the arrival type 1 and 2 call streams. In practical the type 2 call will occupy both of the resources, a link and an operator, if it is setup. Thus, less type 2 call occupy the links on the trunk group. Although the iteration of the Procedure 2 tries to reduce the effect, it very hard to totally eliminate it.

After examining the symmetrical network, an asymmetrical topology is taken into consider. There are 4 and 5 links on $TG0$ and $TG1$, respectively, that is, $N_0 = 4$ and $N_1 = 5$. In the service center of location 0, three operators are allocated to serve type 2 calls and there is only one operator on the other service center, i.e., $S_0 = 3$ and $S_1 = 1$. The arrival rate of type 2 calls is a variable. The arrival rate of type 1 calls on each trunk group is set as: $\lambda_{10} = 4$ calls/min and $\lambda_{11} = 2.5$ calls/min. The average holding time of all calls is 1 min/call. Fig.7 is the comparison of the exact and analytic results under this network configuration. In this diagram the approximate solution is almost the same as the exact solution, but the error gradually increases. From the previous investigation this error is caused by the same reason of the symmetric network case.

6 Conclusion

In the approximate analytic method the network is divided into two separated subnetworks for calculating. Each subnetwork can be divided into two cascade stages. The first stage has two call streams which are independent Poisson processes and one of them will form a IPP call stream as the input of the second stage. Combining the IPP call stream and the overflow IPP call stream from another subnetwork as the inputs, the blocking probability of the destination block and the parameters of the overflow IPP call stream can be obtained. Then, using the gotten blocking probability, calibrate the input arrival rate of the first stage and calculate again. Each time the number of the states depends on the number of the servers. The number of states in the first stage is two times of the number of the links of the trunk group. In second stage the number of states is four times of the operators in each service centre. This algorithm reduces the size of the matrix significantly.

However, in the operation of the real network every type 2 call occupies a link on a certain trunk group and an operator in the service centers. If the call cannot be setup, this call will not occupy any resource in this network. Although the calculation of the IPP parameters of the first stage already exempts the factors of the internal and destination blocks, the error between the exact and approximate solutions is significant when the blocking probability of the destination block is significantly large. It is due to the subnetwork divided into two cascade stages. This approach cannot reflect the behavior that a type 2 call will occupy both of the resources, a link and an operator, or nothing. This algorithm under estimates the effect of the characteristic. Therefore, the convergence of the Procedure 3 is very fast, but the convergence of the Procedure 2 is very slow.

To get the optimization value of the probabilistic parameter the false position method is used. In this method the new probabilistic parameter is obtained from the slope of the blocking probability of the type 2 calls and the old probabilistic parameter. Therefore, every iteration to get a new parameter value needs two times of calculation of the Procedure 1. Every calculation of the Procedure 1 calls the Procedure 2 and the Procedure 3 several times. When the number of the operators in the service centres increases, every increasing operator causes 4 more states of the state space. When the number of the links pf the trunk group increases, every additional link adds 2 more staes of the state space. Therefore, the calculation time is very significant. However, this analytic method lets the calculation of the network model possible.

References

[1] W. D. Ambrosch, A. Maher, and B. Sasscer, *The Intelligent Network*. Berlin: Springer-Verlag, 1989.

[2] P. H. Vapheasky, A. M. Gopin, and R. J. Wojcik, "Advanced intelligent network: Evolution," in *Proc. of IEEE ICC*, pp. 941–947, 1991.

[3] R. B. Robroc II, "The intelligent network – changing the face of telecommunication," *IEEE proc.*, vol. 79, pp. 7–20, Jan. 1991.

134

[4] H.-M. Chen, Z. Tsai, J. Mellor, and P. Mars, "Performace analysis of intelligent network with two interconnected service centres," in *Proceedin of the 3rd International Workshop on Queueing Networks with Finite Capacity* (D. Kouvatsos and Y. Dallery, eds.), pp. 31/1–31/7, 1995.

[5] A. Kuczura, "The interrupted poisson process as an overflow process," *Bell System Tech. J.*, vol. 52, pp. 437–448, Mar. 1973.

[6] A. Kuczura and D. Bajaj, "A method of moments for the analysis of a switched communication networks performance," *IEEE Trans. Commun.*, vol. COM-25, pp. 185–193, Feb. 1977.

[7] D. R. Manfield and T. Downs, "Decomposition of traffic in loss systems with renewal input," *IEEE trans. Commun.*, vol. COM-27, pp. 44–58, Jan. 1979.

[8] K. Akimaru and H. Takahashi, "An approximate formula of estimation individual call losses in overflow systems," *IEEE Tran. Commun.*, vol. COM-31, pp. 808–810, June 1983.

[9] A. A. Fredericks, "Aprroximating parcel blocking via state dependent birth rates," in *Proc. 10th I.T.C.*, (Montreal, P.Q., Canada), 1983.

[10] J. Matsumoto and Y. Watanabe, "Individual traffic characteristics of queueing systems with multiple poisson and overflow inputs," *IEEE Trans. Commun.*, vol. COM-33, pp. 1–9, Jan. 1985.

[11] K. Kawashima, "Trunk reservation models in telecommunications systems," in *Teletraffic Analysis and Computer Performance Evaluation*, Elsevier, 1986.

[12] K. S. Meier-Hellstern, "The analysis of a queue arising in overflow models," *IEEE Trans. Commun.*, vol. 37, pp. 367–372, Apr. 1989.

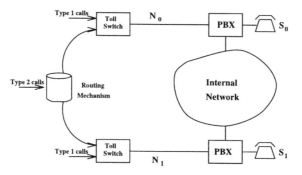

Figure 1: Network topology

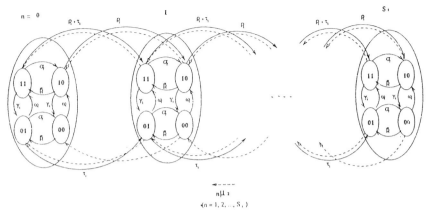

Figure 2: IPP state transition diagram.

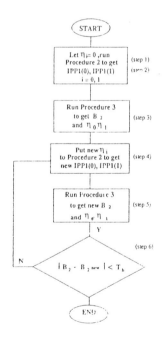

Figure 3: Flow chart of the Procedure 1.

136

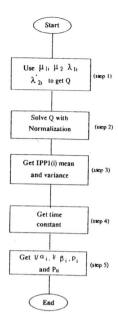

Figure 4: Flow chart of the Procedure 2.

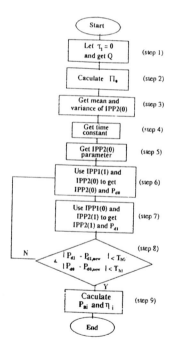

Figure 5: Flow chart of the Procedure 3.

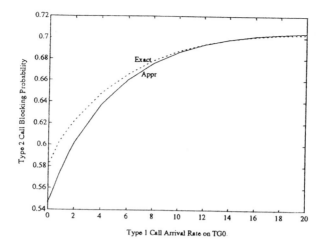

Figure 6: The results of the exact and approximate solutions under a symmetrical network configuration.

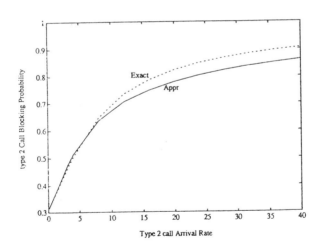

Figure 7: The results of the exact and approximate solutions under an asymmetrical network configuration.

"General" marking-dependent rates and probabilities in *GSPNs*

Isabel Rojas[*]

Department of Computer Science, University of Edinburgh

Scotland, U.K.

Abstract

General marking dependency in Generalised Stochastic Petri Nets (*GSPNs*) has been limited because of the drawbacks that can it can cause, such as *stochastic confusion* and discrepancy with the underlined untimed PN model. Rates of a timed transition can only be made dependent on its input places and probability of an immediate transition on places of its Extended Conflict Set (*ECS*). We show, using a *GSPN* model of the Two-Phase Locking concurrency control mechanism, that in some models, in order to accurately reflect the behaviour of the system modelled, it is useful, if not necessary, to allow the global state of the system to determine transition probabilities and rates.

1   Introduction

When *GSPNs* where originally introduced in [1] no restrictions were established for the characteristics or circumstances under which marking-dependent transition rates and probabilities could be defined. The possibility of constructing models in which a single transition represents a whole complex subsystem whose rate or probabilistic behaviour depend in a complex way on the population inside the subsystem, is considered the main advantage of allowing "general" dependence of rates and probabilities on markings of the Petri net model. Another advantage is that when having probabilities or rates that truly depend on the behaviour of the system in "general", they can be represented in the *GSPN* model, obtaining potentially more accurate representations of the system.

Marking-dependent rates and probabilities, however, have several drawbacks: they permit the behaviour of the model to change with respect to the associated untimed net; and the actual semantics of the model is completely defined only by the reachability graph plus the marking-dependent rates. To overcome these problems, Chiola *et al.* propose a series of restrictions on the type of marking-dependency that is allowed in *GSPNs* [2].

We acknowledge the drawbacks of allowing "general" marking-dependency of the rates and probabilities of timed and immediate transitions, respectively. However, in this paper we show that in some cases it is convenient (if not necessary), to offer these features in order to obtain a model that reflects accurately the behaviour of the system. We present an example of such a system,

---

[*]The author is supported by a grant from the Venezuelan Government Research Council (CONICIT)

showing the differences that marking-dependency can make, suggesting that it is important to consider the need of "general" marking-dependent rates and probabilities, before discarding the notion.

The rest of this paper is structured as follows: in the second section the relevant features of the *GSPN* model and the problems of allowing marking-dependency are presented. Then in Section 3 we present a *GSPN* model of the Two-Phase Locking concurrency control mechanism. We then, in section 4, point out the limitations of the *GSPN* model presented and reflect the improvements that can be made when allowing "global" state marking-dependence of the timed transitions rates and immediate transitions probabilities. Finally, we draw some conclusions and discuss the need for general or extended marking-dependency in *GSPNs*.

2   Marking-dependent rates and probability values of transitions in *GSPNs*

2.1   Relevant *GSPN* concepts

In this section we will assume that the reader is familiar with the Petri net (PN) formalism. We will only refresh the concepts relevant to the problem of general marking dependency in *GSPN*. Readers unfamiliar with PN concepts are referred to [3].

Intuitively a *GSPN* is a Timed Petri net where we distinguish two types of transitions: *timed* and *immediate*. Timed transitions are associated with a rate which determines the delay associated with the transition. This delay is exponentially distributed with mean 1/rate. Immediate transitions have delay zero and have an associated probability.

The rate of a timed transition or the probability of an immediate transition can be a fixed value or a function of the parameters of the model, states of the model or both.

Timed transitions are associated with priority zero, whereas all other priority levels are reserved for immediate transitions. When two or more timed transitions are enabled the transition with the shortest delay will fire first (this does not necessarily mean the transition with the smallest rate). In the case of immediate transitions this is not the case because they have zero delay. The decision of which transition would fire is made according to the probability and priority level of the immediate transitions enabled. If an immediate transition t_i is enabled with another transition t_j of a higher priority, than t_i can not fire. Within the set of higher priority enabled transitions, the choice probability is used to determine which transition will fire.

2.2   Extended Conflict Set (*ECS*)

When the firing of a transition t_j affects the enabling of another transition t_i previously enabled, we say that t_j is in conflict with t_i. The enabled condition of a transition t_i can be affected by a transition t_j, if t_j shares input places with t_i or if its set of output places intersects with the set of inhibiting place of t_i. This is known as structural conflict (*SC*) or direct structural conflict.

The opposite notion of this is the concept of causal connection, which states that the firing of a transition t_j can determine the enabling of another transition t_i that was not previously enabled. In a net with multiple levels of priority, more complex situations of conflict may arise. The enabled condition of a transition t_i may not be affected by the firing of a transition t_j that is in SC with it, but indirectly by a third transition t_k that is causally connected with t_j.

Intuitively we can define the *Extended Conflict Set* (*ECS*) of an immediate transition t_i as an equivalence class formed by the set of immediate transitions that can be concurrently enabled with t_i, have the same priority level of t_i, and that are in SC or in indirect conflict with t_i. In another words, it is the set of immediate transitions that can affect the enabled condition of t_i in a direct or indirect manner (through the firing of other immediate transitions). For formal definition of an *ECS* see [3].

2.3 Confusion

In a general Petri net the resolution of a conflict may depend on the firing sequences of transitions that are not in conflict with each other. This problem is known in the literature as *confusion*, and appears when the firing of a transition t_i, that is not in the *ECS* of a given transition t_k, enables a third transition t_m that belongs to $ECS(t_k)$. This implies that in a confusion free net, the resolution of conflicts is completely determined by single *ECS*s, while in confusion, the specification at the level of *ECS*s may not be sufficient to determine the net's behaviour [4].

2.4 Marking-dependent Rates and Probabilities

Definition 1 *When the rate of a timed transition, or the probability of an immediate transition, is a function of the state or marking of a subset P' of places ($P' \subseteq P$), it is said that it is* marking-dependent.

When *GSPNs* were first introduced [1], marking-dependence was permitted over the markings of any subset of places in the net, without restriction. As mentioned, the main advantage of this feature was considered to be the fact that it permits the compact representation of complex models.

However, the incorporation of marking-dependent rates and probabilities has several drawbacks. As pointed out in [2], the behaviour of the model can change with respect its underlying untimed net. The net can be seen only as an abbreviated description of the state space, and the actual semantics of the model is completely defined only by the reachability graph plus the dependent rates. Chiola *et al.* define two levels of coherence between the stochastic model and the underlying *PN* [2]:

Minimal Coherence: general marking-dependency is allowed, provided that no firing rate or probability is null. With this condition it is guaranteed that all states reachable in the underlying untimed Petri net model can be reached with non-zero probability in the stochastic model.

Complete Coherence: the parallelism of the net is completely reflected in the timing semantics through the use of "infinite-server" semantics. The rates of timed transitions are allowed to depend on their enabling degree through a scaling factor. In this manner the specification of the model can be performed by associating, with the transitions that have marking-dependent behaviours, a attribute that depends only on the state of their input places.

In the case of immediate transitions a general marking-dependency for transitions firing weights inside *ECS*s, may lead to "stochastic confusion", i.e., to improperly defined stochastic models. However, we suggest that in some models, confusion is required or forms part of the model.

In order to prevent stochastic confusion, two sufficient conditions over the definition of marking-dependent probability functions of immediate transitions are proposed in [2]. These are: first, limit dependency to places whose marking cannot be modified by transitions of priority greater than or equal to that of the considered transition; second, allow dependency from the set of transitions that are actually enabled within the considered *ECS*.

Ajmone Marsan *et al.* in [3] consider that in practice the generality allowed by "general" marking-dependency is in contrast with the claim that *GSPN*s are a high-level language for the description of complex systems. Indeed, while the construction of a *GSPN* model requires a local view of the behaviour of the real system, the specification of (general) marking-dependent parameters requires the analyst to be aware of possible (global) states of the system.

An important advantage of allowing "general" marking-dependency is that, in some models a general state of the system is required in order to accurately define the rates and probabilities of the transitions, however, this is not always taken into account. The marking of places that are not inputs to a transition, may affect the value of its rate or probability for many possible reasons, for example resource sharing. The amount of time required to process an element may be affected by the number of other tasks that the processing unit has to perform.

We will present in the following section an example in which we show the importance of allowing "general" marking-dependency, in order to obtain a accurate model of a system's behaviour.

3   A *GSPN* model for the Two-Phase Locking Protocol

3.1   The Two-Phase Locking Protocol

The *Two-Phase Locking protocol (*2PL*)* for transaction database systems is used to ensure the consistency of the database, whose data may be accessed concurrently by a certain number of transactions. A transaction consists of a sequence of access operations to the data (data items). Before accessing a data item the transaction must acquire a lock over it (either read or write lock), according to the operation to be performed. The data items are grouped in granules, which are the units of physical locks. A transaction must release all the locks it owns when it terminates, and cannot acquire any locks once it has started releasing others. This guarantees serialise-ability. An unfortunate property of the *2PL* protocol is that transactions can deadlock.

Deadlocks are detected by most database systems by keeping a *"wait-for-graph" (WFG)*, in which the nodes represent the transactions concurrently active in the system. An arc from a transaction Tr_i to Tr_j, represents the fact that Tr_i is waiting for a lock that is held by Tr_j. Deadlocks are formed by cycles in the graph. A deadlock situation is resolved by aborting a transaction within the cycle. In many systems the transaction selected as a "victim" is the one that finally forms the cycle. Other systems consider as the victim the transaction, within the cycle, who has acquired the lowest number of locks; since it would cost less to restart such a transaction. When a transaction aborts, all the locks that it held are released.

In the *2PL* database model that we present, we consider the *"claim as needed"* locking policy. Under this policy locks are requested by a transaction one by one on demand, and there is no ordering to the granules which the lock requests have to follow. Figure 1 show a diagram of the behaviour of the *2PL* protocol under the "claim-as-needed" policy. We will consider our system to be a Single Site database System (SDBS).

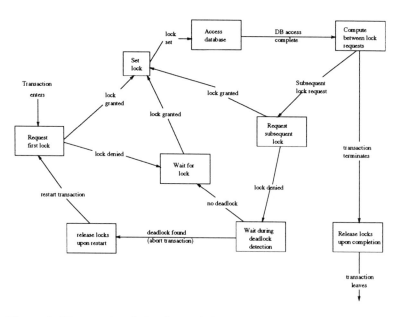

Figure 1: Transaction behaviour of the *2PL* under "claim-as-needed"

With respect to the accessing behaviour of the data items, we will consider the *"well placed"* behaviour, where the data items referenced by a transaction are packed into as few granules as possible. This assumes that the transaction accesses the data items sequentially and that the granule boundaries are optimally located for the transaction. If TS is the transaction size (number of data items referenced), DS and NG the database size and the number of granules in the entire database, respectively, then the number of data items in each granule is $G = \lceil \frac{DS}{NG} \rceil$; and the number of locks required by the transaction is $NL = \lceil \frac{TS}{G} \rceil$. We assume that all transactions require, on average, the same number of locks.

Most of the literature on models for the *2PL* protocol [5, 6, 7], estimate a priori several parameters of the model which are assumed to be constant values:

- the probability that a transaction's first or subsequent lock is not granted,
- the probability that a transaction is deadlocked, and
- the probability that a transaction will require a subsequent lock.

The authors claim that these parameters can be estimated fairly accurately by first estimating the number of locks owned by active transactions in the system and then computing the probability that an event will occur. Nevertheless, these parameters are normally estimated by making assumptions such as: the number of locks per lock owner (transactions that have already acquired locks) is equal to the half of the locks that they are expected to require.

To solve our models we have employed SPNP (Stochastic Petri Net Package), created G. Ciardo, J. Muppala and K. Trivedi at Duke University [8]. This package allows the definition of "general" marking-dependent rates and probabilities.

3.2 A *GSPN* model for the *2PL* protocol

The *GSPN* model presented in this section, is based on the work in [5]. Following the diagram in Fig. 1, we can easily obtain a *GSPN* model of the *2PL* protocol under the "claim-as-needed" policy. Without loss of generality we will make the following assumptions:

- When a transaction terminates or aborts, it is immediately replaced by another transaction, in order to maintain a constant total number of transactions in the system.
- Transactions are indistinguishable
- Each data item in the database occupies exactly one physical block.
- Each lock is an exclusive lock (the model can be extended to distinguish among read and write lock requests).
- The system has infinite resources, i.e., when a transaction is ready to execute it can always go ahead.

The model was designed in order to obtain: the system's throughput, i.e., throughput of successfully completed transactions; and the utilisation of CPU for the useful computations on data items

Fig. 2 shows the *GSPN* model for the *2PL* protocol. Transactions are represented by tokens. When a transaction enters the system (represented by a token in place P_1), it requests its first lock. This request is processed (transition T_1), and the lock is granted with probability P_{g1} or the transaction is blocked to wait for the lock with probability $1\text{-}P_{g1}$.

The value P_{g1} is estimated using the scheme proposed by Pun & Belford in [5]. They assume when a transaction requests a lock: 1) each lock owner has on average acquired half of its locks ($\lfloor NL/2 \rfloor$ locks; denoted HL), and 2) the number of lock owners (NLO) in the system is the maximum possible limited only by N (number of transactions in the system) and NG ($NLO = \min(N -$

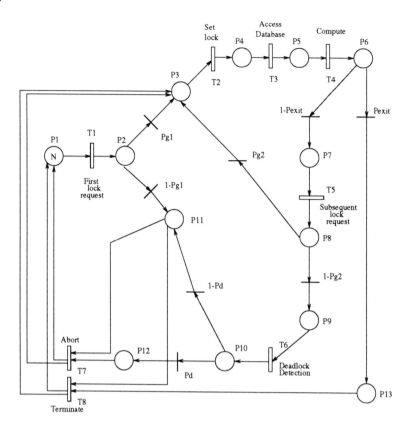

Figure 2: Petri Net model of the 2 Phase Locking Protocol

DS	Number of data items in the database
TS	Number of data items in a transaction
G	Number of data items controlled by a lock
NG	Total number of locks in the system ; $NG = \lceil \frac{DS}{G} \rceil$
$S_{cpu,lreq}$	CPU time required for processing a lock request.
$S_{cpu,lset}$	CPU time required for setting a lock.
$D_{cpu,deadlock}$	CPU time required for deadlock detection algorithm.
$D_{cpu,lrel}$	CPU time required for releasing locks.
S_{cpu}	CPU time required for processing a data item.
S_{disk}	Disk drive time required required per lock.

Table 1: GSPN model Input Variables

$1, \frac{NG}{[NL/2]}$)). To calculate P_{g1} they then have: $P_{g1} = 1 - \frac{NLO*HL}{NG}$. $NLO*HL$ is the number of locks that have been allocated, under the assumption that each lock owner has half of its locks. Dividing this by NG (the number of granules in the database) gives us the percentage of locks assigned. The probability that the lock requested is free is 1 minus this quantity.

If a transaction is granted its lock, it will proceed to process the data item (set lock, access database and CPU processing) represented by transitions T_2, T_3 and T_4. If the lock is denied, it will go to place P_{11} where it will wait until a transaction finishes or aborts, thus liberating the lock required.

Once the lock has been processed, the transaction may or may not require a subsequent lock. This probability (P_{exit}) is considered in [5] to be fixed. They assume that the probability that the transaction requires no more locks is $1/NL$. This is based on the fact that the lock just processed could be any of the NL locks that a transaction will require. If the transaction does not require a subsequent lock, then the transaction will terminate (T_8) and liberate its locks, placing a blocked transaction (if any) from P_{11} into P_3, thus converting it into a lock owner; and by placing a token in P_1 the number of transactions in the system is preserved.

If the transaction requires a subsequent lock (place P_7), it has a probability of P_{g2} that this lock will be assigned. The value of this probability is also considered by [5] to be fixed and calculated by making the same assumption as for P_{g1}, i.e. the number of locks per lock owner is $NL/2$, thus: $P_{g2} = 1 - \frac{NLO*HL}{NG-HL}$.

If the lock is assigned the transaction follows the data processing tasks. If the lock is denied, the transaction is studied to see if blocking it will cause a deadlock in the system. The probability that a deadlock is formed (P_d), is determined by the probability that a cycle is formed in the "wait-for-graph". This is also considered to be fixed. We will not go into detail of how this factor is estimated, and interested readers to [5]. If a deadlock is detected a transaction is aborted. Which transaction depends on the "victim selection" policy. Since transactions are indistinguishable and the number of locks assigned is unknown, it is irrelevant which transaction is taken. Following the assumptions made, the "victim" will have half of the total number of locks it requires ($NL/2$). When a transaction aborts it liberates its locks, consequently enabling a blocked transaction, which is converted into a lock owner by placing a token in place P_3. As in the case of a terminating transaction, a new transaction now enters the system.

Table 1 shows the input variables for the $GSPN$ model. The rate functions of the timed transactions are presented in Table 2. The values of the rates are determined by the size of the transaction and the number of locks a transaction will require (See table 4).

4 Limitations of the $GSPN$ model proposed

One factor which is not reflected in the $GSPN$ model presented, is the fact that the CPU time must be shared among transactions executing different tasks:

- transactions requesting locks,
- transactions setting locks,

Transition	Rate Function
T_1	$1/S_{cpu,lreq}$
T_2	$1/S_{cpu,set}$
T_3	$1/S_{disk}$
T_4	$1/S_{cpu}$
T_5	$1/S_{cpu,lreq}$
T_6	$1/D_{cpu,deadlock}$
T_7	$1/D_{cpu,lrel}$
T_8	$1/D_{cpu,lrel}$

Table 2: Timed Transition Rates in the *PN* model with no marking-dependency

- the execution of the deadlock detection algorithm and
- releasing the locks associated to a transaction.

Without allowing "general" marking-dependency, this factor cannot be accurately reflected in the model, because we can only make the rates of these tasks dependent on the state of their input places and/or the parameters and input values of the system. When one of the timed transitions associated with one of the tasks is enabled, the only elements that can reflect the level of CPU sharing would be N (the number of transactions in the system) and the number of transactions (tokens) that are in the input places of the transition. In order to estimate the level of CPU sharing we would have to assume either that: all other transactions (N - number of tokens in the input place of the transition) are executing or requesting CPU processes, none of them are, or a certain proportion of them are. Depending on these assumptions the results could be quite different. Moreover, none of them accurately reflects the situation.

Another possible way of incorporating the CPU sharing factor could be by adding an extra transition representing the CPU service, having an input place for each type of task requesting CPU time or a single input place and a colour set to distinguish among the different types of tasks. In this way the rate of the transition could be dependent on the input places. The problem would come in determining in which order the tasks would execute. We would have to be assume that several transactions can execute simultaneously (multiple server) and determine in some way the rate of the combined process or sequentially ordered (one server). This does not reflect the interleaving behaviour obtained when several transitions are enabled simultaneously and the delay is calculated to see which transition will fire.

4.1   Improving the *GSPN* model of the *2PL* protocol

When we allow "general" marking-dependent rates, the CPU sharing factor can be incorporated very easily and accurately, knowing at each state the number of transactions requesting CPU service. We introduce an adjustment factor to each of the transitions associated with the CPU, which is dependent on the state of all the other transactions sharing CPU time. For example the rate of T_1 would be multiplied by $\#P_1/\#(P_1 + P_3 + P_5 + P_7 + P_9 + P_{12} + P_{13})$ (where

$\#P_1$ represents marking of place P_1 and $\#(P_1+P_3+P_5+P_7+P_9+P_{12}+P_{13})$ the markings of all the mentioned places), see Fig. 3.

Transition	Rate Function
T_1	$\frac{1}{S_{cpu,lreq}}*\#(P_1)/\#(P_1+P_3+P_5+P_7+P_9+P_{12}+P_{13})$
T_2	$\frac{1}{S_{cpu,set}}*\#(P_3)/\#(P_1+P_3+P_5+P_7+P_9+P_{12}+P_{13})$
T_3	$\frac{1}{S_{disk}}$
T_4	$\frac{1}{S_{cpu}}*\#(P_5)/\#(P_1+P_3+P_5+P_7+P_9+P_{12}+P_{13})$
T_5	$\frac{1}{S_{cpu,lreq}}*\#(P_7)/\#(P_1+P_3+P_5+P_7+P_9+P_{12}+P_{13})$
T_6	$\frac{1}{D_{cpu,deadlock}}*\#(P_9)/\#(P_1+P_3+P_5+P_7+P_9+P_{12}+P_{13})$
T_7	$\frac{1}{D_{cpu,lrel}}*\#(P_{12})/\#(P_1+P_3+P_5+P_7+P_9+P_{12}+P_{13})$
T_8	$\frac{1}{D_{cpu,lrel}}*\#(P_{13})/\#(P_1+P_3+P_5+P_7+P_9+P_{12}+P_{13})$

Table 3: Timed Transition Rates considering general marking-dependency

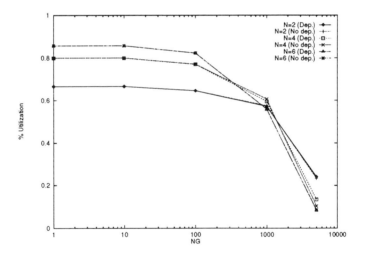

Figure 3: Results obtained for SZtr=250 and SZdb=5000

To show the difference that this factor has on the model's behaviour, we have run the model presented in the previous section (non-dependent model), along with the modified version proposed in this section (dependent model). For both models we made runs employing a small transaction size ($TS=5$) and a large transaction size ($TS=250$). Figures 3 and 4 show these cases with a database size of 5000 granules ($DS=5000$), comparing the results obtained for both models. The values used for the input parameter are shown in Table 4.

$DS = 5000$ and 10000
$TS = 5$ and 250
$S_{cpu,lreq} = 0.005\ sec$
$S_{cpu,lset} = 0.005\ sec$
$D_{cpu,deadlock} = 0.01 * N\ sec$
$D_{cpu,lrel} = 0.01 * N\ sec$
$S_{cpu} = (0.05 * TS)/NL\ sec$
$S_{disk} = (0.05 * TS)/NL\ sec$

Table 4: Model Input Values

The main difference can be observed when the transaction size is small. This is because the number of locks required is also smaller in general, thus the quantities are more sensitive to the factor incorporated. Studying the model with DS=10000 and TS=5, we observe (Fig. 5) that the influence of the dependent factor is clearly visible. The non-dependent model implies that as the number of locks per transaction increases the CPU utilisation decreases at a constant rate, independent of the number of active transactions in the system. However, the model with marking-dependent factors, shows how the decreasing rate is more acute while the number of transactions (N) increases, coinciding with the results obtained by [5, 6, 7].

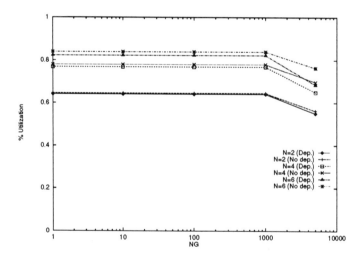

Figure 4: Results obtained for SZtr=5 and SZdb=5000

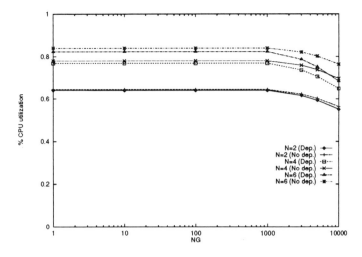

Figure 5: Results obtained for SZtr=5 and SZdb=10000

4.2   Incorporating marking-dependent probabilities

In the *GSPN* models presented so far, several assumptions are made in order to estimate the probabilities that determine the behaviour of the system. For example, it is assumed that when a transaction requests a lock all other transactions that have locks have half of the total number they will require. This assumption is taken in many analytical models of *2PL* [5, 6, 7], based on the fact that the number of locks required by a transaction is very small with respect the number of locks in the system. In most analytical models it is not possible to determine how many locks have been assigned or how many lock owning transactions are blocked. The number of lock owners is estimated by $min(N - 1, NG/\lfloor NL/2 \rfloor)$.

We can precisely know how many lock owners there are in the system at any state by incorporating a place (P_{15}) into the *GSPN* model proposed (see Fig. 6). Here we will keep the transactions that were blocked when requesting a subsequent lock. In this way we can distinguish among the transactions blocked when requesting the first lock (not lock owners) and transactions blocked in subsequent lock requests (lock owners). Adding the number of tokens in P_{15} to the ones in the input places of $T_2, T_3, T_4, T_5, T_6, T_7$ and T_8 we obtain the precise number of lock owners (NLO) in the system in any given state.

In order to use this value as an input to the probability functions that require it, we need to allow "general" marking-dependence in the net. This dependency on NLO breaks both the conditions proposed in [2] and outlined in Section 2. Dependency of transitions will be on places whose marking can be modified by transitions with priority equal to the one of the considered transition. Moreover, the dependency is over places modified by transitions in different *ECS*s. For example, to calculate the probability P_{g1} we require the value NLO, which is calculated using the marking of place P_9 that is determined by the firing of the transition with probability $1 - P_{g2}$, which is in a different *ECS* and has the same probability as the transition with probability P_{g1}.

With the introduction of place P_{14} (see Fig. 6), we can register the number of locks that are busy in the system (TNL). A token is placed when a transaction is granted a lock, and the average number of locks per lock owner is retrieved when a transaction terminates or aborts, thus P_{14} acts like a lock table (only registering number of locks assigned).

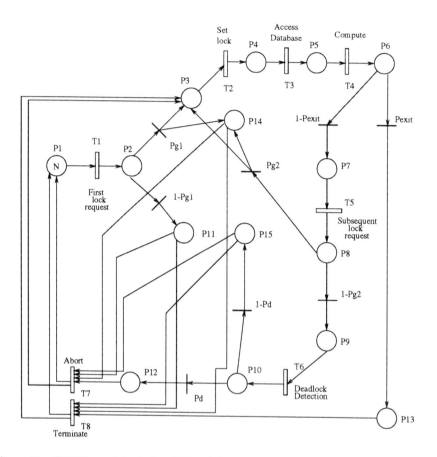

Figure 6: *GSPN* model of the *2PL* with "general" marking-dependent rates and probabilities

Knowing the number of locks assigned and the number of lock owners we can obtain the average number of locks per transaction. This has similar implications as the estimation of NLO, the probability of the immediate transitions will depend of the marking of places that have been changed by the firing of transitions within another *ECS*. We must, however, be aware of possible null probabilities, this can be avoided by incorporating an enabling function [8] over the immediate transitions, such that no transition with probability 0 can be enabled.

Using the values described, P_{g1} can be calculated by $P_{g1} = 1 - \frac{TNL}{NG}$, and P_{g2} would be redefined as $P_{g2} = 1 - \frac{TNL}{NG-(TNL/NLO)}$; where TNL/NLO gives an approximation of the number of locks estimated per lock owner.

The probability that a transaction has required all the locks it needs (P_{exit}), is considered to increase while the average number of locks per lock owner increases, thus $P_{exit} = \frac{TNL}{NLO*NL}$.

The probability that a blocked transaction forms a deadlock, is estimated following the same approach as in [5], but using the values and probabilities proposed. It is considered as sufficient to estimate the probability of creating a cycle of length two [9].

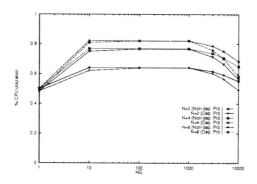

Figure 7: Results obtained when including marking-dependent immediate transitions probabilities (DS=10000 and TS=5)

The main disadvantage of this modifications is that the number of markings in the reachability graph of the $GSPN$, increases dramatically when incrementing the number of transactions in the system and the number of locks required per transaction. The improvement in the results obtained can be seen in Fig.7. Although the difference is very small in the NL=1 region, when NL increases the results diverge. The results obtained with the marking-dependent probabilities approximate better the simulation results obtained in [5].

5   Conclusions

We have observed how in some systems it desirable to allow marking dependent probabilities and rates of immediate and timed transitions, respectively, in order to obtain a more accurate description of the system's behaviour. It can be argued that the improvements obtained by adding marking-dependent rates and probabilities do not justify the means, especially in the case of marking-dependent probabilities, where the size of the net grows dramatically. However, in some cases the results obtained with a model that does not allow "general" marking dependency can lead to false conclusions. In the non "general" marking-dependent $GSPN$ model of the $2PL$ protocol it seems reasonable to conclude that the rate of decrease of the CPU utilisation is constant, independent of the number of transactions in the system. However, the marking-dependent modification shows, as other models do [5, 6], that the rate of decrease of the CPU utilisation increases with the number of transactions in the system.

The explosion of the state space when allowing "general" marking-dependent immediate transition probabilities, could be reduced by employing techniques such as concurrent firing of *ECS*s as proposed in [4]. To avoid stochastic confusion we can assume that transitions of different *ECS*s that are concurrently enabled can view a common state of the system and then fire simultaneously and independently within their *ECS*s, without seeing the intermediate firing steps of other *ECS*s.

More study needs to be devoted to determining under which circumstances this dependency is appropriate rather than directly limiting the dependency over the set of input places. Marking-dependency allows an easy incorporation of factors that depend on the "overall" state of the system, but as pointed out may cause several drawbacks.

References

[1] M. Ajmone Marsan, G. Balbo, and G. Conte. A class of generalized stochastic Petri nets for the performance analysis of multiprocessor systems. *ACM Trans. on Comp. Sys.*, 2(1), May 1984.

[2] G. Chiola, M. Marsan Ajmone, G. Balbo, and G. Conte. Generalized Stochastic Petri Nets: A Definition at the Net Level and Its Implications. *IEEE Trans. on Soft. Eng.*, 19:89–107, February 1993.

[3] M. Ajmone Marsan, G. Balbo, G. Conte, S. Donatelli, and G. Franceschinis. *Modeling with Generalized Stochastic Petri Nets.* John Wiley & Sons Ltd., 1994.

[4] G. Balbo, G. Franceschinis, and G. Molinar Roet. On the Efficient Construction of the Tangible Reachability Graph of Generalized Stochastic Petri Nets. In IEEE Comp. Soc. Press, editor, *Proc. 2nd Int. Workshop on Petri Nets and Performance Models*, pages 136–145, Madison, U.S, 1987.

[5] K.H. Pun and G. Belford. Performance Study of Two Phase Locking in Single-Site Database Systems. *IEEE Trans. on Soft. Eng.*, SE-13(12):1311–1328, December 1987.

[6] I.K. Ryu and A. Thomasian. Analysis of Database Performance with Dynamic Locking. *J. Assoc. of Comp. Machinery*, 37(3):491–523, July 1990.

[7] A. Thomasian. Two-Phase Locking Performance and Its Thrashing Behavior. *ACM Trans. on Database Systems*, 18(4):579–625, December 1993.

[8] G. Ciardo, J. Muppala, and K.S. Trivedi. Manual for the SPNP package version 3.0. Duke University, Durham, NC - U.S.A., May 1991.

[9] C. Devor and C. Carlson. Structural locking mechanisms and their effect on database management system performance. *Info. Sys.*, 7(4):345–358, 1982.

Negative Customers Model Queues with Breakdowns

Peter G. Harrison [*]

Imperial College

London

Edwige Pitel [†]

IRISA, Rennes

France

Naresh M. Patel

Tandem Computers Inc., Cupertino, CA

U.S.A.

Abstract

Negative customers in queueing networks are used to model break-downs at a server which cause some customers to be lost. We consider an unreliable M/M/1 queue with both instantaneous and exponential repairs, and derive expressions for the Laplace transform of the sojourn time density. We apply the model to approximate the performance of a bank of parallel, unreliable servers by modifying the arrival process to a Markov Modulated Poisson Process and considering an approximate decomposition of the underlying Markov chain under the assumption that arrivals and service occur much faster than breakdowns and repairs. Validation is by simulation.

1 Introduction

We apply the theory of G-queues—queues with negative customers introduced by Gelenbe [1]—to approximate the behaviour of networks of queues with break-downs. A breakdown is represented by a negative customer arriving at a queue and removing a random number of customers, if any are present. Performance is characterised in terms of the rate at which tasks are lost, together with the response time of successful tasks.

In the next section, we model an M/M/1 queue with breakdowns and instantaneous repairs which is relevant in systems such as telephone networks, where a "failure" is interpreted as the loss of a call, or the packet-switched network discussed above. The analysis is then extended in section 3 to the case of exponential repair times which provides a more conventional model of unreliable systems and a variation on the contemporary notion of G-queues. In section 4, this model is applied to predict the performance of a bank of

---

[*]Research partially supported by the European Commission under ESPRIT BRA QMIPS n° 7269 and the EPSRC of the UK under Research Grant n° GR/H 46244.

[†]Research supported by the European Commission under ESPRIT bursary n° ERBCH-BICT920179 during the author's PhD at Imperial College.

parallel, unreliable processors in which work sent to a broken processor is redirected randomly to an operational one, if any, and lost otherwise. Each of these queues has arrivals from a Markov Modulated Poisson Process, the underlying continuous time Markov chain describing the evolution of the number of (other) broken servers. We solve this queue approximately, under the assumption that arrivals and service occur much faster than breakdowns and repairs, by decomposition. In section 5, numerical validation with respect to simulation suggests that the approximation is accurate, especially for non-heavily utilised servers. We conclude in section 6 where we indicate how to extend our approach to arbitrarily connected Markovian queueing networks with breakdowns.

2 Breakdowns with instantaneous repair times

We model an M/M/1 queue with breakdowns and instantaneous repairs via an M/M/1 G-queue with batch killings. The batch size, a positive integer-valued random variable, represents the number of tasks lost upon failure. Tasks are killed at the front of the queue under RCH killing strategy ("Remove the Customer at the Head") and at the rear under RCE killing strategy ("Remove the Customer at the End"). One example involving instantaneous repairs is provided by telephone lines which are subject to call losses without any physical breakdown, when they are all busy, since they continue to transmit calls without delay. Both RCH and RCE strategies are relevant. In a communication network in which messages are transmitted in a packet-switching mode, suppose one server fails during a transmission, losing one part of a message. Some other packets have already been transmitted successfully to the next queue, others are left behind in the broken queue. The partial loss of this message makes it unsuitable for further use and we decide therefore to delete what is left in both queues. For the broken server, we use RCH and for the downstream server, RCE.

Consider therefore the M/M/1 G-queue with first-come-first-served (FCFS) queueing discipline, service rate μ, positive arrival rate λ^+, negative arrival rate λ^- and both RCH and RCE killing strategies. We abbreviate the combined arrival rate $\lambda^+ + \lambda^-$ by λ. Additionally, we define Π_k, $k > 0$, as the probability that k positive customers are killed simultaneously by a negative arrival. If the number of queueing customers is less than k, the queue becomes empty.

In [3], it is shown that the stationary probability mass function of the random variable N denoting the number of customers in the queue (including the one in service, if any) is geometric:

$$p_n = (1 - \rho)\rho^n \quad (n \geq 0) \tag{1}$$

where the load, ρ, is the solution of the fixed point equation

$$r = \frac{\lambda^+}{\mu + \lambda^- h(r)} \tag{2}$$

and h is defined by $h(y) = \frac{1-H(y)}{1-y}$ where H is the probability generating function of the batch size of the killings, $H(y) = \sum_{k=1}^{\infty} \Pi_k y^k$. Notice that if only one customer is killed, $h(y) = 1$ and ρ reduces to the familiar $\lambda^+/(\mu+\lambda^-)$ which can be interpreted as "negative customers helping the server", i.e. the service rate in a classical M/M/1 queue is increased to $\mu + \lambda^-$.

With batch killings, as far as the queue length alone is concerned, the queue behaves as an M/M/1 queue with arrival rate λ^+ and service rate $\mu + \lambda^- h(\rho)$, independently of the killing strategy. The queue is stable if and only if $\rho < 1$. However, the response time distribution of customers that are not killed *is* sensitive to killing strategy and we give an expression for the Laplace-Stieltjes transform $S^*(s)$ of the sojourn time distribution function $S(t)$ for a surviving positive customer. Note that the probability that a positive customer is not killed is the (marginal) probability $S(\infty) = S^*(0)$. However, we can obtain an expression for this quantity directly using probabilistic arguments. First, we observe that customers are stochastically identical and so, for a given queue, all new arrivals have the same equilibrium probability of being killed. Since the stationary queue length random variable N is identical in every M/M/1 model, the killing rate (of anonymous customers) is the same and so we conclude that the killing probability is the same for every customer in every model. Specifically, the expected proportion of surviving customers out of all positive customers arriving during any given interval at equilibrium is the ratio of the effective throughtput $\rho\mu$ over the positive arrival rate λ^+. Thus the equilibrium probability of not being killed is $\rho\mu/\lambda^+$.

2.1 Sojourn times

Let S be the sojourn time random variable, $S(t)$ the probability distribution function of S and $S^*(s)$ its Laplace transform. Then we have (see [5] for the proof):

Theorem 2.1 *The Laplace transform of the sojourn time density for an M/M/1 FCFS-RCE G-queue with* batch killings *of probability mass function Π_k for batch size, jointly with the probability of not being killed, is*

$$S^*(s) = (1 - \rho)\frac{\mu}{\lambda^+}\, \frac{y_1(\rho, s)}{1 - y_1(\rho, s)} \tag{3}$$

where $y_1(x, s)$ is the unique fix-point of the function $\mathcal{F}_{x,s}$ defined by

$$\mathcal{F}_{x,s}(y) = \frac{\lambda^+}{s + \lambda + \mu(1 - x) - \lambda^- H(y)}$$

inside the unit polydisc with $H(y) = \sum_{k=1}^{\infty} \Pi_k y^k$ and ρ the fix-point of the function f defined by $f(r) = \frac{\lambda^+}{\mu+\lambda^- h(r)}$ with $h(r) = \frac{1-H(r)}{1-r}$.

It is easy to check that the above expression simplifies to the simple killing case by setting $\Pi_1 = 1$ and $\Pi_k = 0$ for all $k > 1$. In other words $H(y) = y$ and $h(y) = 1$ as required.

The main problem computationally is determining $y_1(\rho, s)$. However, it is not always necessary to compute this quantity at every s, for example when determining mean response time. The following property will prove useful:

Lemma 2.1 $y_1(\rho, 0) = \rho$

Using Lemma 2.1, we easily verify that the probability that the tagged customer is not killed is (as anticipated)

$$S^*(0) = \frac{\mu\rho}{\lambda^+} = \frac{\mu}{\mu + \lambda^- h(\rho)}$$

The mean value of response time, conditional on a customer not being killed, is then obtained from the (negative) derivative of the Laplace transform at 0 divided by $S^*(0)$:

Corollary 2.1

$$E(S) = \frac{1}{(1-\rho)(\mu + \lambda^-(h(\rho) - \rho H'(\rho)))}$$

It is easier to see (by modifying the service time distribution to combine the effect of the batch killings at the head of the queue with that of the server— see [4]) that, for RCH killing, we have:

Theorem 2.2 *The sojourn time density for an M/M/1 FCFS-RCH G-queue with batch killings, conditioned on the probability of not being killed, is exponential with mean*

$$E(S) = \frac{1}{\mu + \lambda^- h(\rho) - \lambda^+}$$

3   Breakdowns with exponential repair times

We now consider an M/M/1 queue with breakdowns and exponential repair times, in which again some but not all of the outstanding tasks are lost upon failure. Both RCE and RCH batch killings will be considered. To do this, we extend the state space of previous models to two-dimensions to account for the two possibilities that a server is either working or broken. We then obtain the queue length distribution and sojourn time density.

3.1   Steady-state probabilities

In this model neither positive nor negative arrivals are permitted when the server is broken: no jobs are accepted during repair times (of average $1/\eta$ (inoperative period)) and the server cannot break down when already down. However, the server can fail even when the queue is empty. This is in contrast to previous G-models where negative arrivals had no effect on an empty system: here a negative arrival with any batch size causes the empty operative queue to become empty and broken.

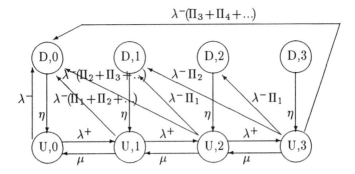

Figure 1: State-space diagram for breakdowns with exponential repair times

Proposition 3.1 *Let the random variables U and D respectively denote the active (up) and broken (down) states of the system and let ρ be the fix-point solution for utilisation defined in the previous section. The joint steady state probability distribution of the state of the system and the queue length are then given by*

$$P(U,n) = \frac{\eta}{\eta + \lambda^-}\,(1-\rho)\,\rho^n,\; n \geq 0$$

$$P(D,n) = \frac{\lambda^- H(\rho)}{\eta + \lambda^-}\,(1-\rho)\,\rho^n\,H(\rho),\; n > 0$$

$$P(D,0) = \frac{\lambda^-(1+\rho\,h(\rho))}{\eta + \lambda^-}\,(1-\rho)$$

$$\rho = \frac{\lambda^+}{\mu + \lambda^-\,h(\rho)}$$

and the marginal steady state probability distribution of the queue length is

$$P(n) = (\frac{\eta}{\eta + \lambda^-} + \frac{\lambda^- H(\rho)}{\eta + \lambda^-})\,(1-\rho)\,\rho^n,\; n > 0$$

$$P(0) = (\frac{\eta}{\eta + \lambda^-} + \frac{\lambda^-(1+\rho\,h(\rho))}{\eta + \lambda^-})\,(1-\rho)$$

Proof By direct solution of the balance equations $\qquad\qquad\square$

3.2 Sojourn times

In this section, we give the results corresponding to those of the previous section, but with exponential, rather than instantaneous repair times. The proofs are again omitted, being available in [5]. We begin with the case of RCE killing.

Theorem 3.1 *The Laplace transform of the sojourn time density for an M/M/1 FCFS-RCE G-queue with* batch killings *and* exponential repairs, *conditioned on the probability of not being killed, is*

$$S^*(s|not\ killed) = \frac{1-\rho}{\rho}\ \frac{y_1(\rho,s)}{1-y_1(\rho,s)} \tag{4}$$

where $y_1(x,s)$ *is the unique fix-point of the function* $\mathcal{F}_{x,s}$ *defined by*

$$\mathcal{F}_{x,s}(y) = \frac{\lambda^+}{s+\lambda+\mu(1-x)-\lambda^-\frac{\eta}{\eta+s}H(y)}$$

inside the unit polydisc with $H(y) = \sum_{k=1}^{\infty}\Pi_k y^k$ *and* ρ *the fix-point of the function* f *defined by* $f(r) = \frac{\lambda^+}{\mu+\lambda^- h(r)}$ *with* $h(r) = \frac{1-H(r)}{1-r}$.

The probability of not being killed in this system and the average sojourn time conditioned on not being killed are given by the following Corollary (using Lemma 2.1):

Corollary 3.1

$$S^*(0) = \frac{\eta}{\eta+\lambda^-}\ \frac{\mu}{\mu+\lambda^- h(\rho)} = P(not\ killed)$$

and

$$E(S) = \frac{\eta+\lambda^- H(\rho)}{\eta}\ \frac{1}{(1-\rho)(\mu+\lambda^-(h(\rho)-\rho H'(\rho)))}$$

For the case of RCH killing strategy (from the front of the queue), the corresponding result for the Laplace transform of sojourn time density is as follows.

Theorem 3.2 *The Laplace transform of the sojourn time density for an M/M/1 FCFS-RCH G-queue with* batch killings *and* exponential repairs, *conditioned on the probability of not being killed, is*

$$S^*(s|not\ killed) = \frac{\mu+\lambda^- h(\rho)-\lambda^+}{s+\mu(1-\rho)+\lambda^-(1-\frac{\eta}{\eta+s}H(\rho))} \tag{5}$$

Note that we no longer have an exponential distribution for $S^*(s|not\ killed)$ as opposed to the case of single RCH killing and instantaneous repairs. Mean sojourn time is given by the following Corollary.

Corollary 3.2

$$E(S) = \frac{\eta+\lambda^- H(\rho)}{\eta}\ \frac{1}{\mu+\lambda^- h(\rho)-\lambda^+}$$

4 Application to a bank of parallel servers

Consider N alternative gateways (servers) modelled as parallel M/M/1 queues. When one breaks down, a random number of jobs at that queue are lost and its incoming stream is redirected to active queues whilst it is repaired. This redirection introduces dependencies amongst the queues.

To model this system we consider (any) one of the queues in isolation— queue N, say. At any instant, it receives Poisson arrivals from its own source (at a known, fixed rate) as well as superposed Poisson streams comprising redirected input from other, broken servers. The net arrival process is therefore Poisson, but the rate of the redirected traffic depends on the state of the rest of the system, specifically on which servers are broken. The binary vector of length $N-1$ with a "1" (respectively "0") in the ith position indicating that server i, of the others, is broken (respectively operational), $1 \leq i \leq N-1$, clearly follows a continuous time Markov chain over the state space $\mathcal{V} = \{(s_1, ..., s_{N-1}) | s_i \in \{0,1\}\}$, each state of which determines the instantaneous arrival rate to the server we are considering. Thus we need to solve a further extended G-queue, with repairs *and* Markov Modulated Poisson Process (MMPP) arrivals.

To solve this problem exactly for $N > 2$ is a complex problem — it is a reformulation of that of [6] where $N = 2$ — and we seek an approximate solution. We assume that the rates at which breakdowns and repairs occur are at least an order of magnitude smaller than the arrival and service rates, which reflects the usual situation in practice. Consider the two-dimensional state characterising queue N, with "horizontal" transitions (on the second component) representing positive and negative arrivals for each possible fixed set of operational, other servers (the first component of the state), and with "vertical" transitions representing changes in operational status of the other servers. Thus vertical transitions are much slower than horizontal. We apply a decomposition in which each row of horizontal transitions is aggregated into a single state, representing the operational status of the other servers. The resulting "vertical" Markov chain is then solved (very cheaply) for the steady state distribution of the operational status of the other servers, given by the binary vector. Let the equilibrium probability that this status is given by the vector s be denoted by π_s, $s \in \mathcal{V}$. Then, under random redirection, the net positive arrival rate to queue N is

$$\lambda_s^+ = \lambda_N^+ + \frac{\sum_{i=1}^{N-1} \lambda_i^+ s_i}{1 + Z(s)} \tag{6}$$

where λ_i^+ is the fixed positive arrival rate to node i $(1 \leq i \leq N)$ and $Z(s)$ is the number of zero components in the vector s. Of course, any redirection strategy dependent on the current operational stati is equally simple to model.

Within each aggregate state s we can now use our existing results for the single extended G-queue with arrival rate λ_s. The net result for performance measure P (e.g. throughput, cell loss probability, sojourn time distribution) at queue N is then approximated by

$$P = \sum_{s \in \mathcal{V}} \pi_s P_s \tag{7}$$

where P_s is the corresponding performance quantity for the extended G-queue with fixed arrival rate λ_s^+, as derived in previous sections. This approximation is only valid when:

- the extended G-queue is stable when its arrival rate is λ_s^+ for all s;

- the assumption about the disparity in rates holds so that the system can be expected to "settle down" before each of the operational stati can change.

Of course, we can always guarantee the former condition by choosing a sufficiently large value for the number killed by a negative arrival—more precisely choosing a batch size probability mass function yielding $\rho < 1$.

Now, the operational status of each server is independent of that of every other and queue i has equilibrium probability $\xi_i/(\xi_i + \eta_i)$ of being broken where ξ_i and η_i are respectively the parameters of the negative exponential distributions of the times to failure and repair. Thus,

$$\pi_s = \prod_{i=1}^{N-1} \frac{s_i\xi_i + (1 - s_i)\eta_i}{\xi_i + \eta_i}$$

4.1 Symmetrical queues

We consider first the simplest case of N parallel *symmetrical* queues with exponential service rate μ, breakdown rate $\xi \equiv \lambda_i^-$ and repair rate η. An external Poisson arrival stream of rate Λ splits equally amongst all queues. In this case, all queues are identical with a common modified arrival rate, λ_s^+ in operational state s, given by:

$$\lambda_s^+ = \Lambda \left(\frac{1}{1 + Z(s)} \right)$$

There are therefore only N different values that λ_s^+ may take, corresponding to the number of zeros in the operational status vector.

The total average number of jobs lost per unit time in this system is given by:

$$
\begin{aligned}
L_s &= N\lambda_s^+(1 - P(not\ killed)) + \Lambda P(All\ broken) \\
&= N\lambda_s^+ \left(1 - \frac{\mu}{\mu + \xi h(\rho)} \right) + \Lambda \left(\frac{\xi}{\xi + \eta} \right)^N
\end{aligned}
$$

by Corollary 3.1. We use RCH batch killings to model a failure, which is the most realistic case since the task in service would normally be lost, and hence (Corollary 3.2) to obtain the average response time of a successful job:

$$S_s = \frac{\eta + \xi H(\rho)}{\eta} \frac{1}{\mu + \xi h(\rho) - \lambda_s^+}$$

Since all queues experience the same arrival rate in every operational status, we can simplify equation 7 to sum only over aggregate states with the same number of zeros (and hence same λ_s^+), i.e. from 0 to $N-1$. Then we have

$$P = \sum_{n=0}^{N-1} \left(\sum_{s:Z(s)=n} \pi_s \right) P_n$$

where P_n is the value of the relevant performance quantity when there are n other servers operational, i.e. when the queue has positive arrival rate λ_s^+ for *any* s with $Z(s) = n$. Since, in this symmetrical case, the equilibrium probability of being broken is the same for every server, say $b = \xi/(\xi + \eta)$, the binomial theorem immediately yields

$$P = \sum_{n=0}^{N-1} \binom{N-1}{n} b^{N-1-n}(1-b)^n P_n$$

5 Numerical validation

In this section, we consider the loss rate and mean response time for successful jobs under various configurations, as evaluated by the analytical model and as estimated by simulation. In all cases, the simulation was run long enough so that the 90% confidence interval was within 5% of the mean response time, using the batch-means method. We first consider a bank of three servers (N=3) and then a larger bank of eight servers (N=8). Results for completely symmetrical cases, as considered in section 4.1, are somewhat uninteresting, allowing few parameter variations, and we do not include them.

In an asymetric model, we consider an arbitrary one of the queues, G-queue 1 say, with parameters $(\lambda_1^+, \xi_1, \mu_1, \Pi_1, \eta_1)$. We use Corollary 3.2 to obtain the average response time of a successful job at queue 1 in operational state s:

$$S_{1s} = \frac{\eta_1 + \xi_1 H(\rho_{1s})}{\eta_1} \frac{1}{\mu_1 + \xi_1 h(\rho_{1s}) - \lambda_{1s}^+}$$

where λ_{1s}^+ is the net arrival rate of tasks in operational state s and ρ_{1s} is the corresponding load given by equation 2. The average number of jobs lost per unit time at queue 1 in operational state s is obtained from Corollary 3.1:

$$L_{1s} = \lambda_{1s}^+(1 - P(not\ killed_{1s})) = \lambda_{1s}^+ \left(1 - \frac{\mu_1}{\mu_1 + \xi_1 h(\rho_{1s})} \right)$$

Assuming all the other queues are statistically identical to queue 2, the total average number of jobs lost per unit time in an N-queue system is therefore

$$L = \sum_{s\in\mathcal{V}} \pi_s \{L_{1s} + (N-1)L_{2s}\} + \frac{\Lambda\xi_1(\xi_2)^{N-1}}{(\xi_1 + \eta_1)(\xi_2 + \eta_2)^{N-1}}$$

where the last term corresponds to arrivals lost when all servers are broken. Similarly, the average response time of a successful job is

$$S = \sum_{s\in\mathcal{V}} \pi_s S_s$$

where

$$S_s = \frac{\lambda_{1s}^+ S_{1s} + (N-1)\lambda_{2s}^+ S_{2s}}{\lambda_{1s}^+ + (N-1)\lambda_{2s}^+}$$

We again have symmetrical arrival processes and can use the version of equation 7 $P = \sum_{n=0}^{N-1} \left(\sum_{s:Z(s)=n} \pi_s \right) P_n$, but we cannot simplify the inner summation using the binomial theorem as before, except when considering queue 1. Computationally, however, it is still most efficient to precompute each of the N sums, to give σ_n say, and then use the formula $P = \sum_{n=0}^{N-1} \sigma_n P_n$ at each queue with arrival rate λ_n^+ equal to one of the λ_s^+ above, $0 \le n \le N-1$.

5.1 Three-server bank

For the $N=3$ case, the arrival rate is 0.9 and the service rate is 1, so that even when all but one server is down, the queueing system would remain stable even without losses through failures. We have assumed a batch killing of one or two jobs with equal probability. For larger batch killings, higher arrival rates are possible without loss of stability.

In a given configuration we alter two parameters: the breakdown rate of server 1 and the proportion of jobs that are directed to server 1. All the other servers receive the same proportion of work. In the symmetrical case, all servers have a breakdown rate of 0.01 and a repair rate of 0.1. More generally, we increase the breakdown rate of server 1 from 0.01 to 0.05, 0.1, and 0.5. The proportion (q) of arriving jobs that are directly to server 1 is varied from 0 to $2/N$ (so that twice as much traffic is routed to server 1 as in the even loading case).

The following 2 figures show plots for the metrics of interest. Each figure contains plots for 4 breakdown rates: 0.01 (symmetrical), 0.05, 0.1, and 0.5. Since we are fixing the repair rate, these rates correspond to server 1 availability percentages of 90.9%, 66.7%, 50%, and 16.7%, respectively. Thus we have to be careful in interpreting the value of q, since the actual offered traffic to server 1 is reduced considerably by the fact that it is not available much of the time.

Figure 2 shows how the mean response time varies as we vary q. The minimum response time occurs at $q = 1/3$ for the symmetrical case. For higher breaking rate curves, the mimimum occurs at higher values of q. When the breakdown rate for server 1 is very high (0.5), the response time is almost a straight line.

In Figure 3, we find that the lowest loss rate occurs at $q = 1/3$ for the symmetrical case. For other cases, there is no mimimum and the losses mount as more traffic is routed to the least reliable server.

5.2 Eight-server bank

For the $N=8$ case, we consider first a moderately loaded system with arrival rate 4.0 and loss of either one or two jobs with equal probability upon failure, and secondly a heavily loaded system with arrival rate 7.2. The other parameters are the same as in the previous subsection (when N was 3). In either case, when some of the servers are down, the arrival rate is higher than the available servicing power. In the latter case, we assume that all jobs in the queue are

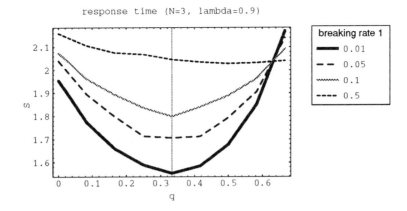

Figure 2: 3-server bank - response time

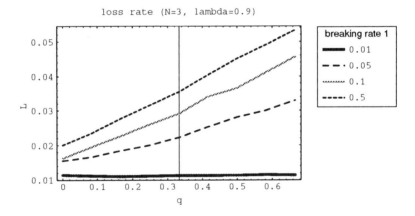

Figure 3: 3-server bank - loss rate

killed on the arrival of a negative customer, so that overall stability is assured for any negative arrival rate. This is achieved in our model by choosing a sufficiently large killing batch size, as determined by successively increasing it until there is no change in predicted performance. For the lower arrival rate, there is a stability condition, of course, satisfied by our choice of batch size distribution. In this case, the system will rarely be in a state with higher arrival rate than total service capacity. Thus we can expect a better approximation from our model than in the heavily loaded case where the approximate decomposition of the MMPP queueing model is suspect.

We again study the effects of varying the breakdown rate of server 1 and the proportion of jobs that are directed to server 1.

5.2.1 Moderate loading: arrival rate 4.0

Figure 4 shows that the mean response time increases rapidly after $q = 0.2$ as saturation is approached. The curve for a breaking rate of 0.5 remains relatively flat which indicates that the system is not saturated.

In figure 5, the lowest loss rate occurs for breaking rate of 0.01 and $q = 1/8$. For the higher breaking rate curves, the loss rate increases linearly as more jobs are routed to the unreliable server.

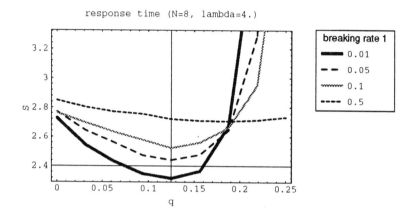

Figure 4: 8-server bank - response time

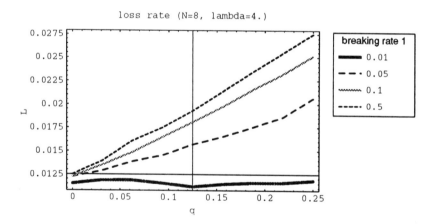

Figure 5: 8-server bank - loss rate

5.2.2 Heavy loading: arrival rate 7.2

Figure 6 shows that the mean response time decreases for all curves except at a breaking rate of 0.01, when it begins to increase after $q = 1/8$ as server 1 becomes more utilised. This is as expected since response time is conditioned on a task not being lost. Thus, a task known to survive at any server is aided by disposing of other tasks at other servers where, if they become lost, do not contribute to net mean response time.

Figure 7 shows that the lowest loss rate occurs at breaking rate 0.01 and $q = 1/8$. When the breaking rate is 0.05 or 0.1, the lowest loss rate occurs when slightly more jobs are routed to the less reliable server (i.e. $q > 1/8$)! Although, perhaps, counter-intuitive, a possible explanation for this may be that it prevents too many tasks being lost upon failure of the more reliable servers which may accumulate very long queues in heavy traffic.

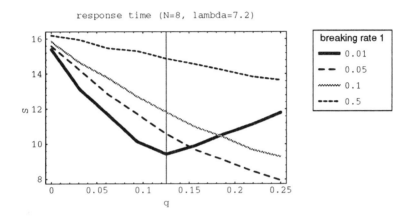

Figure 6: 8-server bank - response time

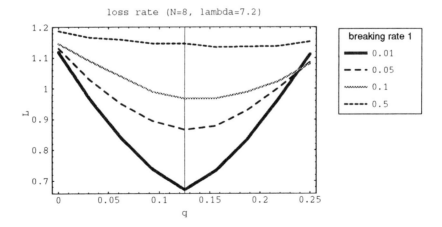

Figure 7: 8-server bank - loss rate

6 Conclusion

We have applied an extended theory of the G-queue to approximate queues with breakdowns by introducing Markov modulated Poisson arrivals relating to the operational state of the network. Our method is efficient computationally and is tractable for much larger examples than those which we validated by simulation.

Almost the same problem was considered by Mitrani and Wright in [6] who solved it exactly in the case of a two-server bank by the method of spectral analysis. The difference is that in their model a breakdown causes all work in a queue to be lost, so that the system is *always* stable. In their numerical examples, they concentrated on a system where the queue at one server built up rapidly when the other was broken, the arrival rate then being greater than the service rate. Stability in this situation depended on occasional breakdowns. Our model has to be applied carefully in this situation, i.e. with sufficiently large batch sizes killed by a negative arrival. Otherwise, between breakdowns, the system may not come close to stability due to an ever-growing queue, invalidating our assumptions about decomposability of the G-queue with MMPP arrivals. This is an inherent problem with the method in that we assume that the system is approximately in equilibrium between breakdowns and yet depends on a breakdown locally to achieve such stability. However, this mode of operation, with massive queues forming, is also uncommon in real systems.

The obvious future development of the research is to extend it to more general networks of queues with breakdowns. The same approach, *viz* considering each queue separately with Markov modulated Poisson arrivals, could be applied, but more complex traffic equations would have to be solved. This could lead to an efficient, approximate method for solving arbitrary queueing networks of unreliable servers for sojourn time distributions as well as the more commonly derived resource-based performance measures. However, in large, strongly connected networks where the input process at each node is a mixture of the output processes from several others, the assumption that net arrivals are Poisson is typically a valid approximation, both theoretically under appropriate assumptions and also as supported by simulation tests. This results in a much simpler model.

References

[1] GELENBE, E. (1991) Product form networks with negative and positive customers. *J. Appl. Prob.* **28**, 656-663.

[2] GELENBE, E. AND SCHASSBERGER, R. (1992) Stability of product form G-networks. *Probability in the Engineering and Informational Sciences*, **6**, 271-276.

[3] GELENBE, E. (1993) G-Networks with signals and batch removal. *Probability in the Engineering and Informational Sciences*, **7**, 335-342.

[4] HARRISON, P.G. AND PITEL, E. (1993) Sojourn times in single server queues with negative customers. *Journal of Applied Probability* **30**, 943-963.

[5] HARRISON, P.G., PATEL N.M. AND PITEL, E. (1995) Reliability Modelling Using G-queues. *Research Report, Imperial College, London, 1995*

[6] MITRANI, I. AND WRIGHT, P.E. (1993) Routing in the presence of breakdowns. *Performance '93*, Iazeolla, G. and Lavenberg, S. (eds), Elsevier Science publishers.

MEM for Closed Exponential Queueing Networks with RS-Blocking and Multiple Job Classes*

Demetres D. Kouvatsos and Irfan-Ullah Awan[†]

Abstract

A new product-form approximation, based on the method of entropy maximisation (MEM), is characterised for arbitrary closed exponential queueing networks with multiple classes of jobs, mixed service disciplines and repetitive-service (RS) blocking with random destination (RS-RD). The ME approximation implies decomposition of the network into individual multiple class M/M/1/N queues, satisfying constraints on population and flow conservation, and it is, in turn, truncated and efficiently implemented by a convolution type recursive procedure. Numerical validation results against simulation are included to demonstrate the credibility of typical ME performance metrics.

1 Introduction

Queueing network models (QNMs) are widely used for the performance evaluation and modelling of discrete flow systems such as computer system, communication networks and production systems. However, not much work has been done so far on the analysis of closed finite capacity queueing networks with blocking and multi-classes of jobs. Comprehensive reviews on open and closed QNMs under various blocking mechanisms have been carried out by Onvural[1] and Perros [2,3]. Akyildiz and Von Brand [4-6] analysed open, closed and mixed QNMs with finite capacity and multi-classes under reversible routing and repetitive service (RS) blocking. Choukri [7], and Akyildiz and Huang [8] solved exactly special types of multiple class QNMs with finite capacity and reversible routing.

In this paper, the method of entropy maximisation (MEM) is applied to analyse arbitrary closed exponential queueing networks with multiple classes of jobs under repetitive service (RS) blocking and mixed service disciplines (FCFS, LCFS, PS, LCFS-NONPR). The work is based on earlier applications of MEM for (i) single class open and closed QNMs (cf., Kouvatsos and Xenios [9], Kouvatsos and Denazis [10]) and (ii) multiple class open QNM with mixed non-priority based service disciplines, (cf., Kouvatsos and Denazis [11]),

---

*Supported in part by the Engineering and Physical Sciences reseach Council (EPSRC), UK, under grants GR/H/18609 and GR/K/67809 and in part by British Council

†Computer Systems Performance Modelling Research Group, Department of Computing, University of Bradford, Bradford, BD7-1DP, U.K.

under repetitive service blocking with random destination (RS-RD). A new ME product-form approximation for the joint queue length distribution of the closed exponential networks is characterised, subject to marginal mean value constraints. The M/M/1/N queue with multiple classes of jobs -which has been analysed in [11]- is used as an efficient building block in the solution process.

The ME analysis of multiple class M/M/1/N queue is introduced in Section 2. The ME product-form approximation for arbitrary exponential multi-class closed QNMs and the associated ME algorithm are presented in Section 3. Numerical results are given in Section 4. Conclusions follow in Section 5.

Remark: RS-RD blocking: RS blocking occurs when a job upon service completion at queue i attempts to join a destination queue j and immediately receives another service at queue i. In the RS-RD case, each time the job completes service at queue i, a downstream queue is selected independently of the previously chosen destination queue j.

2 The multiple class M/M/1/N queue

This section reviews the ME analysis of a stable M/M/1/N queue with:

R classes of jobs,

Fist Come First Served (FCFS), Processor Sharing (PS), Last Come First Served (LCFS), Last Come First Served with Non Preemptive (LCFS-NPR) scheduling disciplines, and

censored arrival process cf., [11].

Notation
Let

S $= (c_1, c_2, \ldots, c_n)$ $n \leq N$ where c_i is the class of ith job in the queue and c_1 is the class of the job in service

Q be the set of all feasible states of **S**

k $= (k_1, k_2, \ldots, k_R)$ be a system state seen by a random observer

k_i be the number of jobs of class i, $i = 1, 2, \ldots, R$, present in the queue

For each state **S** let

$n_i(\mathbf{S})$ be the number of class i customers and $s_i(\mathbf{S}), f(\mathbf{S})$ be auxiliary functions defined by

$$s_i(\mathbf{S}) = \begin{cases} 1, & \text{if the job in service is of class i} \\ 0, & \text{otherwise} \end{cases}$$

$$f(\mathbf{S}) = \begin{cases} 1, & \text{if} \sum_{i=1}^{R} n_i(\mathbf{S}) = N, \\ 0, & \text{otherwise} \end{cases}$$

Λ_i be the arrival rate

μ_i be the service rate

π_i be the blocking probability that an arrival of class i finds the queue full

$p(\mathbf{S})$ be the stationary state probabaility.

Suppose that the following mean value constraints about the state probability $p(\mathbf{S})$ are known to exist:

(i) Normalisation, $\displaystyle\sum_{\mathbf{S}\in\mathbf{Q}} p(\mathbf{S}) = 1$ (1)

(ii) Utilisation, $\displaystyle\sum_{\mathbf{S}\in\mathbf{Q}} s_i(\mathbf{S})p(\mathbf{S}) = U_i, 0 < U_i < 1, i = 1, 2, \ldots, R.$ (2)

(iii) Mean queue length,

$$\sum_{\mathbf{S}\in\mathbf{Q}} n_i(\mathbf{S})p(\mathbf{S}) = \langle n_i \rangle, \ U_i < \langle n_i \rangle < N, \ i = 1, 2, \ldots, R.$$ (3)

(iv) Full buffer state probability, $\displaystyle\sum_{\mathbf{S}\in\mathbf{Q}} f(\mathbf{S})p(\mathbf{S}) = \phi, 0 < \phi < 1,$ (4)

and also the flow balance equations are satisfied

$$\Lambda_i(1 - \pi_i) = \mu_i U_i, i = 1, 2, \ldots, R.$$ (5)

The choice of mean values (1) - (4) is based on the type of constraints used for the ME analysis of stable single class FCFS G/G/1/N queue [12]. The form of the state probability distribution, $p(\mathbf{S}), \mathbf{S} \in \mathbf{Q}$, can be characterised by maximizing the entropy function $H(\mathbf{p}) = -\sum_s p(\mathbf{S}) \log p(\mathbf{S})$, subject to constraints (1) - (4). By employing Lagrange's method of undetermined multipliers the following solution is obtained

$$p(\mathbf{S}) = \frac{1}{Z} \prod_{i=1}^{R} g_i^{s_i(\mathbf{S})} x_i^{n_i(s)} y^{f(\mathbf{S})}, \forall \mathbf{S} \in \mathbf{Q},$$

where Z is the normalizing constant and $g_i, x_i(i = 1, 2, \ldots, R)$ and y are the Lagrangian coefficients corresponding to constraints (2) - (4), respectively. Defining the sets

$$\begin{aligned}
\mathbf{S_0} &= \{\mathbf{S}/\mathbf{S} \in \mathbf{Q} : s_i(\mathbf{S}) = 0, i = 1, 2, \ldots, R\}, \\
\mathbf{Q_i} &= \{\mathbf{S}/\mathbf{S} \in \mathbf{Q} : s_i(\mathbf{S}) = 1\}, i = 1, 2, \ldots, R, \\
\mathbf{Q_{i;k}} &= \{\mathbf{S} \in \mathbf{Q_i} : n_j = k_i \geq 1, j = 1, 2, \ldots, R\}, i = 1, 2, \ldots, R,
\end{aligned}$$

where $\mathbf{k} = k_1, k_2, \ldots, k_R$.

It is implied, after some manipulation, that the ME joint state probability distribution is given by

$$P(\mathbf{S_0}) = p(0, \ldots, 0) = \frac{1}{Z}$$ (6)

$$p(\mathbf{k}) = \sum_{i=1}^{R} Prob(\mathbf{Q}_{i;\mathbf{k}})$$

$$= Z^{-1} \frac{\left(\sum_{j=1}^{R} k_j - 1\right)!}{\prod_{j=1}^{R} k_j!} \left(\prod_{j=1}^{R} x_j^{k_j}\right) \left(\sum_{i=1}^{R} k_i g_i\right) y^{\delta(\mathbf{k})} \qquad (7)$$

where $\delta(\mathbf{k}) = 1$, if $\sum_j k_j = N$, or 0, otherwise. The Lagrangian coefficients x_i and g_i can be approximated analytically by making asymptotic connections to an infinite capacity queues. Assuming x_i, g_i and y are invariant to the buffer capacity size N, it can be established that

$$x_i = \frac{\langle n_i \rangle - \rho_i}{\langle n \rangle} \qquad (8)$$

$$g_i = \frac{(1 - X)\rho_i}{(1 - \rho)x_i} \qquad (9)$$

and, using flow balance condition (5),

$$y = \frac{1 - \rho}{1 - X} \qquad (10)$$

where $X = \sum_{i=1}^{R} x_i$, $\langle n \rangle = \sum_{i=1}^{R} \langle n_i \rangle$ and $\langle n_i \rangle$ is the asymptotic marginal mean queue length of a multi-class M/M/1 queue. Note that statistics $\langle n_i \rangle, i = 1, 2, \ldots, R$ can be determined by

$$\langle n_i \rangle = \begin{cases} \rho_i + \frac{\Lambda_i}{1-\rho} \sum_{j=1}^{R} \frac{\rho_j^2}{\Lambda_j}, & \text{if FCFS or LCFS-NPrule} \\ \frac{\rho_i}{1-\rho}, & \text{if LCFS-PR rule} \\ \rho_i \left\{ 1 + \frac{1}{1-\rho} \sum_{j=1}^{R} \frac{h_j}{h_i} \rho_j \right\}, & \text{if PS rule} \end{cases} \qquad (11)$$

where $\rho_i = \Lambda_i / \mu_i$, $\rho = \sum_{i=1}^{R} \rho_i$ and $h_i, i = 1, 2, \ldots, R$, is a set of discriminatory weights that impose different treatment to different classes of jobs.

3 ME and closed queueing network

Consider a closed queueing network with the following arbitrary configuration:

M queueing stations each with a single server

R distinct classes of jobs

Fist Come First Served (FCFS), Processor Sharing (PS), Last Come First Served (LCFS), Last Come First Served with Non Preemptive (LCFS-NPR) scheduling disciplines

censored arrival process

repetitive service blocking with random destination (RS-RD)

Let

L be the total number of customers in the system with $L \leq N$, where N is the total capacity of the system

N_k be the total capacity of queue k, $k = 1, 2, \ldots, M$, such that

$$\sum_{k=1}^{M} N_k = N$$

$\mathbf{L}_c$ be an integer valued vector defined as:

$$\mathbf{L}_c = (L_{c1}, L_{c2}, \ldots . L_{cR})$$

where L_{ci}, represents the total number of jobs of class i, $i = 1, 2, \ldots, R$, in the network and

$$\sum_{i=1}^{R} L_{ci} = L \leq N$$

Moreover, the maximum effective capacity, L_k, is defined as

$$L_k = min(N_k, L)$$

and minimum effective capacity, K_k, as,

$$K_k = max(0, L - \sum_{k \neq j = 1}^{M} L_j)$$

for each queue, k, $k = 1, 2, \ldots, M$. K_k, represents the minimum number of jobs always present at queue k.

3.1 The joint queue length distribution

Let

n_{ik} be the number of jobs of class i at queue k

$\mathbf{S}_k$ represent the state of queue k and defined as:

$$\mathbf{S}_k = (n_{k1}, n_{k2}, \ldots, n_{kR})$$

$\mathbf{S}$ be the state of the network and can be described by an integer valued vector $\mathbf{S} = (\mathbf{S}_1, \mathbf{S}_2, \ldots, \mathbf{S}_M)$
and finally

$\mathbf{S}(\mathbf{L}_c, M)$ be the set of all posssible states of the network.

The form of the ME solution $p(\mathbf{S})$, subject to normalization and marginal constraints of the type (1) - (4), can be clearly established in terms of the product-form approximation

$$p(\mathbf{S}) = \frac{1}{Z(\mathbf{L}_c, M)} \prod_{k=1}^{M} \omega_k(\mathbf{S}_k) \tag{12}$$

where

$$\omega_k(\mathbf{S}_k) = \begin{cases} 1 & \sum_{i=1}^{R} n_{ki} = K_k \\ \prod_{i=1}^{R} g_{ki} x_{ki}^{n_{ki}} y_{ki}^{f_{ki}(S_k)} & K_k < \sum_{i=1}^{R} n_{ki} \leq L_k \end{cases}$$

where g_{ki}, x_{ki} and y_{ki}, $k = 1, 2, \ldots, M$, $i = 1, 2, \ldots, R$ are the Lagrangian coefficients corresponding to the utilization, U_{ki}, mean queue length, $\langle n_{ki} \rangle$ and the joint full buffer state probability, ϕ_{ki} constraints respectively and are given by equations (1)- (4). $Z(\mathbf{L}_c, M)$ is the normalizing constant and given by

$$Z(\mathbf{L}_{cR}, M) = \sum_{\mathbf{S} \in \mathbf{S}(\mathbf{L}_{cR}, M)} \prod_{k=1}^{M} \omega_k(\mathbf{S}_k).$$

3.2 Computational technique

The ME approximation (12) cannot be implemented directly since the performance measures U_{ki}, $\langle n_{ki} \rangle$ and ϕ are not known a-priori and subsequently no closed form expressions are available for the corresponding Lagrangian multipliers. However, approximate estimates for these coefficients can be obtained by making use of psuedo open multiple class network with no external arrival or departure processes and almost identical topology to that of closed network (i.e. both networks have the same number of queues and servers, service-time characteristics and transition probabilities satisfying the principles of:
i) the conservation of flow, expressed by the job flow-balance equations:

$$\lambda_{ki}^{*} = \sum_{k=1}^{M} \hat{a}_{kim} \lambda_{ki}^{*}, k = 1, 2, \ldots, M, i = 1, 2, \ldots, R, \tag{13}$$

ii) the conservation of population, represented by the fixed population mean constraints:

$$\sum_{k=1}^{M} \langle n_{ki} \rangle^{*} = L_{ci}, i = 1, 2, \ldots, R \tag{14}$$

where λ_{ki}^{*} and $\langle n_{ki} \rangle^{*}$, are the effective throughputs and the mean queue lengths, respectively, for each queue k, $k = 1, 2, \ldots, M$, of the psuedo open network.

The introduction of constraints (13) and (14) accounts, in a mathematical sense, for the absence of the external arrivals. In particular, constraint (13) is not sufficient to determine values for the effective throughputs, since only $(M - 1) * R$ out of $M * R$ equations are independent. Thus the system of equations (13) has a unique solution up to a multiplicative constant, and therefore, one of the components of the vector, $(\lambda_{1i}, \lambda_{2i}, \ldots, \lambda_{Mi}), i = 1, 2, \ldots, R,$

should be chosen arbitrarily. This motivates the introduction of the second constraint (14), which will play the role of additional non-linear equation in the set of non-linear equations and will give us an accurate estimate for the multiplicative constatnt.

The performance measures U_{ki}^*, $\langle n_{ki} \rangle^*$ and ϕ^* of the psuedo open network are determined by assuming that the intervals between class-i customers, $i = 1, 2, \ldots R$, to and from centre k, $k = 1, 2, \ldots, M$, are renewal processes. These quantities can be determined by solving an open network under repetitive service blocking and random destination with additional contraints given by (14). Having solved the psuedo open network, the Lagrangian coefficients can then be determined and subsequently the ME solution (12) can be used to approximate the joint steady state of the network. However, unless the network is separable [13] the throughputs λ_{ki}, obtained from the truncated ME solution of the psuedo open network does not satisfy the job flow-balance equations

$$\lambda_{ki} = \sum_{k=1}^{M} \hat{a}_{kim} \lambda_{ki}, k = 1, 2, \ldots, M, i = 1, 2, \ldots, R, \tag{15}$$

In fact, the routing matrix $\hat{A} = (\hat{a}_{kim})$, obtained from the solution of the psuedo open network, does not correspond to the exact routing behaviour of the throughputs λ_{ki}^* and thus equation (13) can only be used as an approximation. Consequently the throughput λ_{ki}^* may be approximated by

$$\lambda_{ki}^* = \hat{\mu}_{ki} U_i,$$

where $\hat{\mu}_{ki}$ is the mean effective service rate of station k of the psuedo open network.

To obtain, now, a set of throughputs, λ_{ki}, $k = 1, 2, \ldots, M$, $i = 1, 2, \ldots, R$, that satisfy (15), the ME probabilities are corrected by introducing a new coefficient g_{0ki}, to adjust the Lagrangian coefficients, g_{ki}, which correspond to the ME utilization constraints. The adjustment is carried out by substituting coefficients g_{ki} by $\tilde{g}_{ki}$, using the relation, $\tilde{g}_{ki} \leftarrow g_{0ki} g_{ki}$, $k = 1, 2, \ldots, M$, $i = 1, 2, \ldots, R$. The correction factors, g_{0ki} are initially set equal to 1, and are evaluated iteratively until the relation below is satisfied;

$$\frac{\hat{\rho}_{ki}}{U_{ki}} = \frac{\hat{\rho}_{mi}}{U_{mi}} = \text{constant} \, \forall k \neq m, \, k, m = 1, 2, \ldots, M, \, i = 1, 2, \ldots, R,$$

where $\hat{\rho}_{ki}$ is the utilization of station k for customers of class i, of the psuedo open network.

The following heuristic formulae have been proposed:

(i) $$g_{0ki} \leftarrow \frac{\hat{\rho}_{ki} L_{ci}}{\rho_{ki} \sum_{m=1}^{M} \frac{\langle n_{mi} \rangle \hat{\rho}_{mi}}{\rho_{mi}}} g_{0ki}, \tag{16}$$

Kouvatsos [14]

(ii) $$g_{0ki} \leftarrow \left[\frac{1 - \rho_{ki}}{\rho_{ki}} \right] \frac{\hat{\rho}_{ki} L_{ci}}{\sum_{m=1}^{M} \frac{\langle n_{mi} \rangle \hat{\rho}_{mi}}{\rho_{mi}} - \hat{\rho}_{ki} L_{ci}} g_{0ki}, \tag{17}$$

Walstra [15]

where $\rho_{m,i} \neq 0, \forall m = 1, 2, \ldots, M$.

Both expressions have been thoroughly tested and generally give convergence. However, the speed of convergence of each approach varies with the configuration of the network and as a consequence, it is very difficult to know before hand which one is better. On balance expression (16) has been found experimently more credible and therefore is used in the ME algorithm.

3.3   Convolution algorithm for product form Network

To reduce the time and space complexities of the ME algorithm, the maximum entropy solution of a closed queueing network given by equation (12) can be obtained more efficiently by using convolution type formulae analogous to the ones used in the standard convolution algorithm [16].

Sum of all state probabilities must equal 1

$$
1 = \sum_{\mathbf{S} \in \mathbf{S}(\mathbf{L}_{cR}, M)} p(\mathbf{S}) \tag{18}
$$

$$
= \frac{1}{Z(\mathbf{L}_{cR}, M)} \sum_{\mathbf{S} \in \mathbf{S}(\mathbf{L}_c, M)} \prod_{k=1}^{M} \omega_k(\mathbf{S}_k) \tag{19}
$$

Therefore, from above equation,

$$
Z(\mathbf{L}_{cR}, M) = \sum_{\mathbf{S} \in \mathbf{S}(\mathbf{L}_{cR}, M)} \prod_{k=1}^{M} \omega_k(\mathbf{S}_k) \tag{20}
$$

An auxiliary function is essential for the derivation of a computational procedure

$$
z(\mathbf{n}_R, k) = \sum_{\mathbf{S} \in \mathbf{S}(\mathbf{n}_R, k)} \prod_{\ell=1}^{k} \omega_\ell^*(\mathbf{S}_\ell) \tag{21}
$$

where $\mathbf{n}_R = (n_1, n_2, \ldots, n_R)$. Let $\mathbf{k}_R = \mathbf{n}_R$ and

$$
n_i = K_k, K_k + 1, \ldots, L_{ci}, \forall k = 1, 2, \ldots, M, \forall i = 1, 2, \ldots, R,
$$

$$
\omega_\ell^*(\mathbf{S}_\ell) = \begin{cases} 0 & \begin{cases} 0 \leq \sum_{i=1}^{R} n_{ki} < K_\ell \\ L_\ell < \sum_{i=1}^{R} n_{\ell i} \leq L \end{cases} \\ \omega_\ell(\mathbf{S}_\ell) & K_\ell \leq \sum_{i=1}^{R} n_{\ell i} \leq L_\ell \end{cases} \tag{22}
$$

equation (21) can be written as:

$$
z(\mathbf{n}_R, k) = \sum_{\substack{\mathbf{k}_R = 0}}^{\mathbf{n}_R} \sum_{\substack{\mathbf{S} \in \mathbf{S}(\mathbf{n}_R, k) \\ \mathbf{n}_k = \mathbf{k}_R}} \prod_{\ell=1}^{k} \omega_\ell^*(\mathbf{S}_\ell) \tag{23}
$$

The normalizing constant $Z(\mathbf{L}_c, M)$,can be computed using the following recursive formula obtained after some manipulation from eq. (23):

$$z(\mathbf{n}_R, k) = \sum_{\mathbf{k}_R=\mathbf{K}_k}^{min(\mathbf{L}_c, \mathbf{n}_R - \sum_{m=1}^{k-1} K_m)} \omega_k^*(\mathbf{k}_R) z(\mathbf{n}_R - \mathbf{k}_R, k-1), \tag{24}$$

where $n_i = \sum_{m=1}^{k} K_m, \ldots, L_{ci}$ $k = 2, \ldots, M, i = 1, 2, \ldots, R$ with the initial conditions,

$$z(\mathbf{n}_R, 1) = \begin{cases} \omega_1(\mathbf{n}_R) & n_i = K_1, K_1 + 1, \ldots, L_{ci}, \\ 0 & n_i = L_{ci} + 1, \ldots, L \end{cases}.$$

3.4 An efficient calculation of the Marginal Queue Length Distribution

The marginal proabability per station of finding $\mathbf{n}_R$ customers, where $\mathbf{n}_R = (n_1, n_2, \ldots, n_R)$ at queue k is given by

$$
\begin{aligned}
p_k(\mathbf{n}_R) &= \sum_{\mathbf{S} \in \mathbf{S}(\mathbf{L}_{cR}, M)} p(\mathbf{S}) \\
&= \sum_{\mathbf{S} \in \mathbf{S}(\mathbf{L}_{cR}, M)} \frac{1}{Z(\mathbf{L}_{cR}, M)} \prod_{k=1}^{M} \omega_k(\mathbf{S}_k) \\
&= \frac{\omega_k(\mathbf{n}_R)}{Z(\mathbf{L}_{cR}, M)} \sum_{\mathbf{S} \in \mathbf{S}(\mathbf{L}_{cR}, M)} \prod_{k \neq j=1}^{M} \omega_j(\mathbf{S}_j)
\end{aligned}
\tag{25}
$$

Focusing on the sum that appears in eq. (25), which can be viewed as a normalization constant of a network related to the original network but with station k removed together with its $\mathbf{n}_R$ jobs,[17], and, hence, there are $L - \sum_i n_i$ jobs left within the network. Hence, we define the following auxiliary functions:

$$\zeta^k(\mathbf{n}_R, M) = \sum_{\mathbf{S} \in \mathbf{S}(\mathbf{L}_{cR}, M)} \prod_{k \neq j=1}^{M} \omega_j(\mathbf{S}_j), \forall n_i = K_k, \ldots, L_{ci} \tag{26}$$

Now, eq. (25) becomes

$$p_k(\mathbf{n}_R) = \frac{\omega_k(\mathbf{n}_R)}{Z(\mathbf{L}_{cR}, M)} \zeta^k(\mathbf{L}_{cR} - \mathbf{n}_R, M), \forall n_i = K_k, \ldots, L_{ci} \tag{27}$$

The derivation of an iterative algorithm to compute the auxilary function relies on the marginal probabilities. The marginal probabilities $p_k(\mathbf{n}_R)$ sum up to 1.

$$
\begin{aligned}
1 &= \sum_A p_k(\mathbf{k}_R), 1 \leq k \leq M, \\
Z(\mathbf{L}_{cR}, M) &= \sum_{\mathbf{n}_R=0}^{\mathbf{L}_c} \omega_k(\mathbf{n}_R) \zeta^k(\mathbf{L}_{cR} - \mathbf{n}_R, M)
\end{aligned}
$$

Solving with respect to $\zeta^k[\mathbf{L}_{cR} - \mathbf{k}_k, M]$, we have

$$\zeta^k[\mathbf{L}_{cR} - \mathbf{k}_k, M] = Z(\mathbf{L}_{cR}, M) - \sum_{\mathbf{n}_R = \mathbf{K}_k + 1}^{\mathbf{L}_k} \omega_k(\mathbf{n}_R)\zeta^k(\mathbf{L}_{cR} - \mathbf{n}_R, M) \quad (28)$$

which, in fact, defines the recursive scheme

$$\zeta^k(\mathbf{j} - \mathbf{k}_k, M) = Z(\mathbf{j}, M) - \sum_{\mathbf{n}_R = \mathbf{K}_k + 1}^{min(\mathbf{j}, N_k)} \omega_k(\mathbf{n}_R)\zeta^k(\mathbf{j} - \mathbf{n}_R, M) \quad (29)$$

where $\mathbf{j} = \sum_{\ell=1}^{M} \mathbf{K}_\ell, \ldots, \mathbf{L}_\ell$, with initial conditions

$$\zeta^k(\mathbf{J} - 1, M) = 0, \text{ and } \zeta^k(\mathbf{J}, M) = 1, \text{ where } \mathbf{J} = \sum_{k \neq \ell = 1}^{M} \mathbf{K}_\ell$$

3.5 The ME Algorithm

INPUT

- $L, M, R,$

- $\mathbf{L}_c = (L_{c1}, L_{c2}, \ldots, L_{cR}),$

- for each station k

  - type of service discipline,
  - $\mu_k = (\mu_{k1}, \mu_{k2}, \ldots, \mu_{kR})$

- $\{A_{kim}\}$, the transition probability matrix

PART A {* Solve the pseudo open network *}

Step 1 Feedback correction
For each queue k, $k = 1, 2, \ldots, M$, and class i, $i = 1, 2, \ldots, R$, with $a_{kim} > 0$, substitute

$$\mu_{ki} \leftarrow \mu_{ki}(1 - a_{kim})$$

$$a_{kim} \leftarrow \begin{cases} 0 & , k = m \\ a_{kim}/(1 - a_{kik}) & , m \neq k \end{cases}$$

Step 2 Initialize $\pi_k \leftarrow$ any valu in (0,1), $k, = 1, 2, \ldots, M$;

Step 3 Solve the system of non-linear equations $\{\pi_k\}$, $k = 1, 2, \ldots, M$, and $\{L_{ci}\}$, $i = 1, 2, \ldots, R$,

$$\pi_k \leftarrow \frac{1 - \rho_k}{1 - \rho_k^2 X_k^{N_k - 1}} \rho_k X_k^{N_k - 1}, \text{ and}$$

$$L_{ci} \leftarrow \sum_{k=1}^{M} \langle n_{k,i} \rangle_*,$$

using Newton-Raphson method:

Step 3.1 Calculate effective flow transition probabilities $\{\hat{a}_{kim}\}$:

$$\hat{a}_{kim} \leftarrow a_{kim} * (1 - \pi_m)/(1 - \pi_{cki})$$

Step 3.2 Calculate effective job flow balance equations:

$$\lambda_{k,i} \leftarrow \sum_{k \neq m=1}^{M} \hat{a}_{kim}\lambda_{mi}, \forall k, m, i;$$

Step 3.3 Calculate blocking probabilities $\{\pi_{cki}\}$

$$\pi_{cki} \leftarrow \sum_{k \neq m=1}^{M} (\sum_{j=1}^{R} a_{kim})\pi_m, \forall k, m, i;$$

Step 3.4 Calculate effective service rates $\{\hat{\mu}_{ki}\}$

$$\hat{\mu}_{ki} \leftarrow \mu_{ki}(1 - \pi_{cki}), \forall k, m, i;$$

Step 3.5 Calculate overall arrival rates $\{\Lambda_{ki}\}$

$$\Lambda_{ki} \leftarrow \lambda_{ki}/(1 - \pi_k), \forall k, i;$$

PART B {* Solve closed queueing network *}

Step 4 Introduce new Lagrangian coefficients, $\{g_{0ki}\}$, and initialize them by setting

$$g_{0ki} \leftarrow 1, \forall k, i;$$

Step 5 Set $\tilde{g}_{ki} \leftarrow g_{0ki}g_{ki}$

Step 6 Use the convolution algorithm of section 3.3 to obtain marginal statistics:

step 7 Adjust the values of $\{g_{0ki}\}$ using the formula

$$g_{0ki} \leftarrow \frac{\hat{\rho}_{ki}/\sum_{m=1}^{M} \hat{\rho}_{mi}}{U_{ki}/\sum_{m=1}^{M} U_{m,i}} g_{0ki}$$

Step 8 Return to step 4 until eqs. $\hat{\rho}_{ki}/U_{ki} = \hat{\rho}_{mi}/U_{mi} = $ constant $\forall k \neq m$, are satisfied.

4 Numerical results

The credibility of the proposed ME algorithm is demonstrated via typical numerical experiments presented in Tables 1 and 2, and Figures 1-4. Table 1 shows the input data for a closed exponential queueing network with 3 station and 2 classes of jobs whilst Table 2 contains typical performance measures obtained from ME analysis and simulation (obtained via QNAP-2 at 95% confidence interval [18]). For validaton purpose the following tolerances (TOL) and relative differences (RD) have been used:

For the marginal mean queue lengths and queue length distribution TOL is defined by

$$\left| \frac{\text{SIM}(\langle n_{ki} \rangle) - \text{MEM}(\langle n_{ki} \rangle)}{\sum_{k=1}^{M} \text{SIM}(\langle n_{k,i} \rangle)} \right|, \text{ and}$$

$\left| \mathrm{SIM}(p_{ki}(0)) - \mathrm{MEM}(p_{ki}(0)) \right|$, respectively

for each class i and station k. For throughputs RD is defined by

$$\left| \frac{\mathrm{SIM}(\lambda_{ki}) - \mathrm{MEM}(\lambda_{ki})}{\mathrm{SIM}(\lambda_{k,i})} \right|,$$

Figures 1-2 represent the tolerances at each station for mean queue length $\langle n_{k,i} \rangle$, empty buffer probabilities, $p(0)$ and full buffer probabilities per class whilst figures 3-4 show the relative throughput differences obtained for each class i at all stations. It can be observed that the ME solutions are very comparable to those obtained by corresponding simulation models with maximum error tolerance 0.012 and maximum error RD 0.064. **Experiment**

Table 1

Raw data:	$M = 3,\ R = 2,\ L_1 = 3,\ L_2 = 3,$ FCFS
Queue 1:	$N = 3,\ \mu = (5,4),\ C_s^2 = (1,1),$
	$a_{112} = 0.5,\ a_{113} = 0.5,\ a_{212} = 0.5,\ a_{213} = 0.5,$
Queue 2:	$N = 3,\ \mu = (6,5),\ C_s^2 = (1,1),$
	$a_{121} = 0.5,\ a_{123} = 0.5,\ a_{221} = 0.5,\ a_{223} = 0.5$
Queue 3:	$N = 3,\ \mu = (4,6),\ C_s^2 = (1,1),$
	$a_{131} = 0.5,\ a_{132} = 0.5,\ a_{231} = 0.5,\ a_{232} = 0.5$

Table 2
Marginal Statistics

Queue	Class	Measures	MEM	SIM	TOL
1	1	$\lambda_{1,1}$	1.469	1.400	0.049
		$\langle n_{1,1}\rangle$	1.030	1.026	0.001
		$u_{1,1}$	0.414	0.410	0.004
		$p_{1,1}(0)$	0.258	0.257	0.001
		$p_{1,1}(5)$	0.019	0.018	0.001
	2	$\lambda_{1,2}$	1.464	1.410	0.038
		$\langle n_{1,2}\rangle$	1.130	1.120	0.002
		$u_{1,2}$	0.516	0.517	0.001
		$p_{1,2}(0)$	0.214	0.226	0.012
		$p_{1,2}(5)$	0.027	0.029	0.002
2	1	$\lambda_{2,1}$	1.561	1.503	0.038
		$\langle n_{2,1}\rangle$	0.872	1.876	0.001
		$u_{2,1}$	0.389	0.391	0.002
		$p_{2,1}(0)$	0.353	0.347	0.006
		$p_{2,1}(5)$	0.015	0.014	0.001
	2	$\lambda_{2,2}$	1.556	1.533	0.015
		$\langle n_{2,2}\rangle$	0.945	1.946	0.001
		$u_{2,2}$	0.466	0.469	0.003
		$p_{2,2}(0)$	0.314	0.317	0.003
		$p_{2,2}(5)$	0.019	0.020	0.001
3	1	$\lambda_{3,1}$	1.504	1.463	0.028
		$\langle n_{3,1}\rangle$	1.099	1.098	0.0002
		$u_{3,1}$	0.541	0.545	0.004
		$p_{3,1}(0)$	0.235	0.242	0.008
		$p_{3,1}(5)$	0.027	0.029	0.002
	2	$\lambda_{3,2}$	1.500	1.410	0.064
		$\langle n_{3,2}\rangle$	0.925	0.933	0.002
		$u_{3,2}$	0.359	0.359	0.000
		$p_{3,2}(0)$	0.317	0.305	0.012
		$p_{3,2}(5)$	0.015	0.013	0.002

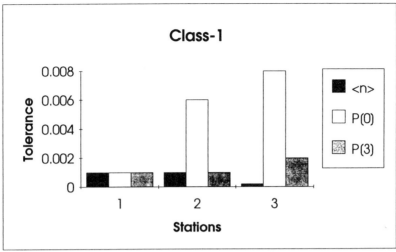

Figure 1: TOLERANCES of typical performance
measures for class-1 jobs in the network

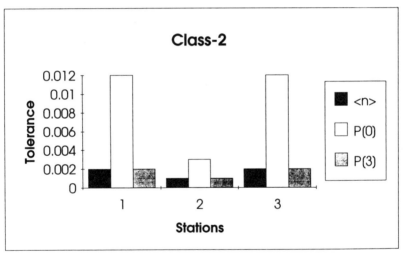

Figure 2: TOLERANCES of typical performance
measures for class-2 jobs in the network

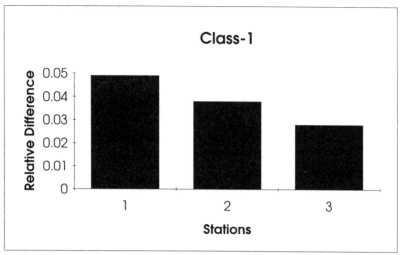

Figure 3: Relative Differences of the marginal throughputs
among all queues of network for calss-1 jobs

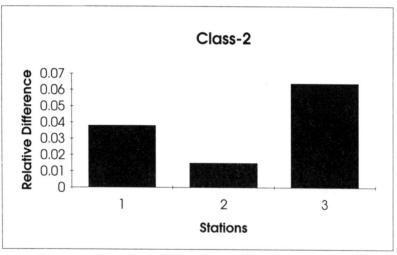

Figure 4: Relative Differences of the marginal throughputs
among all queues of network for calss-2 jobs

5 Conclusions

A new ME approximation algorithm is proposed for obtaining the queue length disrtibutions and typical performance metrics for arbitrary multiple class closed exponential QNMs with finite capacity, complete buffer sharing, RS-RD blocking and mixed service disciplines. An efficient convolution algorithm for the computation of nomalizing constant is presented. Numerical experiments against simulation indicate the credibility of the ME algorithm. Future work will extend the ME solution to GE (Generalised Exponential) and GGeo (Generalised Geometric) type QNMs.

References

1. R.O.Onvural, Survey of Closed Queueing Networks with Blocking, *ACM Comput. Surv.* Vol. 22(2), 83-121, 1990.

2. H.G.Perros, Approximation Algorithms for Open Queuing Networks with Blocking, *Stochastic Analysis of Computer and Communication Systems*, H. Takagi (ed.), North Holland, pp. 451-494, 1990.

3. H.G.Perros, Queueing Networks with Blocking. *Oxford University Press*, 1994.

4. I.F.Akyildiz and H. Von Brand, Exact Solutions for Open, Closed and Mixed Queueing Networks with Rejection Blocking, *Theoret. comput. Sci.* Vol. 64, 203-219, 1989.

5. I.F.Akyildiz and H. Von Brand, Computational Algorithm for Networks of Queues with Rejection Blocking, *Acta Inform.* Vol. 26, 559-576, 1989.

6. I.F.Akyildiz and H. Von Brand, Central Server Models with Multiple Job Classes, State Dependent Routing and Rejection Blocking, *IEEE Trans. Software Eng.* Vol. 15, pp. 1305-1312, 1989c.

7. T. Choukri, Exact Analysis of Multiple Job Classes and Different Types of Blocking, R.O. Onvural and I.F..Akyildiz, eds., *Proc. 2nd International Workshop on Queueing Networks with Finite Capacity*, Research Triangle Park, pp.25-40, May 1992.

8. I.F.Akyildiz and C.C.Huang, Exact Analysis of Multi-Job Class Networks of Queues with Blocking-After Service, *Proc. 2nd International Workshop on Queueing Networks with Finite Capacity*, R.O. Onvural and I.F..Akyildiz, (eds.), Research Triangle Park, pp. 258-271, May 1992.

9. D.D. Kouvatsos and N.P. Xenios, MEM for Arbitrary Queueing Networks with Multiple General Servers and Repetitive-Service Blocking, *Performance Evaluation*, Vol. 10(3), 1989.

10. D.D. Kouvatsos and S.G. Denazis, MEM for Arbirary Queueing Networks with Blocking: A Revision, *Technical Report* CS-18-91, University of Bradford, May 1991.

11. D.D. Kouvatsos and S.G. Denazis, Entropy Maximised Queueing Networks with Blocking and Multiple Job Classes, *Performance Evaluation*, Vol. 17, pp. 189-205, 1993.

12. D.D. Kouvatsos, Maximum Entropy and the G/G/1/N Queue, *Acta Informatica*, Vol. 23, pp. 545-565, 1986.

13. F. Baskett, K.M. Chandy, R.R. Muntz and F.G. Palacios, Open, Closed and Mixed Networks of Queues with Different Classes of Customers, *J. ACM*, Vol. 22(2), pp. 248-260, 1975.

14. D.D. Kouvatsos, A Universal Maximum Entropy Algorithm for the Analysis of General Closed Networks, *Computer Networking and Performance evaluation*, Hasegawa, T. et al (eds.), North Holland, pp. 113-124, 1986.

15. R. Walstra, Iterative Analysis of Networks of Queues, *Ph.D. Thesis*, Tech. Rep. CSE1-166, Toronto Univ., 1984.

16. S.C. Bruell and G. Ballbo, Computational Algorithm for Closed Queueing Networks, North Holland, New York, 1980.

17. J. Labetoulle, and G. Pujolle, Isolation Method in a Network of Queues, *IEEE Transaction on Software Eng.*, Vol SE-6 pp. 373-381, 1980.

18. M. Veran and Potier D. QNAP-2: A Portable Environment for Queueing Network Modelling Techniques and Tools for Performance Analysis, Potier D. (ed.), North Holland, pp. 25-63, 1985.

Formal Methods for Performance

A Stochastic Process Algebra Based Modelling Tool

Holger Hermanns and Vassilis Mertsiotakis

University of Erlangen-Nürnberg, IMMD VII, Martensstr. 3,
91058 Erlangen, Germany

Abstract

The incorporation of time into classical process algebras aiming at the integration of functional design and performance analysis has become very popular recently. There are many ways to include time in process algebras. Current research in this area concentrates mainly on the annotation of actions with exponentially distributed random variables. This allows us to make use of a large repertoire of analysis algorithms. This paper presents some first results of ongoing work which aims at providing a tool for the efficient performance evaluation and functional analysis of computer and communication systems based on the stochastic process algebra paradigm. It provides facilities for model creation, reachability analysis, as well as several numerical algorithms for the solution of the underlying Markov chain and the computation of characteristic performance measures.

1 Introduction

Stochastic process algebras (SPA) have been introduced as an extension of classical process algebras, like CCS [21] or CSP [17], with timing information aiming mainly at the integration of functional design and quantitative analysis of computer systems [8]. Time is represented by associating exponentially distributed random variables to activities in the model.

As in classical process algebras, the main characteristic of SPA is constructivity, i.e. the ability to describe complex systems as a composition of several smaller ones. This feature together with a powerful abstraction mechanism to hide internal structures form the basis of the successful application of SPA for describing and analysing systems of various domains [11, 12, 13].

TIPP-tool (*timed processes and performability evaluation*) is a prototype modelling tool for creating and evaluating stochastic process algebra models of parallel and distributed systems. It supports a LOTOS-oriented input language [1] and apart from facilities to apply functional analysis based on reachability analysis, it comes up with a few numerical solution modules for the stationary as well as transient analysis of the Continuous Time Markov Chain (CTMC) underlying a TIPP-specification. The input language and the emphasis on numerical evaluation rather than analytical evaluation represent the main difference to the *PEPA Workbench*, another modelling tool for SPA that relies upon symbolic evaluation of the underlying Markov chain [6].

The structure of the TIPP-tool is shown in Fig. 1. Model descriptions are syntactically and semantically analysed using a parser written in the functional programming language *Standard ML*. The parser has been generated using the tools *ML-YACC* and *LEXGEN* from the New Jersey Standard ML toolkit. The use of this programming language has shown to be especially useful for implementing the formal semantics of TIPP, the calculus the tool is based on [14].

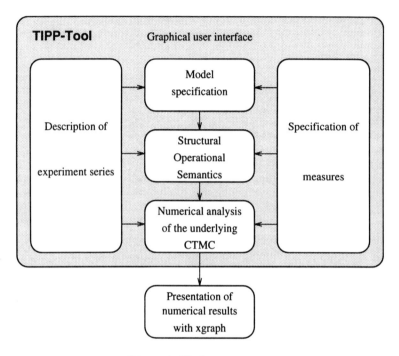

Figure 1: Tool components

The state space that has been derived formally serves as a base for further reduction into a CTMC. For the steady state analysis of the underlying CTMC we are using a variety of numerical algorithms. TIPP-tool supports also transient analysis by providing methods to compute the mean time to absorption of an absorbing Markov chain or the transient state probabilities. For the latter, a refined randomization scheme is provided, similar to the method adopted by Lindemann in [20]. All methods work on a sparse matrix data structure [18].

The result of numerical analysis is usually a vector with state probabilities. In order to obtain more sophisticated and expressive results the user can specify measures. This is done via weights that are assigned to states that match a certain regular expression the user has to specify. Experiment series are also supported by allowing rates to be symbolic variables. The specification of the model as well as the measures and experiments is supported through a Tcl/Tk-graphical user interface.

The model specification language of the tool is explained in detail in the following section. Section 3 describes the analysis modules. We conclude the paper with a discussion on intended extensions to our tool.

2 Model Description Language

In this section we introduce the specification language of our tool designed to be suitable for describing the behaviour of a system on a high level [22]. This description can be refined step by step, in order to obtain a more accurate model of a system.

The language is closely related to LOTOS, but enriched with stochastic timing information. This is achieved by the convention that a certain random distribution function is associated with every activity specifying its duration. The set of random distributions that can be handled with TIPP is restricted to exponentially distributed ones. This implies that every term in our language can be interpreted as a high level description of a Continuous Time Markov Chain (CTMC). Exponential distributions can be characterized by a single parameter, drawn from the set of positive reals. Therefore, each activity of the system possesses a specific name and a parameter, called rate, characterizing its duration.

The major building blocks of a system specification are behaviour expressions that perform actions. Behaviour expressions can be composed out of smaller ones by combining them via a language operator. The set of operators that can be used in the input language is listed in Table 1. Their intuitive meaning is as follows:

Inaction: The process `stop` denotes an inactive process which cannot perform any action.

Action Prefixing: The process $(a, \lambda); P$ (or $(a, \lambda).P$) carries out the action with name a after a duration which is exponentially distributed with parameter λ. Thus the mean time of the delay until the action occurs is $1/\lambda$. The process subsequently behaves as process P. Internal actions, denoted by (i, λ), are unobservable for the environment, as far as functionality is concerned. A delay with mean time $1/\lambda$, however, can be witnessed. Actions that happen without delay will be discussed in Section 4.2.

Name	Syntax
inaction	`stop`
action prefix	
– observable	$(a, r); P$ or $(a, r).P$
– unobservable	$(i, r); P$ or $(i, r).P$
choice	$P\ []\ Q$ or $P + Q$
parallel composition	
– pure interleaving	$P\ |||\ Q$
– full synchronization	$P\ ||\ Q$
– general case	$P\ |\ [a_1, \ldots, a_n\ |]\ Q$
hiding	`hide` $a_1, \ldots, a_n$ `in` P
process instantiation	$P[a_1, \ldots, a_n](r_1, \ldots, r_m)$

Table 1: Syntax for behaviour expressions. P and Q are place-markers for any behaviour expressions. $a, a_1, \ldots, a_n$ stand for action names. The rates $r, r_1, \ldots, r_m$ can be instantiated by arbitrary positive real numbers.

190

Choice: The process P [] Q (or $P + Q$) represents a system which may behave either as P or as Q. The decision is taken as soon as either P or Q performs any action. The duration until this choice is made is obviously the minimum of the concerned random distributions. As a matter of fact, due to our restriction to exponential distributions, the according random variable is given by the sum of the involved rates.

Parallelism: Concurrency and cooperation of behaviour expressions can be modelled by specific operators. P ||| Q denotes that P and Q proceed independently. In contrast P || Q implies that both proceed in complete synchrony, except for possible internal actions. The general case, where parts of the behaviour must proceed synchronously and other parts are independent is specified by P |[$a_1, \ldots, a_n$]| Q. The processes P and Q proceed independently and concurrently with any action whose action name is not contained in the list of synchronizing actions $a_1, \ldots, a_n$. However, for action names a_j, the processes P and Q must cooperate in order to perform the action. Such actions can be completed only when they are enabled in both P and Q. Thus both partners have to synchronize on this action before proceeding. The rate of a synchronized action is given by the multiplication of the rates of the involved actions. This implies that each of the partners has a scalable influence on the mean duration of a synchronizing action. In practice many cooperations consist of a passive and an active component (e.g. workload/machine). Passivity is expressed by the rate 1 — the neutral element of multiplication.

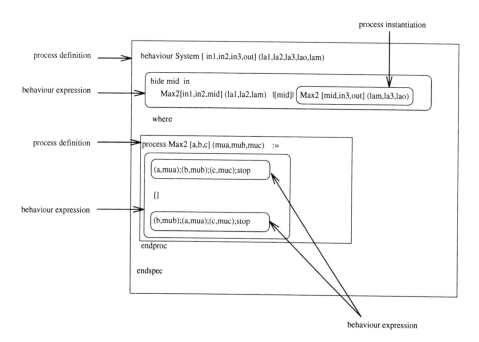

Figure 2: Process definition

Hiding: The process hide $a_1, \ldots, a_n$ in P behaves as the process P except that any actions within the list $a_1, \ldots, a_n$ are *hidden*, meaning that their name is not visible. Instead they appear with the name i and can be regarded as an internal delay by the component. Hiding is important for a structured design methodology: At every structural level the actions that do not interact with the environment can be hidden in order to be invisible for external components.

Process Instantiation: Process instantiations $P[a_1, \ldots, a_n](r_1, \ldots, r_m)$ may occur within behaviour expressions and resemble the invocation of procedures in procedural programming languages such as PASCAL. A process instance is generated by replacing the list of formal action names and rates in the corresponding process definition process $P[\ldots](\ldots) := \ldots$ endproc by actual action names $a_1, \ldots, a_n$ and rates $r_1, \ldots, r_m$.

A complete specification is constituted by a number of process definitions, each describing a process scheme with generic action names and action rates within a behaviour expression. The typical structure of a process definition is given in Figure 2. It defines a behaviour expression that itself is built out of smaller behaviour expressions by applying any of the language operators.

The behaviour of a specification is formally defined in a structural operational style (see [14] for more details). A Markovian Labelled Transition System (MLTS) is used as domain for the semantic model, consisting of nodes representing processes and arcs between them. These arcs are labelled with pairs of actions and rates[1]. Each of these arcs $\square \xrightarrow{\,a,\lambda\,} \square$ symbolizes that the first process can evolve to the second by carrying out the action a with rate λ. Figure 3 for example shows an MLTS, that is the semantic model of the specification in Figure 2.

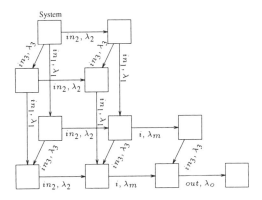

Figure 3: Markovian Labelled Transition System

---

[1] An additional third label is used for some subtle management purposes, cf. [14].

3 Analysis of the Underlying CTMC

Quantitative analysis is based on the MLTS derived formally from a process specification as shown in the last section. Here, both labels of an arc, denoting the action name and rate, are regarded as being significant, as are the arc multiplicities. For any SPA model with finite state space the underlying CTMC can be derived directly by associating a state with each node of the MLTS. The transitions of the CTMC are the amalgamation of all the arcs joining the nodes, and the transition rate is the sum of the individual activity rates. The action name information is not incorporated into the CTMC representation but it is often crucial in defining the measures to be extracted from a model.

Characteristic performance and dependability measures based on equilibrium behaviour are extracted via the steady state solution of the CTMC, i.e. solving the matrix equation

$$\pi Q = 0 \qquad \text{subject to} \qquad \sum_i \pi_i = 1$$

using standard numerical techniques, where Q denotes the infinitesimal generator matrix of the CTMC. The steady state probabilities π_i can be interpreted as the relative frequency with which the behaviour corresponding to each process state i is exhibited by the system. From these values it is straightforward to obtain high-level measures such as throughput, or utilization. Unfortunately the only sufficient condition for a unique equilibrium solution is strong connectivity of the MLTS.

If the latter condition is not fulfilled or steady state measures do not give a sufficient insight into the system behaviour, it is possible to carry out transient analysis. Standard algorithms like randomization or calculation of higher moments may be used to find time-related properties of systems. Transient analysis is central to the integration of qualitative and quantitative aspects of model analysis as it allows quantitative implications of qualitative properties of a system to be investigated [11].

3.1 Solution Methods

Our tool offers a small set of solution algorithms for steady state analysis as well as transient analysis of the underlying CTMC. For the former we provide the following methods: *LU-factorization*, *power method* and *Gauß-Seidel* method. The first two methods can be used efficiently for systems with less than 1000 states, while the latter method enables us to analyse chains with up to 100000 states. All the algorithms are implemented in C and work on a data structure for sparse matrices that came along with the Sparse1.3 package of K.S. Kundert [18]. Apart from basic matrix subroutines this package contains a LU-factorization solver. We extended this library with two iterative solution methods for steady state analysis.

For transient analysis we implemented the moment method for the computation of the mean time to absorption. Time-related state probabilities can be computed using the randomization method that we integrated into this library as well. The main problem with the randomization method is the computation of Poisson-probabilities for stiff differential equation systems. Therefore, we adopted a similar refined randomization method based on a numerically stable subroutine for the computation of Poisson-probabilities as proposed in [4] and has been used by several other tool developers [2, 20].

3.2 Specification of Measures

The result of numerical analysis is the state probability distribution vector. This is usually a too detailed measure and not well-suited for getting insight into the system behaviour. More high-level measures can be computed out of this vector. For this, the user has to specify a number of measures using the dialog box shown in Fig. 4.

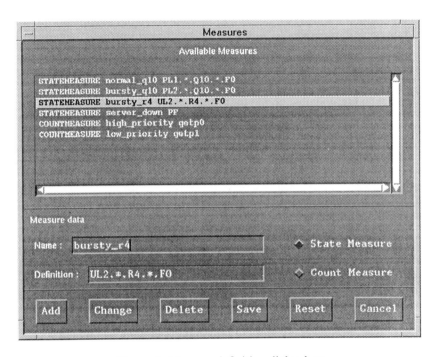

Figure 4: Measure definition dialog box

TIPP-tool supports two types of measures. *Count measures* that are typically throughputs of actions the user has to specify and *state measures*. The latter ones are specified using regular expressions the user has to supply for each state measure. The measure is computed by summing over all states that match the specified expression. Typical state measures are: blocking probabilities, probability of deadlock states, probability of being in a repair phase, etc.

3.3 Description of Experiment Series

It is very often desirable to use symbolic names for activity rates instead of fixed numerical values. In that case, the user has to describe an experiment using the dialog box shown in Fig. 5. Here, the user can specify the actual numerical values of the activity rates. In each experiment there can be one activity rate which may be varied. The user needs to specify the first value, the last value, and a stepwidth.

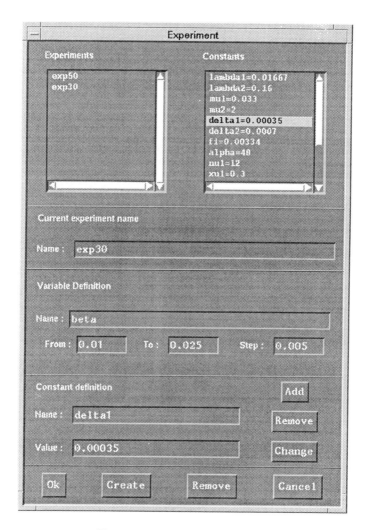

Figure 5: Experiment dialog box

3.4 Presentation of Numerical Results

If the user has specified an experiment using a variable rate, the tool automatically analyses the model for each value within the variable range and the results can afterwards be presented graphically. For this purpose we employ an existing plotting program called `xgraph` that is freely available from UC Berkeley. TIPP-tool transforms the numerical results into an appropriate input format (see Fig. 6).

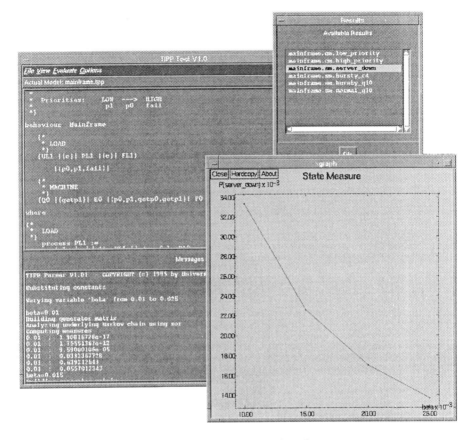

Figure 6: Graphical user interface

3.5 Case Studies

Even though the present paper is our first publication about the TIPP-tool, it has already been used successfully for a number of case studies that have been published in various conference proceedings and journals. The case studies we investigated include:

- data transfer with multiple send-and-wait communications [9]

- multiprocessor and task graph modelling [8]

- electronic mail system [11]

- alternating bit protocol [13]

- robot control system [7]

- multiprocessor mainframe with software failures [12, 16]

All these models were built up out of small modules that were composed to obtain the entire system. The multiprocessor mainframe example additionally contains concepts for the decomposition of the generated state space along the structure given by the modular process description.

We computed various types of performability measures such as response time, availability of components, failure probabilities, mean execution times and many more.

4 Extensions

In this chapter we will outline some extensions that shall be included in the future in order to enhance the modelling convenience. Although these extensions take place mainly on the level of the input language they have a strong impact on the underlying theory.

4.1 Value Passing

The concept of process instantiation that is already integrated in our tool allows us to parameterize processes over action names. Value passing can be regarded as a generalization of this concept. Here, values are attached as parameters to processes as well as actions in the system description. Thus, it gets a lot more easier to describe systems at a lower level of abstraction, where very often buffers have to be described and their respective scheduling strategy. This can become even in the simplest case of a FIFO-buffer a hard and errorprone task if value passing is not provided, as the following example should demonstrate:

```
process Q0[in,out] :=
       (in,1) ; Q1[in,out]
endproc
process Q1[in,out] :=
       (in,1) ; Q2[in,out] [] (out,1) ; Q0[in,out]
endproc
process Q2[in,out] :=
       (in,1) ; Q3[in,out] [] (out,1) ; Q1[in,out]
endproc
    ⋮
process Qmax[in,out] :=
       (out,1) ; Qmax-1[in,out]
endproc
```

Such processes can become arbitrarily complex as soon as more sophisticated scheduling strategies as FIFO have to be implemented. Often the insertion of jobs into the queue is performed by different processes possibly using different action names. This results in process descriptions with a huge number of choices. For these cases the concept of parametric processes and value passing is a very attractive means to help the user in specifying such systems.

Our current language has to be extended by a number of constructs in order to realize value passing, among them are [1]:

data types The precondition for value passing is the support of at least one basic data type, like `integer` or `boolean`. Abstract data type specification languages like ACT ONE that is used in Full LOTOS may be employed to represent more abstract data structures.

value declarations A value declaration has the form ! (*value*) and is usually attached to an action. *value* may be a concrete value, a variable or a value expression if its data type provides any operators.

variable declarations The counterpart to value declarations are variable declarations, which have the form ? (*variable* : t). *variable* is the name of the variable and t its data type. Actions a attributed with variable declarations describe a set of actions a<v> for all values in the domain of type t.

interprocess communication This is achieved by composing two or more processes in parallel that offer the same structured action on which the participating processes are required to cooperate. Three different types of interaction may occur (cf. [1] for more details):

- **value matching** The synchronizing actions are combined with value declarations. If the supplied values are equal synchronization is possible.

- **value passing** Combination of value declaration and variable declaration results in value passing. The effect is that a value is transmitted from one process to the other.

- **value generation** The only remaining case is the one where actions are combined with variable declarations of the **same type** attached to them. Superposing another process in parallel that offers a value of the required type yields a form of multicast communication.

conditional constructs The flexibility of the description is increased if it is possible to describe behaviours that depend on conditions on values, e.g. an action should be enabled if a certain value (buffer size) is below a threshold. Such conditions are usually in square brackets and attached to actions as this example shows:

```
(send,1)!(5) ; stop [] (recv,1)?x:nat [x>0] ; stop
```

generalized choice With this construct it is possible to specify choice among behaviours B(v) for all values of v by writing

```
choice x:nat [] B(x)
```

parametric processes Processes can be parameterized by a list of variable declarations that is attached to the formal action name list. The syntax for such a parameterized process specification is:

$$\text{process } P[a_1,\ldots,a_n](x_1 : t_1,\ldots,x_n : t_n) \ := \ \ldots\text{endproc}$$

Example

```
process Queue[in,out](n:nat, max:nat) :=
  (in,1)?x:nat [n+x lt max+1] ; Queue[in,out](n+1,max)
[]
  (out,1)!n [n gt 0] ; Queue[in,out](n-1,max)
endproc
```

The binary operators lt and gt denote "less than" and "greater than" respectively. We can see clearly how elegant it becomes to describe components like queues using value passing and parametric processes.

4.2 Immediate Actions: Disabling and Enabling

Currently, the algebraic foundations our tool strictly rely on exponentially distributed durations. Recently, TIPP has been extended with *immediate actions* that happen instantaneously if enabled [15]. Their key feature is that they can be eliminated out of the transition system under certain circumstances identified in the paper. Therefore, the extended expressiveness of the language does not influence the treatment of the underlying stochastic process. By virtue of this independence, immediate actions can be integrated into the TIPP-tool without affecting the analysis modules.

The elimination procedure for immediate actions can be implemented in two different ways. First, special graph algorithms for equivalence checking can be used for this purpose operating on the transition system. Another possibility is the syntactic manipulation of the process description in order to automatically eliminate immediate actions. The advantage of the latter is that the elimination happens on the fly, during the construction of the transition system. Both methods rely on algebraic properties that have been established for the enriched calculus.

Once immediate actions have been integrated in our tool it should become easy to enrich our language with two constructs taken from LOTOS, namely *process enabling* and *disabling*. The former construct, denoted by $P \gg Q$, models the sequential execution of two processes P and Q. By means of the disabling operator $P \; [> Q$ the preemption of process P by Q is modelled. The process P can perform arbitrary actions as long as Q, typically representing a timer, does not perform any action. But as soon as Q executes an action, P is stopped and control is handed over to process Q.

The reason why we need immediate actions first before adding these operators to our language lies in the semantics of these operators (cf. [1]). Both make use of internal control activities that are meant to have no impact on the temporal behaviour of the system. Immediate, internal actions allow us to model those control activities in a proper way.

4.3 Measure Definition

As mentioned above, in order to extract more expressive results out of the state probabilities the user has to provide information about what he is interested in. The implemented method associates weights with states that match a certain regular expression specified by the user. This relies on the fact that the formal semantics identifies process terms with states of the transition system. Therefore it is possible to use a regular expression like

```
.*.Q3[in,out].*.
```

in order to refer to the states of the above example where three places of the queue are in use. In the future this will not be as easy as it is nowadays. The inclusion of value passing as sketched in Section 4.1 is one reason for this development. In addition, various research groups are working on different techniques to reduce the state space of the semantic model, cf. [5]. Most of them try to exclude redundant information from the model. As a consequence, groups of states are amalgamated to form a single macro state. Unfortunately, the automation of these techniques implies that those macro states are no longer identifiable by regular expressions. Therefore, new ways to define measures have to be developed.

One approach to solve this problem arises from the fact that process algebras describe the *behaviour* of a system whereas the labelling of states is not significant. States are considered to be different if their subsequent behaviour differs. *Temporal logic* is an adequate means to classify states according to their subsequent behaviour. In [10] the temporal logic CTL^* [3] is used to define classes of states by means of their behaviour. Measures are defined by associating weights with each of these classes.

Even though it is not clear to the authors if the expressive power of CTL^* is really necessary for this purpose, the future development seems to force a shift from state based definition of measures towards behaviour oriented definition. This implies that efficient techniques for investigating the behaviour have to be integrated into the tool. For this purpose it is necessary to take a close look at algorithms used for model checking of temporal logic formulae.

5 Conclusions

In this paper we presented the stochastic process algebra based modelling tool TIPP-tool. Models of parallel and distributed systems can be described in a text-oriented process algebraic specification language. The model description language that is currently supported by our implementation is Basic LOTOS oriented. The main difference to classic process algebra based specification languages is that apart from pure functional behaviour the user can specify timed behaviour by attaching exponentially distributed durations to each activity in the model. This allows the integration of qualitative and quantitative system analysis into one comprehensive design methodology.

We described briefly the various modules of our current implementation. It contains facilities for syntactic and semantic checking of the specification and reachability analysis of the model based on the formal semantics of our process algebra. The resulting state space can be checked for deadlocks or reduced into a CTMC, suitable for further numerical analysis. The various tool components – written partially in C and Standard ML – are integrated by a graphical user interface that uses `Expectk`, an enhanced version of the command language Tcl/Tk which supports dialog with existing interactive applications [19].

One of our main research aims is to integrate some concepts of Full LOTOS to make the input language more comfortable. In Section 4 we showed some first ideas to this direction and outlined some conceptual difficulties arising from this extension. It is obvious that the development of our tool is by far not yet finished and a lot of new features will be added in the near future. Nevertheless, the experience we made already with our tool has been documented in several publications and helped in getting new ideas for future extensions.

References

[1] T. Bolognesi and E. Brinksma. Introduction to the ISO Specification Language LOTOS. In P.H.J. van Eijk, C.A. Vissers, and M. Diaz, editors, *The Formal Description Technique LOTOS*, pages 23–73, Amsterdam, 1989. North-Holland.

[2] J. Couvillion, R. Freire, R. Johnson, W. D. Obal, A. Qureshi, M. Rai, W. H. Sanders, and J. E. Tvedt. Performability Modeling with UltraSAN. *IEEE Software*, 8(5):69–80, 1991.

[3] E.A. Emerson and J.Y. Halpern. "sometimes" and "not never" revisited: on branching versus linear time temporal logic. *Journal of the ACM*, 33:151–178, 1986.

[4] B.L. Fox and P.W. Glynn. Computing Poisson Probabilities. *Communications of the ACM*, 31(4):440–445, 1988.

[5] S. Gilmore and J. Hillston (editors). Special issue on Process Algebras and Performance Modelling. *The Computer Journal*, 1995. to appear.

[6] S. Gilmore and J. Hillston. The PEPA Workbench: A Tool to Support a Process Algebra-Based Approach to Performance Modelling. In G. Haring and G. Kotsis, editors, *7th Int. Conference on Modelling Techniques and Tools for Computer Performance Evaluation*, pages 353–368, Wien, May 1994.

[7] S. Gilmore, J. Hillston, R. Holton, and M. Rettelbach. Specifications in Stochastic Process Algebra for a Robot Control Problem. *accepted for: International Journal of Production Research*, 1995.

[8] N. Götz, U. Herzog, and M. Rettelbach. Multiprocessor and Distributed System Design: The Integration of Functional Specification and Performance Analysis Using Stochastic Process Algebras. In *Tutorial Proc. of the 16th International Symposium on Computer Performance Modelling, Measurement and Evaluation, PERFORMANCE 1993*. Springer LNCS 729, 1993.

[9] N. Götz, U. Herzog, and M. Rettelbach. *Verteilte Systeme - Grundlagen und zukünftige Entwicklungen aus der Sicht des Sonderforschungsbereichs 182 "Multiprozessor- und Netzwerkkonfigurationen"*, chapter TIPP - Einführung in die Leistungsbewertung von verteilten Systemen mit Hilfe von Prozeßalgebren, pages 509–531. BI-Wiss.-Verlag, 1994.

[10] H. Hermanns. Leistungsvorhersage von Verhaltensbeschreibungen mittels temporaler Logik. In Reinhard Gotzhein and Jan Bredereke, editors, *Formale Beschreibungstechniken für verteilte Systeme*. Universität Kaiserslautern, Fachbereich Informatik, 1995.

[11] H. Hermanns, U. Herzog, J. Hillston, V. Mertsiotakis, and M. Rettelbach. Stochastic Process Algebras: Integrating Qualitative and Quantitative Modelling. Technical Report 11/94, Universität Erlangen–Nürnberg, IMMD VII, Martensstr. 3, 91058 Erlangen, May 1994.

[12] H. Hermanns, U. Herzog, and V. Mertsiotakis. Stochastic Process Algebras as a Tool for Performance and Dependability Modelling. In *Proc. of IEEE International Computer Performance and Dependability Symposium*, Erlangen, April 1995. IEEE Computer Society Press.

[13] H. Hermanns, V. Mertsiotakis, and M. Rettelbach. Performance Analysis of Distributed Systems Using TIPP – a Case Study. In J. Hillston and R. Pooley,

editors, *Proc. of the 10th U.K. Performance Engineering Workshop for Computer and Telecommunication Systems*, pages 131–144. Department of Computer Science, University of Edinburgh, 1994.

[14] H. Hermanns and M. Rettelbach. Syntax, Semantics, Equivalences, and Axioms for MTIPP. In U. Herzog and M. Rettelbach, editors, *Proc. of the 2nd Workshop on Process Algebras and Performance Modelling*, pages 71–88, Erlangen-Regensberg, July 1994. IMMD, Universität Erlangen-Nürnberg.

[15] H. Hermanns and M. Rettelbach. Formal Characterisation of Immediate Actions in an SPA with Non-Deterministic Branching. *The Computer Journal*, special issue on Process Algebras and Performance Modelling, 1995. to appear.

[16] U. Herzog and V. Mertsiotakis. Applying Stochastic Process Algebras to Failure Modelling. In U. Herzog and M. Rettelbach, editors, *Proc. of the 2nd Workshop on Process Algebras and Performance Modelling*, pages 107–126, Regensberg/Erlangen, July 1994. Arbeitsberichte des IMMD, Universität Erlangen-Nürnberg.

[17] C.A.R. Hoare. *Communicating Sequential Processes*. Prentice-Hall, Englewood Cliffs, NJ, 1985.

[18] S.K. Kundert and A. Sangiovanni-Vincentelli. A Sparse Linear Equation Solver. Technical report, University of California, Berkeley, 1988.

[19] D. Libes. *Exploring Expect: A Tcl-based Toolkit for Automating Interactive Programs*. O'Reilly & Associates, 1994.

[20] C. Lindemann. Employing the Randomization Technique for Solving Stochastic Petri Net Models. In A. Lehmann and F. Lehmann, editors, *Proc. 6. GI/ITG Conf. on Modelling, Measurement and Evaluation of Computing Systems (MMB '91)*, pages 306–319, Munich, Germany, 1991. Springer.

[21] R. Milner. *Communication and Concurrency*. Prentice Hall, London, 1989.

[22] R. Studt. Syntactical and Semantical Analysis of TIPP processes. Internal study, Universität Erlangen-Nürnberg, IMMD VII, March 1995.

Specification of Stochastic Properties in Real-Time Systems *

Abderrahmane Lakas, Gordon S. Blair, Amanda Chetwynd

School of Engineering, Computing and Mathematical Sciences,

Lancaster University, Lancaster LA1 4YR, UK.

email: {lakas,gordon,amanda}@comp.lancs.ac.uk

Abstract

In this paper we present a new approach to the formal specification of distributed real-time systems using the formal description technique LOTOS together with a stochastic temporal logic STL. This approach previously presented in the context of LOTOS/QTL, is characterized by a *separation of concerns*. The functional behaviour is described in LOTOS without regard for the time critical constraints. The specification is then extended with precise real-time requirements written in STL. We present a method to generate a timing event scheduler from the requirements in order to monitor the functional behaviour.

1 Introduction

Over the past few years there have been several techniques used for the design and the validation of distributed systems. Traditionally, these techniques fall into two distinct fields: *qualitative validation* and *quantitative evaluation*. The former field is concerned by the formal correctness of the system and usually employs formal description techniques. These techniques are based on a range of paradigms such as finite state machines, Petri nets or process algebras. The latter field is mainly based on simulation models by applying stochastic process theory.

More recently, the rise of multimedia applications and real-time systems required the definition of new techniques including both qualitative and quantitative analysis. Indeed, time critical properties have become the main issue in such applications and hence the validation is not complete until these properties are satisfied. Several methods attempted to integrate qualitative and quantitative information in the same description technique. A Real-time CCS is presented in [18, 9] where agents are associated with constant value delays. A timed bisimulation is then defined in order to compare the systems. In [8, 10] the authors present process algebras where delays are stochastically distributed and systems are compared over a Markovian bisimulation. In a more performance oriented work, [13] presented an extension of LOTOS for performance analysis.

Due to the fact that the qualitative and the quantitative description of systems are of different natures, we opted for a methodology that uses two

---

*This work reported in this paper is funded by the UK EPSRC under grant number GR/J44926

description techniques: one for the functional behaviour and one for the quantitative information.

This paper focuses on the use of the LOTOS specification language [5] for the formal description of functional behaviours, and a stochastic temporal logic for the specification of timing constraints. LOTOS has been proved to be a powerful formal description technique (FDT[1]) with a well defined semantics. Considerable research has been carried out on the validation of LOTOS specifications. The validation techniques are mainly based either on the bisimulation relations [12, 16] or on the non exhaustive simulation techniques as in [7]. We also consider STL as a linear real-time temporal logic used for the description of timing properties. The purpose of this language is not only to specify the temporal ordering between the events occurring during the system's life (e.g, *receive* events are always preceded by *send* events in a link protocol), but also to specify the time interval between these events (e.g, the time distance between *send*'s and *receive*'s is exponentially distributed with mean λ). Time critical constraints in real-time applications are commonly gathered under the concept of *Quality of Service* (QoS) [6] expressed in terms of concepts such as *delay*, *throughput* and *jitter*. A wide range of real-time temporal logics can be found in [1, 11, 15].

Our approach is motivated by the desire to maintain a separation between functional concerns and real-time concerns. This approach first adopted in [3], is used in order to tackle complex systems in a simpler way. This mainly allows us to reuse the qualitatively validated specification and extend it with the time constraint requirements for further timing analysis. Also, there is no need to change the functional specification when timing constraints are changed. Furthermore, it is easier for the designer to tune the whole system by adding more time constraints.

2 Background on LOTOS

LOTOS behaviour expressions are defined by the following rule:

$$B := \mathbf{stop} \mid \mathbf{exit} \mid g; B \mid B\backslash S \mid B_1[\,]B_2 \mid B_1[> B_2 \mid B_1 \gg B_2 \mid B_1|[S]|B_2 \mid P$$

with $P \stackrel{def}{=} B$. A LOTOS specification is interpreted as a *Labelled Transition System* (LTS). A transition $B \stackrel{a}{\rightarrow} B'$ indicates that the system behaving as B, executes an action a and moves to B'. The following presents the structure of labelled transition systems which represent the LOTOS semantics.

Definition 2.1
A labelled transition system is defined by the tuple $LTS = \langle S, \mathrm{Act}, \rightarrow, s_0 \rangle$ where:

- *S is a finite set of states and $s_0 \in S$ is the initial state.*

- *Act is a set of labels*

- *$\rightarrow \subseteq S \times S$ is a transition relation. $\rightarrow = \{\stackrel{a}{\rightarrow} | a \in \mathrm{Act}, \exists (s, s') \in S^2, s \stackrel{a}{\rightarrow} s'\}$*

---

[1]FDT is the term reserved for the three languages standardized by ISO, namely LOTOS, Estelle and SDL

The set of transitions that a LOTOS system may execute is derived from the inference rules defined in table 1. In order to calculate the associated LTS of a LOTOS system, we define the function Der as the smallest set satisfying the following properties:

- $B \in Der(B)$

- if $B' \in Der(B)$ and $\exists a \in Act, B' \xrightarrow{a} B''$ then $B'' \in Der(B)$

Der is finite as we assume that the finiteness conditions holds for the specifications that we consider. A specification is considered finite when the processes are guarded and recursion through parallel operators or in the left of operators $\gg$ and $[>$ is forbidden. A Labelled Transition System of a behaviour expression B is defined by the tuple $LTS(B) = \langle S, Act, \rightarrow, s_0 \rangle$ where:

- $S = Der(B)$ and $s_0 = B$

- $Act = Act(B)$

- $\rightarrow = \{ \xrightarrow{a} | a \in Act(B), \exists s, s' \in Der(B), s \xrightarrow{a} s' \}$.

Action:	$\dfrac{}{g;B \xrightarrow{g} B}$	$\overline{\text{exit} \xrightarrow{\delta} \text{stop}}$
Abstraction:	$\dfrac{B \xrightarrow{g} B',\ g \notin G}{B\backslash G \xrightarrow{g} B'\backslash G}$	$\dfrac{B \xrightarrow{g} B',\ g \in G}{B\backslash G \xrightarrow{i} B'\backslash G}$
Choice:	$\dfrac{B_1 \xrightarrow{g} B_1'}{B_1[\ ]B_2 \xrightarrow{g} B_1'}$	$\dfrac{B_2 \xrightarrow{g} B_2'}{B_1[\ ]B_2 \xrightarrow{g} B_2'}$
Enabling:	$\dfrac{B_1 \xrightarrow{g} B_1'}{B_1 \gg B_2 \xrightarrow{g} B_1' \gg B_2}$	$\dfrac{B_1 \xrightarrow{\delta} B_1'}{B_1 \gg B_2 \xrightarrow{\delta} B_2}$

Disabling:
$$\dfrac{B_1 \xrightarrow{g} B_1'}{B_1[>B_2 \xrightarrow{g} B_1'[>B_2} \qquad \dfrac{B_1 \xrightarrow{\delta} B_1'}{B_1[>B_2 \xrightarrow{\delta} B_1'} \qquad \dfrac{B_2 \xrightarrow{g} B_2'}{B_1[>B_2 \xrightarrow{g} B_2'}$$

Parallel:
$$\dfrac{B_1 \xrightarrow{g} B_1',\ g \notin G}{B_1 |[G]| B_2 \xrightarrow{g} B_1' |[G]| B_2} \quad \dfrac{B_2 \xrightarrow{g} B_2',\ g \notin G}{B_1 |[G]| B_2 \xrightarrow{g} B_1 |[G]| B_2'} \quad \dfrac{B_2 \xrightarrow{g} B_2',\ B_1 \xrightarrow{g} B_1',\ g \in G}{B_1 |[G]| B_2 \xrightarrow{g} B_1' |[G]| B_2'}$$

Table 1: The operational semantics of LOTOS

3 The Stochastic Temporal Logic

3.1 Introducing the logic

STL is a linear time temporal logic featuring logical and modal operators. For the sake of simplicity, we consider here a minimal version of STL where logical operators consist of *true*, the negation $\neg$ and the disjunction $\vee$. The *conjunction* and the *imply* operators are generated as $\psi_1 \wedge \psi_2 \equiv \neg(\neg\psi_1 \vee \psi_2)$ and $\psi_1 \rightarrow \psi_2 \equiv \neg\psi_1 \vee \psi_2$.

The modal operators consist of two operators: the *next* operator, where the formula $\bigcirc\psi$ stands for: the formula ψ will hold next time, and the *until* operator, where the formula $\psi_1 \, \mathcal{U} \, \psi_2$ stands for: the formula ψ_1 holds until the formula ψ_2 holds. Other operators can be generated like the *"eventually"* operator where $\Diamond \, \psi = true \, \mathcal{U} \, \psi$ standing for: ψ will hold some time in the future, and the *"henceforth"* operator where

$\Box \, \psi = \psi \, \mathcal{U} \, true$ and standing for: ψ always holds. This fragment of temporal logic is sufficient to describe qualitative properties such as *liveness* and *safety* properties.

The modalities are now extended with quantitative information on how soon or for how long such formulae will hold. The first extension, in common with other real-time logics, is to allow the expression of quantitative modalities such as:

$$\psi_1 \, \mathcal{U}_{\leq I} \, \psi_2$$

where I is a time constant that bounds the time during which ψ_1 will hold and at the end of which ψ_2 will hold. This formulation is equivalent to:
$\bigvee_{0 \leq \delta \leq I} \psi_1 \, \mathcal{U}_{=\delta} \, \psi_2$ or in more compact and more general form

$$\psi_1 \, \mathcal{U}_{[l,u]} \, \psi_2$$

The time interval between events can also be stochastically distributed as in:

$$\psi_1 \, \mathcal{U}_{=Exp,\lambda} \, \psi_2$$

where (Exp, λ) means that the time interval I is now randomized according to an exponential distribution with a mean interval of $\frac{1}{\lambda}$, i.e, I is following an exponential distribution whose probability density function is $f(I) = \frac{1}{\lambda}.e^{-\frac{1}{\lambda}.I}$ with $I \geq 0$ and $\int_0^\infty f(x) \, dx = 1$.

The temporal logic expressing only qualitative properties represents then a subset of quantified temporal logic where time distance between the situations is undefined, i.e:

$$\Box \, \psi \equiv \Box_{<\infty} \qquad \Diamond \, \psi \equiv \Diamond_{<\infty} \, \psi \qquad \psi_1 \, \mathcal{U} \, \psi_2 \equiv \psi_1 \, \mathcal{U}_{<\infty} \, \psi_2$$

We also consider in a similar way past operators *"previous"* denoted by $\ominus_\lambda \psi$ and *"still"* denoted by $\psi_1 \, \mathcal{S}_\lambda \, \psi_2$. Consequently we have the following derived operators *"has-always-been"* $\boxminus_\lambda \psi$ and *"once"* $\diamondminus_\lambda \psi$. We also define a *"probabilistic or"* which is an *exclusive or* quantified by a probability such that $\psi_1 \overset{p}{\vee} \psi_2$ with $p \in [0,1]$ and standing for either ψ_1 holds with probability p or ψ_2 holds with probability $1 - p$.

3.2 Expressing time critical properties in STL

Quantitative information is introduced in order to express time critical properties such as delays and rates as discussed below.

Delays: This includes the delay between events as well as the delay between situations dated by the time at which associated formulae hold. The formula:

$$\Box \,(send \rightarrow \Diamond_{=Exp,\lambda} \; receive)$$

expresses the fact that an event *send* is always followed by an event *receive* in a time exponentially distributed with a mean λ. The delay is also the basic component to express *jitter* and *drifts*. The following formula:

$$\Box \,(send \rightarrow \bigvee_{l \le I \le u} \Diamond_{=I} \; receive)$$

describe a bounded delay between *send*'s and *receive*'s where l, u are constants.

Rates: This represents the rate at which events are occurring. The following formula expresses the exponential input rate of an event a with a mean of λ:

$$\Box \,(a \rightarrow \bigcirc(\neg a \; \mathcal{U}_{=Exp,\lambda} \; a))$$

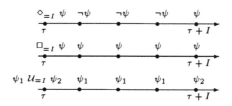

Figure 1: Time distances expressed in modal operators

3.3 Formal presentation

3.3.1 Syntax

The complete syntax of STL is then defined by the following rule:

$$\psi := a \,|\, true \,|\, \neg\psi \,|\, \psi_1 \vee \psi_2 \,|\, \psi_1 \overset{p}{\vee} \psi_2 \,|\, \bigcirc_{=\delta}\psi \,|\, \ominus_{\prec Q}\psi \,|\, \psi_1 \,\mathcal{U}_{\prec Q}\, \psi_2 \,|\, \psi_1 \,\mathcal{S}_{\prec Q}\, \psi_2$$

where $\prec \in \{=, <, \le\}$ and Q a quantification that could be any constant value or an exponential distribution and $p \in [0, 1]$

3.3.2 Semantics

We define a *time sequence* $\vec\tau = \tau_0 \tau_1 ... \tau_i ...$ as a finite or infinite sequence of positive values $\tau_i \in \mathbb{R}_{\ge 0}$ such that $\vec\tau$ increases *monotonically* and *progressively*; that is $\forall i \ge 0 \colon \tau_i < \tau_{i+1}$ [2]. Let $\mathcal{D}$ be the domain of time delays ranging over positive reals $\mathbb{R}_{\ge 0}$. An *interval sequence* over $\mathcal{D}$ is a finite or infinite sequence $\vec{I} = I_0 I_1 ... I_n ...$ such that for some time sequence $\tau = \tau_0 \tau_1 ... \tau_i ...$ we have $I_i = \tau_{i+1} - \tau_i$. We say that an interval sequence is following a stochastic distribution X with parameter λ if the interval values of $\vec{I}$ are also values of (X, λ). We then denote by $\mathcal{D}_{X,\lambda} \subseteq \mathcal{D}$ the interval sequence generated by the distribution X of parameter λ.

Let $\mathcal{P}$ be the finite set of atomic proposition. A *state* is a subset of $\mathcal{P}$. We then denote by $S = 2^{\mathcal{P}}$ the set of all possible states. We define the *timed state* as a pair (s, τ) of state $s \in S$ and a time $\tau \in \mathbb{R}_{\geq 0}$. A *computation* is then a possibly infinite sequence of timed states:

$$\sigma = (s_0, \tau_0), (s_1 \tau_1), ..., (s_i \tau_i), ... \qquad s_i \in S, \tau_i \in \mathbb{R}_{\geq 0}$$

The *evaluation* $I: (S, \mathbb{R}_{\geq 0}) \to 2^{\mathcal{P}}$ mapping each *timed state* (s, τ) where $s \in S$ and $\tau \in \mathbb{R}_{\geq 0}$, to the set of propositions $I(s, \tau) \subseteq \mathcal{P}$ that are true in s at τ. Then, we define the *satisfaction relation* $\models$ between a temporal formula ψ, a computation σ and the time position τ in the computation as follows:

$$
\begin{aligned}
(\sigma, \tau_i) &\models true & & \\
(\sigma, \tau_i) &\models a & \iff & \quad a \in I(\sigma, \tau_i) \\
(\sigma, \tau_i) &\models \neg\psi & \iff & \quad (\sigma, \tau_i) \not\models \psi \\
(\sigma, \tau_i) &\models \psi_1 \vee \psi_2 & \iff & \quad (\sigma, \tau_i) \models \psi_1 \text{ or } (\sigma, \tau_i) \models \psi_2 \\
(\sigma, \tau_i) &\models \psi_1 \overset{p}{\vee} \psi_2 & \iff & \quad (\sigma, \tau_i) \models \psi_1 \text{ with probability } p \\
& & & \quad \text{and } (\sigma, \tau_i) \models \psi_2 \text{ with probability } 1 - p \\
(\sigma, \tau_i) &\models \bigcirc_{X,\lambda}\psi & \iff & \quad \exists I \in \mathcal{D}_{X,\lambda}, (\sigma, \tau_{i+1}) \models \psi \text{ and } \tau_{i+1} = \tau_i + I \\
(\sigma, \tau_i) &\models \psi_1 \, \mathcal{U}_{X,\lambda} \, \psi_2 & \iff & \quad \exists I \in \mathcal{D}_{X,\lambda}, (\sigma, \tau_i + I) \models \psi_2 \text{ and } \forall I' \in \mathcal{D} \\
& & & \quad \text{such that } 0 \leq I' < I: (\sigma, \tau_i + I') \models \psi_1
\end{aligned}
$$

When the time intervals are reduced to constant values, the prefix X is replaced by a constant and λ is omitted.

4 Implications of using LOTOS with STL

A LOTOS system is seen as a black box interacting with its environment through gates. An action is executed when the system and its environment both agree on it. However, the semantics of LOTOS do not specify when these actions are executed or how long they last. States are considered as waiting states. By introducing real-time constraints, the actions are now constrained and specified to happen at precise times. In addition to the ordering causality between events we extend LOTOS with time causality. This extension leads to three consequences: event dating, less non-determinism and timed synchronization.

4.1 Dated events

The introduction of a quantified temporal logic allows us not only to constrain the behaviours with precedence properties but also to mark in time certain actions and to state delays between these time marks. Actions generated from a LOTOS expression are time-stamped with delays as in (a, δ) expressing the fact that action a lasts δ units of time before it is considered to have happened.

If we consider (s, τ) as the current state at the date τ and if a is scheduled to happen with a delay δ then the next state s' will be at the date $\tau + \delta$: $(s, \tau) \xrightarrow{a,\delta} (s', \tau + \delta)$. These delays express the time distances that must conform to the requirements stated in STL. Let us assume that we have the following LOTOS behaviour expression:

```
B[a,b]=run;B[run,quit] [] quit;stop
```

Let us state that the time at which `quit` is performed must be 30 units of time after the beginning. This property would be expressed as:

$$\Diamond_{=30}\ quit$$

During the execution, the first action marks the start of the *countdown*. The intermediate actions which are different from `quit` are enabled on the condition that it is not yet time to perform `quit`. The ultimate action is `quit` which resumes the behaviour (Figure 2).

Figure 2: LOTOS behaviours reduced by timing properties

4.2 Reduced non-determinism

The introduction of dated events also reduces the possible behaviours and hence reduces non-determinism. The non-determinism is reduced by two means:

- time determinism: The introduction of time quantification increases the determinism by stating when an event should occur. As an example, consider the following expression:

$$B = a; B_1 \,[]\, b; B_2$$

If the action a and b are respectively scheduled to happen after δ_1 and δ_2 time-units, then the chosen action is enabled as soon as the associated delay has passed. This property is called the *maximal progress property*. i.e, $a; B_1 \,[]\, b; B_2 \xrightarrow{a, \delta_1} B_1$ if $\delta_1 < \delta_2$ and $a; B_1 \,[]\, b; B_2 \xrightarrow{b, \delta_2} B_2$ if $\delta_2 < \delta_1$. Otherwise, the choice is non-deterministic ($\delta_1 = \delta_2$). In addition, the *persistence property* [18] is required which implies that delaying an action before it happens will not affect the choice alternatives.

- Probabilistic alternatives: The non-determinism can also be explicitly reduced using probability information in the *"exclusive or"* operator:

$$\Box \; (send \rightarrow \Diamond \; (receive \overset{0.9}{\vee} error))$$

which reduce the non-determinism in expressions like $send; (receive[]error)$ using a probabilistic distribution $\{0.9, 0.1\}$. This construction is similar to the probabilistic choice introduced in timed LOTOS [14]: $send; (receive[]_p error)$. Note that the difference between this probability distribution and the stochastic distribution is that the former concerns the choice alternatives of a single event to happen with different delays such that for some event a $\sum_{I \in \mathcal{D}} f(a, I) = 1$. The latter concerns the choice alternatives amongst different events such that $\sum_{a \in Act} f(a) = 1$.

4.3 Timed synchronization

Synchronization through the parallel operator $|[...]|$ in LOTOS is now explicitly or implicitly timed by specifying when it takes place and how long it takes. As actions are constrained with deadlines, their synchronization is assumed to have the same constraints. In addition to the semantics of synchronization in LOTOS we assume that synchronization fails when the synchronized actions have incompatible deadlines. In the expression $a; B_1|[a]|a; B_2$ the synchronization will happen at the same time δ as actions a, i.e: $a; B_1|[a]|a; B_2 \overset{a,\delta}{\longrightarrow} B_1|[a]|B_2$. However, due to the maximal progress property, when the deadline constraint is bounded, the participating agents must engage in the synchronization as soon as they are ready.

5 The underlying model

5.1 Scheduling LOTOS events

The aim of this section is to build up a automaton-based model from the timing properties which will be able to schedule the functional behaviour according to the specified constraints. In other words, every action enabled to happen in the LTS is submitted to the event scheduler which will decide, according to the timing properties, if this action is going to happen, and, if so, will indicate when it will happen.

The resulting model of the event scheduler and the LTS is represented by a *timed labelled transition system* where each labelled transition $\overset{a}{\rightarrow}$ in the LTS is associated with a fixed delay which is either determined from a probabilistic distribution or from a bounded interval, such that we have transitions which are extended to $\overset{a,\delta}{\longrightarrow}$. Based on the model of timed automata[2] [2] and probabilistic automata[17], the event scheduler is seen as an LTS $A = \langle S, \Sigma, E, s_0 \rangle$

---

[2] The main difference between this model and the timed automata [2] resides in the use of timers rather than clocks that increase indefinitely. The reset function is extended to any positive deadline value rather than a zero-initialization.

extended with timers, time constraints, an updating function, and a probability distribution over the elements of Σ. The timers are set to *deadline* values and decreased with the elapsing of time. The transitions are chosen in accordance with the constraints stated on the reading of the timers and the probability distribution associated with the transitions.

Definition 5.1
An event scheduler is defined by the tuple $Sch = \langle S, \Sigma, E, s_0, T, P, R, \Pi \rangle$ where:

- *S is a finite set of states and $s_0 \in S$ is the initial state.*

- *Σ is a finite set of events*

- *$E \subseteq S \times S$ is a finite set of edges. $(s_1, s_2) \in E \Rightarrow \exists a \in \Sigma : s_1 \xrightarrow{a} s_2$.*

- *T is a finite set of timers. The timers are initially set to a deadline value. The value of every timer coupled with the current state represents a generalized state $(s, [T_i]_i)$ where $s \in S$ and $[T_i]_i$ denotes the current reading of all timers $T_i \in T$. The reading of every timer changes automatically with the progress of time. It is decreased with the elapsed delay: $(s, [T_i]) \xrightarrow{a, \epsilon} (s', [T_i \leftarrow T_i - \epsilon])$.*

- *$P : E \to \mathcal{C}_T$ is a function associating with each edge in E a timing constraint from $\mathcal{C}_T$. A transition $s_1 \xrightarrow{a} s_2$ is enabled if the constraint $P(s_1, a, s_2)$ holds. $\mathcal{C}_T$ is constructed from algebraic relation $\prec \in \{=, <, \leq\}$ with the standard form*

$$\mu = \bigwedge [\tau_l \prec T_i \prec \tau_u]$$

where τ_l, τ_u are values representing respectively the lower and upper bound of the timer T_i. Then we say that a value ϵ satisfies a constraint μ and write $\epsilon \models \mu$ if ϵ is a possible solution to μ.

- *$R : E \to V$ is reset function associating a deadline value from $\mathcal{D}^*$ with each edge in E. $V : 2^T \to \mathcal{D}^*$ is a partial function when applied on timers $T' \subseteq T$, it resets the elements of T' to corresponding values $\nu \in \mathcal{D}^*$. $\mathcal{D}^*$ denotes either the set $\mathcal{D}$ of fixed intervals or the set $\mathcal{D}_{X, \lambda}$ of stochastic intervals. When the reset is not defined, the timer is automatically reset according to the time elapsed:*

- *$\Pi : E \to [0, 1]$ is a probability distribution associating a probability with each edge in E such that $\forall (s, s_i) \in E : \sum_{s_i \in S} \Pi(s, s_i) = 1$. When the probabilities of outcoming edges of some state are not specified, the choice is resolved as a non-determinism.*

We denote by $s_1 \xrightarrow{\mu, a, \nu} s_2$ the full transition which moves the event scheduler from a state s_1 to the state s_2 by executing the action a. The action is fired after a delay ϵ determined according to the timing constraint μ; i.e, $\epsilon \models \mu$. The re-initialization of the timers is then carried out according to ν. Timers not affected by the re-initialization are automatically decreased with the elapsing time; i.e, ϵ. The event scheduler behaviour is formally defined by the following rules:

$$\frac{Sch \xrightarrow{\mu, a} Sch' \text{ and } \epsilon \models \mu}{(Sch, [T_i]) \xrightarrow{a} (Sch, [T_i \leftarrow T_i - \epsilon])} \qquad \frac{Sch \xrightarrow{\mu, a} Sch' \text{ and } \epsilon \models \mu \text{ and } Sch \xrightarrow{\nu} Sch'}{(Sch, [T_i]) \xrightarrow{a} (Sch, [T_i \leftarrow \nu(T_i)])}$$

5.2 Generating timing event schedulers from STL

In order to give an operational interpretation of the event scheduler, we define a *derivative* of a formula at some state as the formula that holds in the next state. Intuitively, given a timed state sequence σ, if for some formula ψ, $(\sigma, \tau_i) \models \psi$ then the derivative $D_\alpha \psi$ is such that $(\sigma, \tau_{i+1}) \models D_\alpha \psi$ with $\alpha \in I(\sigma, \tau_{i+1})$.

The event scheduler is then interpreted as a finite state machine where states are defined as a pair (ψ, τ) representing the state and the time at which ψ holds. The transitions from state to state indicate the move in the interpretation of the current formula: $(\psi, \tau_i) \xrightarrow{\alpha} (D_\alpha \psi, \tau_{i+1})$. This approach allows us to build up an event scheduler that permits only appropriate behaviours since the input real-time properties are satisfied by construction. The event scheduler moves from state to state according to the initial property.

In a first stage, we inductively deduce the derivative of a formula according to its structure:

$$
\begin{array}{lcl}
D_\alpha(true, \tau) & = & (true, \tau) \\
D_\alpha(a, \tau) & = & (true, \tau) \text{ if } a = \alpha, \ (false, \tau) \text{ otherwise} \\
D_\alpha(\neg \psi, \tau) & = & \neg D_\alpha(\psi, \tau) \\
D_\alpha(\psi_1 \vee \psi_2, \tau) & = & D_\alpha(\psi_1, \tau) \vee D_\alpha(\psi_2, \tau) \\
D_\alpha(\bigcirc_{=\delta} \psi, \tau) & = & (\psi, \tau + \delta) \\
D_\alpha(\psi_1 \, \mathcal{U}_{=\delta} \, \psi_2, \tau) & = & D_\alpha(\psi_2, \tau + \delta) \bigvee\limits_{0 \leq \tau' < \delta} [D_\alpha(\psi_1, \tau + \tau') \wedge \psi_1 \, \mathcal{U}_{=\delta - \tau'} \, \psi_2]
\end{array}
$$

where Σ represents all the actions belonging to the treated formula. The following function D^* defines the minimal set satisfying the following properties, in order to build the set of event scheduler states:

- $\psi \in D^* \psi$

- if $\phi \in D^*$ then $\forall a \in \Sigma(\psi) : D_a \phi \in \psi$

The set D^* corresponds to the set of states, each representing the state of the formula interpretation and the current time. It is easy to see that the size of the transitive closure D^* is exponential in the size of ψ. However, it is not necessary to worry about the size of D^* since most formulae are not understandable when their size is over 3.

5.3 Example

As an example, consider the formula $\psi = \Diamond_{=30} \, a$. We can generate a three-state event scheduler (Figure 3). We derive ψ with respect to the action a and the event scheduler moves to the state $(true, 30)$ since $D_a(a, 30) = (true, 30)$. We can also derive ψ to $(\Diamond_{=30-\epsilon} \, a, \epsilon)$, where $0 \leq \epsilon < 30$, by executing any action different from a with a delay within $[0, 30[$; i.e, $(*, \epsilon)$ ("$*$" means any other action than a). The timer is decreased with the elapsing time, i.e, $30 - \epsilon$. The event scheduler may either keep executing any action after any delay $\epsilon \in [0, T]$ so long the timer reading is positive, or move to the state $(true, 30)$ by executing an action a in the remaining time, i.e, (a, T).

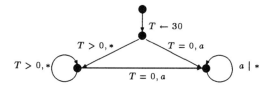

Figure 3: The associated event scheduler to $\Diamond_{=30}\, a$

6 Composing LOTOS with STL

As explained previously, the aim of this approach is to provide a model able to be executed according both to the functional description and the timing properties. The functional specification is provided by the generated LTS and the timing properties are represented by the event scheduler. The behaviour of the event scheduler is to synchronize with the LTS for each step and to monitor their actions. Informally, every possible transition of the LTS is submitted to the event scheduler which will first check if its associated label is also a label of a possible transition within the event scheduler. If so, the event scheduler will check, according to the current date, if its prefixing predicate holds and chose a delay such that the predicate holds. The LTS transition is then enabled for this label with the associated delay. The formal semantics of the composition of the LOTOS associated LTS and the STL event scheduler is defined by the following rules:

$$\frac{B \xrightarrow{\alpha} B' \quad,\quad Sch \xrightarrow{\mu,\beta,\delta} Sch' \quad \wedge \quad \epsilon \models \mu \quad \wedge \quad \beta \in \{\alpha, *\}}{(B, Sch, \{T_i\}_i) \xrightarrow{\beta,\epsilon} (B', Sch', \{T_i \leftarrow \nu'(T_i)\}_i)}$$

where

$$\nu'(T_i) = \left\{ \begin{array}{ll} \nu(T_i) & \text{if } Sch \xrightarrow{\nu} Sch' \\ T_i - \epsilon & \text{otherwise} \end{array} \right.$$

7 The multimedia stream example

As an illustration of the approach presented above, we present the case of a multimedia stream (first introduced in [4]). In this, example we consider a structure linking a multimedia data source and a data sink. The data source and the data sink are assumed to be communicating asynchronously over an unreliable channel. The functional description is given by LOTOS. Figure 4 illustrates the associated LTS generated by the tool SMILE [7] from its LOTOS specification below.

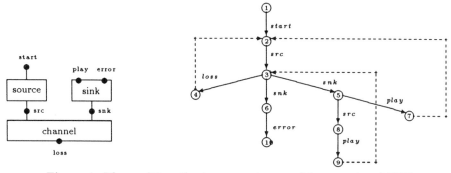

Figure 4: The multimedia stream system and its associated LTS

```
specification stream [start,play,error]:noexit
behaviour
        start;( hide src, snk, loss in
                ( ( source[src] ||| sink[snk,play,error])
                |[src,snk]|
                channel[src,snk,loss] ))
where
        process source[src]:noexit:=
                src; source[src]
        endproc (* source *)
        process sink[snk,play,error]:noexit:=
                snk;(    play;sink[snk,play,error]
                        [] error;stop )
        endproc (* sink *)
        process channel[src,snk,loss]:noexit:=
                src;(    snk; channel[src,snk,loss]
                        [] loss;channel[src,snk,loss] )
        endproc (* channel *)
endspec (* stream *)
```

The above functional description is now extended with real time constraints. These constraints are described with STL formulae. Each formula is assumed to describe a local property for each component of the functional specification.

- The data source arrivals are modelled as Poisson process with an average rate of one frame every $50ms$. This implies that the waiting time I between successive **src** events follows an exponential distribution with a probability density function $X(I) = 0.02.e^{0.02.I}$ with $I \geq 0$.

With this constraint, whenever an **src** event occurs at a time τ we will fix its next occurrence at a time $\tau + I$ such that d is randomly generated from an exponential distribution. The next occurrences are then decided in advance in a deterministic way. The This property is specified below by the STL formula with the associated event scheduler in Figure-5.a.

$$\Box \ [src_i \rightarrow (\Diamond_{<e,0.02} \neg src_{i+1} \land \Diamond_{=e,0.02} \ src_{i+1})]$$

- Another constraint states that successfully transmitted frames arrive at the data sink between $80ms$ and $90ms$ after their transmission. This property represents the channel latency. It fixes the occurrence time of **snk** within $[80ms, 90ms]$ after a **src**. A jitter of $10ms$ is then permitted. However, some frame loss is permitted; a rate of 10% of the sent frames are lost. This property is described as:

$$\Box\ [src_i \rightarrow (\bigcirc loss_i \overset{0.1}{\vee} \Diamond_{[80,90]}\ snk_i)]$$

The associated event scheduler is illustrated is Figure-5.b.

- The next property contributes to making the decision on which of **play** and **err** events to enable when a frame arrives in the data sink. The choice is now decided in a deterministic way by looking at the arrival time of **snk**. If the arrival rate at the data sink is not within $[15, 20]$ frames per second, then an error should be reported. Next is the corresponding formula with its associated event scheduler in Figure-5.c.

$$\Box\ [snk_i \wedge (\Diamond_{\leq 1000}snk_{i-21} \vee \Diamond_{\leq 1000}\neg snk_{i-15}) \rightarrow \bigcirc error]$$

- The final property states that if **play** is selected then it happens exactly $5ms$ after a frame is received. This is described by:

$$\Box\ [snk_i \wedge (\Diamond_{\leq 1000}snk_{i-15} \wedge \Diamond_{\leq 1000}snk_{i-20} \wedge \Diamond_{\leq 1000}\neg snk_{i-21}) \rightarrow \Diamond_{=5}\ play]$$

The associated event scheduler is illustrated is Figure-5.d.

These constraints stated above with STL and interpreted as event scheduler in Figure 5 are composed with the LOTOS-associated LTS and used to construct a real-time model corresponding to the multimedia stream system (Figure 6). This model conforms both to the LOTOS functional specification, and to the STL real-time properties.

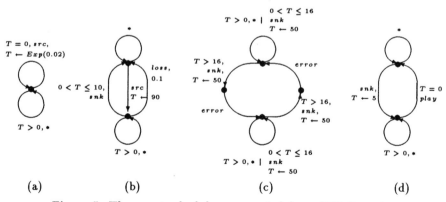

Figure 5: The event schedulers generated from STL formulae

8 Conclusion

In this paper we have introduced a new approach that uses the standardized LO-TOS specification language together with stochastic real-time temporal logic, STL. Timing modalities have been defined to extend LOTOS specifications with time critical and stochastic constraints. The constraints have the effect of reducing non-determinism, denoting time distances between actions and fixing process synchronizations in time. The main direction for future work is the development of verification methods based on both performance analysis and formal validation.

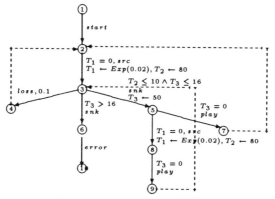

Figure 6: The timed stream model

References

[1] R. Alur, C. Courcoubetis, and D.L. Dill. Model-checking for probabilistic real-time systems. In *Automata, Languages and Programming: Proceedings of the 18th ICALP*, Lecture Notes in Computer Science 510, pages 115–136, 1991.

[2] R. Alur and D. Dill. A theory of timed automata. *Theoretical Computer Science*, 126, 1994.

[3] G. Blair, L. Blair, H. Bowman, and A. Chetwynd. A framework for the formal specification and verification of distributed multimedia systems. In *Workshop on Quality of Service and Performance, Aachen, Germany*, 1994.

[4] L. Blair. *The Formal Specification and Verification of Distributed Multimedia Systems*. PhD thesis, Lancaster University, 1994.

[5] T. Bolognesi and E. Brinksma. Introduction to the ISO specification language LOTOS. *Computer Networks and ISDN Systems, North-Holland*, vol. 14, No. 1, 1988.

[6] A. Campbell, G. Coulson, F. Garcia, D. Hutchinson, and H. Leopolod. Integrated quality of service for multimedia communications. Technical report, Lancaster Univerity Internal Report MPG-92-34, Nov 1992.

[7] H. Eertink and D. Wolz. Symbolic execution of LOTOS specifications. In Michel Diaz and Roland Groz, editors, *5th International Conference on Formal Description Techniques*, Lannion, France, October 13-16 1992.

[8] S. Gilmore and J. Hillston. The PEPA workbench: A tool to support a process algebra-based approach to performance modelling. In *Proceedings of the 7th Int. Conf. on Modelling Techniques and Tools for Computer Performance Evaluation*, Vienna, 1994.

[9] H. Hansson and B. Jonsson. A calculus for communicating systems with time and probabilities. *Proceeding of the 11th IEEE Real-Time Systems Symposium, Florida*, December 1990.

[10] H. Hermanns, V. Mertsiotakis, and M. Rettelbach. Performance analysis of distributed systems using TIPP - a case study. In *10th UK Computer and telecommunications Performance Engineering Workshop*, Edinburgh, UK, September 1994.

[11] R. Koymans. Specifying real time properties with metric temporal logic. *Journal of Real Time Systems*, 2, 1992.

[12] E. Madelaine and D. Vergamini. AUTO: A verification tool for distributed systems using reduction of finite state automata networks. In *2nd International Conference on Formal Description Techniques*, Vancouver, Canada, May 1989.

[13] M. A. Marsan, A. Bianco, L. Ciminiera, R. Sisto, and A. Valenzano. A LOTOS extension for the performance analysis of distributed systems. *IEEE/ACM Transaction on Networking*, 2:151–165, April 1994.

[14] C. Miguel. Extended LOTOS definition. Technical Report OSI95/DIT/B5/8/TR/R/V0, OSI'95, Depto. Ingeniería de Sistemas Telemáticos, Univbersidad Politécnica de Madrid, Spain, November 1991.

[15] J. S. Ostroff. *Temporal Logic for Real Time Systems*. Research Studies Press, 1989.

[16] B. Steffen R. Cleaveland, J. Parrow. The concurrency workbench: A semantics-based verification tool for finite-state systems. In *Proceedings of the Workshop on Automated Verification Methods for Finite-state Systems, Lecture Note in Computer Science*. Springer-Verlag, 1989.

[17] R. Segala and N. A. Lynch. A compositional trace-based semantics for probabilistic automata. In *Proceedings of CONCUR'95*, Philadelphia, PA, USA, August 1995.

[18] W. Yi. Real-time behaviour of asynchronous agents. In *LNCS, CONCUR'90 Theories of Concurrency: Unification and extension*, volume 458, pages 502–520. Springer-Verlag, 1990.

Detecting reversibility in Markovian Process Algebras

Madhu Bhabuta*   Peter Harrison[†]

Kamyar Kanani[‡]

Department of Computing
Imperial College
London SW7 2BZ

Abstract

A methodology is developed to determine when a process defined in terms of a Markovian Process Algebra is a reversible Markov process and so has a product form solution. In particular, we show how to detect birth-death processes, tree-like extensions of these and examine an approach to detecting more general reversible processes through the Kolmogorov criteria applied to a minimal state transition graph. Finally, we indicate how to exploit the compositionality of process algebras to detect more complex reversible, and other product-form, processes formed from interactions of simpler processes already known to be reversible.

1 Introduction

Recent advances in extending process algebras to include time and probablistic behaviour e.g. in [11], [5] have made it possible to use them as a performance modelling tool. Markov models have for long been widely used for this purpose but their direct specification is cumbersome and error-prone. However, when a Markov model has a *product form solution*, closed form results can be used efficiently. In particular, product-form solutions exist for Markov models which are reversible. In this paper we present a methodology to determine when the stationary Markov process associated with a given process algebra program is reversible and so has a product form solution.

We first introduce in the next section the algebra that we use although we emphasise that the methodology we present is applicable to any Markovian Process Algebra (MPA). Section 3 gives a brief introduction to reversibility in general and sketches how we may recognise systematically whether or not the Markov Process associated with a MPA program is reversible. Such detection can be straightforward when an MPA program is isomorphic to the definition of a reversible Markov process. However, when there are also "internal" instantaneous transitions—such as τ-actions in CCS but not only these—the structure

---

*mb3@doc.ic.ac.uk, funded by a bursary from Esprit Basic Research Action Qmips
[†]pgh@doc.ic.ac.uk, work partly supported by Esprit Basic Research Action Qmips
[‡]kk2@doc.ic.ac.uk, supported by a research grant from the EPSRC

of the underlying Markov chain becomes obscured and detection can be difficult. Such problems are addressed in section 4 and a detection algorithm is prescribed. The paper concludes by considering how the methodology can contribute to recognising more general product-form processes, for example those that are quasi-reversible.

2 Markovian Process Algebra

The algebra we use is based on CCS [10], with stochastic, timed extensions very like those in PEPA [5]. The methodology presented to recognise reversibility in Markovian Process Algebras is, however, generic and applicable to most MPAs.

The syntax and structural operational semantics we detail below, under the assumption that the reader is familiar with the concepts of process algebra. A detailed introduction is given in [10].

2.1 BNF notation for MPA

$$
\begin{aligned}
< \text{Vis} > \; &= \; \text{nil} \mid \text{a} \mid \text{b} \mid \text{c} \mid \cdots \\
< \text{Acts} > \; &= \; \tau \cup < \text{Vis} > \\
< \text{Rate} > \; &= \; \top \mid \alpha \mid \beta \mid \gamma \mid \cdots \\
< \text{MActs} > \; &= \; (\text{Acts, Rate}) \\
< \text{Process} > \; &= \; 0 \mid A \mid B \mid C \mid \cdots \\
&\mid \; < \text{MActs} > . < \text{Process} > \\
&\mid \; < \text{Process} > + < \text{Process} > \\
&\mid \; < \text{Process} > \mid < \text{Process} > \\
&\mid \; < \text{Process} > [\text{Vis}/\text{Vis}] \\
&\mid \; < \text{Process} > \backslash \{\text{Vis}\}
\end{aligned}
$$

$\top$ is an undefined rate and 0 is the stopped process.

2.2 Operational Semantics for MPA

ActionPrefixing

$$\overline{(a, \lambda).P \xrightarrow{(a,\lambda)} P}$$

Choice

$$\frac{P \xrightarrow{(a,\lambda)} P'}{P + Q \xrightarrow{(a,\lambda)} P'} \qquad \frac{Q \xrightarrow{(b,\mu)} Q'}{P + Q \xrightarrow{(b,\mu)} Q'}$$

Restriction

$$\frac{P \xrightarrow{(a,\lambda)} P'}{P \backslash L \xrightarrow{(a,\lambda)} P'} \quad (a, \lambda) \neq L$$

Relabelling

$$\frac{P \stackrel{(a,\lambda)}{\rightarrow} P'}{P[f] \stackrel{f(a,\lambda)}{\rightarrow} P'[f]}$$

Parallel

$$\frac{P \stackrel{(a,\lambda)}{\rightarrow} P' \quad Q \stackrel{(\bar{a},\top)}{\rightarrow} Q'}{P \mid Q \stackrel{(\tau,\lambda)}{\rightarrow} P' \mid Q'} \qquad \frac{P \stackrel{(a,\top)}{\rightarrow} P' \quad Q \stackrel{(\bar{a},\mu)}{\rightarrow} Q'}{P \mid Q \stackrel{(\tau,\mu)}{\rightarrow} P' \mid Q'}$$

$$\frac{P \stackrel{(a,\lambda)}{\rightarrow} P' \quad Q \stackrel{(\bar{a},\mu)}{\rightarrow} Q'}{P \mid Q \stackrel{(\tau,\min(\lambda,\mu))}{\rightarrow} P' \mid Q'}$$

$$\frac{P \stackrel{(a,\lambda)}{\rightarrow} P'}{P \mid Q \stackrel{(a,\lambda)}{\rightarrow} P' \mid Q} \qquad \frac{Q \stackrel{(b,\mu)}{\rightarrow} Q'}{P \mid Q \stackrel{(b,\mu)}{\rightarrow} P \mid Q'}$$

The | operator is associative and commmutative. When 2 processes are placed in parallel, it is unclear what is meant by the rate at which the communication takes place. We have chosen a semantics where the rate of synchronisation between two actions is that of the slower one [6].

3   Reversible Markov Processes

The Markov process is the basis of many modelling methodologies and central to it is the concept of state and state transitions. The state of a physical system at any given time may be described by a set of values of measurable properties of that system. State transitions describe the evolution of the system over time, e.g. a queue can often be expressed in terms of its length and transitions between the different lengths are determined by events—arrivals of customers and service by the server. The state and state transitions thus fully describe the system.

A Markov Process is reversible if and only if the probability flux between any pair of states is the same in both directions. More formally, there must exist a collection of positive numbers, $\pi(j)$ summing to unity that satisfy the *detailed balance condition*:

$$\pi(j)\, q_{jk} = \pi(k)\, q_{kj}$$

where q_{jk} is the instantaneous transition rate from state j to state k. Kolmogorov's criterion states that a Markov process is reversible iff the product of transition rates, q_{ij} between states i, j satisfy

$$q_{12}q_{23}\cdots q_{k-1,k}q_{k,1} = q_{1k}q_{k,k-1}\cdots q_{32}q_{21}$$

for all state cycles $1, \ldots, k, 1$. A more complete account of reversibility can be found in [7].

3.1 Detection strategy

We propose to recognise reversible MPA programs, i.e. those with associated Markov chains that are reversible, by first defining a mapping for processes to states and then determining the state transition matrix via the given MPA programs. The methodology we will use is :-

1. Specify a MPA program along with a function that maps a process to its associated Markov state, if there is one, else returns 'undefined'.

2. Construct the derivation tree of the processes in the given MPA program, that is, unfold it.

3. Apply the process → state function to the (process) nodes in the derivation tree to determine the state transition graph of the Markov process.

4. Analyse the state transition graph and test it for reversibility as follows:-

   (a) Check that all the transitons are bidirectional, i.e. if there is an arc i → j there is also an arc j → i.

   (b) If there are no cycles (of length more than two) in the graph, then the graph is

   • a linear structure *or*
   • a tree structure

   and is reversible trivially.

   (c) If there are cycles in the graph then, following the approach of [8]:

   i. Create the spanning tree of the graph.

   ii. Identify the edges in the graph that are not in the spanning tree; these form the *fundamental circuits* of the graph.

   iii. Check Kolmogorov's criterion for each of the fundamental circuits.

   The fundamental circuits form a basis vector for the graph, so if each of the fundamental circuits is found to satisfy Kolmogorov's Criterion, then the graph represents a reversible process.

The state transition graph obtained could be either finite or infinite. In the finite case, the above method would detect a linear, tree-shaped or cyclic reversible process. In the infinite case some form of inductive analysis is necessary to identify simple unbounded reversible processes. Alternatively, from a definition given in terms of recursion equations, an equivalent definition could be sought in terms of higher-order functions. These can then be analysed more simply for reversibility, analogously to algebraic program transformation techniques [3]. In this way a pool of "basic" reversible MPA templates can be assembled.

The important property of compositionality of process algebras can now be employed. Successively larger reversible models can be created from smaller ones, starting with the pool of processes that have been recognised as reversible, and linked in such a manner as to preserve the reversibility property. In figure 1, graphs of the three reversible processes A, B, C are combined such that the detailed balance condition must hold on the arcs connecting each to the state

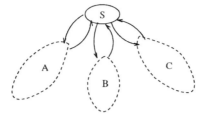

A, B and C are reversible processes

Figure 1: Composition of simple reversible structures

S—only one in each direction—by considering an appropriate contour across which in- and out-flux must balance. Thus this more complex process, built from 3 reversible sub-processes, is also reversible since Kolmogorov's criterion is preserved; every cycle is built from smaller cycles satisfying the criterion. In this way, we can recognise ever more reversible processes and possibly add them to the "pool". Indeed, we can apply this methodology with just linear birth-death processes in the pool to deduce that single step, tree-like processes are also reversible.

For more complex combinations of reversible sub-processses, i.e. when a new cycle is created which is not formed solely out of known reversible cycles, the composite program has to be checked using the above analysis of fundamental circuits. However, given knowledge of the (reversible) fundamental circuits of the sub-processes, the analysis of those in the whole new graph can be simplified for particular types of interaction between these sub-processes. We will not address this topic further here, however.

Ideally, we would like to be able to recognise classes of MPA programs *syntactically* by working directly with the process algebra. Having identified a suitable pool of basic processes—at source level, but via semantic analysis at the level of the Markov graph—compositionality will facilitate this.

4 Analysis

4.1 The Derivation Graph

The semantics for MPA defines a labelled transition system that enables one to describe the evolution of an MPA program in the form of a *derivation tree*. A derivation tree is a directed, connected, acyclic structure. Each node represents a process. The children are those processes that the parent process can evolve into by one action, visible or otherwise. The arc connecting a parent and child node is labelled with the action that caused the corresponding process transition. This tree may be collapsed into a more general *derivation graph* if all syntactically identical nodes are grouped together. This provides a more compact notation for the derivation tree and in some instances leads to a graph that is actually isomorphic to the required Markov chain. We can formally define the derivation tree as the triple (V, E, L) where V is the set of vertices, representing processes, of the graph, E is a relation of type *Process* × *Process*

222

and L is a labelling function of type $Process \rightarrow Process \rightarrow Event$ such that

$$EPQ \quad \Leftrightarrow \quad P \xrightarrow{(a,r)} Q$$
$$L(PQ) = (a,r) \quad \Leftrightarrow \quad P \xrightarrow{(a,r)} Q$$

and $\mid E^{-1}Q \mid \leq 1$ where $E^{-1} : Process \longrightarrow \{Process\}$

The final constraint, that E^{-1} is a function, ensures that each node has only a single parent. This ensures the tree structure. Relaxing this constraint produces the definition for a derivation graph.

For example, if we define an M/M/1 queue by:

$$Q_0 \quad = \quad (arr, \lambda).Q_1$$
$$Q_n \quad = \quad (arr, \lambda).Q_{n+1} + (dep, \mu).Q_{n-1}, n > 0$$

we obtain the derivation graph shown in Figure 2

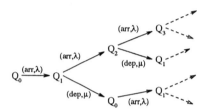

Figure 2: Derivation tree for simple queue

We can identify identical states which we can use to fold the tree to give us a derivation graph as in Figure 3.

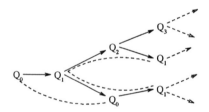

Figure 3: Identify identical states

Thus the tree can be folded along the dotted lines to give the following (familiar looking) graph as in Figure 4.

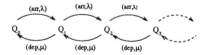

Figure 4: Resulting derivation graph

The size of the derivation tree may be unbounded, such as for the queue example above. This would be the case as well for any other non-terminating or recursive process. This may present us with a problem in implementation of an algorithm that relies on the direct analysis of the derivation tree. There are ways around this problem; for example one could choose a lazy functional programming language for the basis of the implementation. The graph corresponding to an unbounded tree may itself be bounded and instead contain a cycle. However, the finite nature of the graph would simplify its analysis if we ensure that nodes that have been analysed are marked as such.

4.2 The Mapping Function

We now introduce a function F of type $Process \rightarrow num$ that maps a process algebra program to an associated (numerical) state of some underlying continuous time Markov chain with countable state space. It is a partial function in that some processes may not have a defined Markov state associated with them. As we have already discussed, this is typically true of processes involved in sequences of immediate τ actions. These represent internal, instantaneous states in the Markov chain, and are therefore not wanted when analysing steady-state behaviour. Moreover, there may be some detail in the definition of a process that is of no interest in the Markov analysis. For example, the explicit occupancy of each of a set of buffers, where we are really only interested in the *number* of occupied buffers.

The process of choosing the function F could be automated in an MPA with an extended form of weak bisimulation equivalence. This would have to equate states not only up to τ-actions but also *any* internal actions which occur at infinite rate. Moreover, the resulting set of equivalence classes would have to be isomorphic to the desired state space. This would not necessarily be the case, as noted above.

By applying the function to a node in the tree and to its immediate children, we can determine the states that are the immediate transitions out of any current Markovian state, i.e we can define the associated Markov chain.

Take as an example, a section of the derivation tree for our simple queue shown in Figure 5. Here we have $F(Q_n) = n \; for \; all \; n \geq 0$.

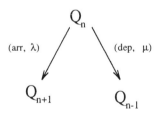

Figure 5: Section of derivation tree for simple queue

We can apply the state function to all nodes in this tree segment. The result is that if we are in a Markovian state n, upon the event (arr, λ) occuring, we move to Markovian state $n + 1$ and upon event (dep, μ) we move to state $n - 1$ (for n > 0). We rely on the fact that actions in our MPA are indivisible. They are either single actions or synchronisations. This precludes the possibility that our system could evolve through several Markov states during the course of a single MPA transition.

More formally, the Markov transition graph corresponding to the derivation tree (V,E,L) is given by (V_M, E_M, L_M) where V_M is the set of states defined by $V_M = \{s \mid s = Fp, p \in V\}$, E_M is the edge relation of type $state \times state$ defined by $E_M RS \Leftrightarrow EPQ \wedge FP = R \wedge FQ = S$ and L_M is the labelling function of type $state \rightarrow state \rightarrow rate$ defined by $L_M RS = \sum_{i:(a,x_i) \in \{LPQ \mid FP=R, FQ=S\}} x_i$. Here we have to sum all of the rates going from one state into another to get the total flow between the two states. Since we are dealing with exponentially distributed rates we can just sum them due to the superposition property.

This definition is actually not complete since during analysis we may be faced with the situation that a given node has an undefined Markov state associated with it. This means that the application of F to that process should also be undefined. In this situation we can conclude that the system is in an *implementation specific transient state*. The process is evolving through a stage which is needed to allow implementation of the system. Such a transient state can be discarded and replaced by its children, provided its holding time is zero.

Consider an implementation of a birth-death process:

$$
\begin{aligned}
C &= (arr, \lambda).(C^\frown C) + (dep, \mu).D \\
D &= (\overline{d}, \top).C + (\overline{z}, \top).B \\
B &= (arr, \lambda).(C^\frown B) + (z, \top).B \\
System &= C^\frown C^\frown \ldots C^\frown B
\end{aligned}
$$

where $P^\frown Q$ represents $(P \mid (Q[d \backslash dep, \top \backslash \mu]) \backslash \{arr\}) \backslash \{d, z\}$. This defines a system which is the same system as the 'stack' defined in [10] of C processes linked together to form a chain tailed off by B processes. Let us assume we are interested in the number of C processes linked together. Informally looking at the system we see that upon an (arr, λ) event, the lead C process (the only process that can accept the event) splits into two, increasing the length of the chain by one. However, upon a (dep, μ) event the leading C process becomes a D process which can now synchronise with the second C process in the chain to swap places and so on down the chain until it comes to a B process which will absorb it, effectively decreasing the total number of C processes in the chain by one. We therefore define the state function thus :

$$
\begin{aligned}
F(C^\frown Rest) &= 1 + F Rest \\
F(B^\frown Rest) &= 0 \\
FB &= 0
\end{aligned}
$$

Thus the function recursively counts the number of C processes in the chain. Note that there is no mention of process D in the definition for F. Whilst the D process is rippling down the chain after a (dep, μ) event the state of the system is undefined. Therefore, whilst the D process is rippling down the chain, we would like to think that our system is moving instantaneously through a number of implementation specific transient states. In an ideal definition for the system, the length of the chain would decrease by one in just one transition but this is not the case here.

Moreover, situations can arise within the system that we do not address here. For example, whilst the D processes are rippling down the chain, the leading C process can still continue to accept (arr, λ) or (dep, μ) events. This confuses the system beyond our interest but one should be aware of this short-coming in the definition. [1]

The definitions we have thus given for the Markov graph in terms of the definition graph therefore have to be modified to take these transient states into account.

Thus, we give new definitions for V_M, E_M and L_M

$$V_M = \{s \mid s = Fp, p \in V\} \backslash \{\perp\}$$
$$E_M RS \Leftrightarrow EPQ \wedge FP = R \wedge FQ = S$$
$$\vee E^* PQ \wedge \forall Z[E^* PZ \wedge E^* ZQ \Rightarrow FZ = \perp]$$
$$L_M RS = \sum_{i:(a,x_i)\in\{LPQ \mid FP=R, FQ=S\}} x_i$$

where E^* is the transitive closure of E, i.e. $E^* PQ \Leftrightarrow P \longrightarrow^* Q$. Thus V_M has the undefined state removed from it and two states in the Markov graph have an edge between them if there is a path from one process, corresponding to the origin state, to a process with corresponding target state, in the derivation tree such that either the path is single step or that all intermediate processes have no state associated with them.

5 Topological structures

5.1 Linear structure

The canonical example of a system that has a linear structure is the birth-death process, in which the population increases on a birth, at rate λ and decreases on a death, at rate μ. This example has been covered by the M/M/1 queue specification given in section 4.1.

Of course, many linear systems are not birth-death processes since they have multi-step jumps. In this case the Markov graph contains cycles and the Kolmogorov criteria must be checked. In simple cases, e.g. where the jumps follow a regular pattern specified inductively along with their rates, such checking is straightforward, involving only one or two generic cycles.

5.2 Tree-shaped structure

Consider a model of a network distribution protocol. Messages are sent to each child node from the root. Replies are recieved from each of the child nodes upwards. The permormance measure of interest in the model is the load at each of the nodes in the network. The rate at which messages are sent from the root node to its children is λ and the rate at which the results are received back is μ. The process algebra for the system would look like:

$$\text{Node}(1,1) = (\text{send}, \lambda).(\text{Node}(2,1) + \text{Node}(2,2))$$
$$\text{Node}(n,x) = (\text{send}, \lambda).(\text{Node}(n+1, 2x-1) + \text{Node}(n+1, 2x))$$
$$+ (\overline{\text{resend}}, \mu)(\text{Node}(n-1, \frac{1}{2}x + \frac{1}{2}), \text{if } x \bmod 2 \neq 0$$

---

[1] Actually this can be ignored if we employ the rule that $\top$ rated events occur in preference to ones with defined rates; similar to immediate and stochastic transitions in GSPNs

$$(\overline{\text{resend}}, \mu)(\text{Node}(n-1, \tfrac{1}{2}x), \text{otherwise}$$

The function mapping process to state is defined in the obvious way:

$$F(\text{Node}(n, x)) = (n, x)$$

since the Markov state is again isomorphic to the MPA process. The resulting state transition graph is then Figure 6. Obviously this definition can

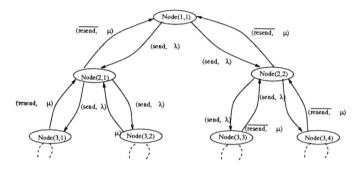

Figure 6: Tree-shaped graph

be extended recursively to yield various tree-structures of arbitrary size, and reversibility is immediate since all transitions are single step.

5.3   Cyclic structure

Consider now a ring-based network protocol, in which messages can travel to and fro, from sender to receiver. We present 2 alternative process algebras [2] to represent the process.

1. The communication is illustrated by the following definition for a single cycler:

$$
\begin{aligned}
C &= a.C' + c.C' \\
C' &= b.C + d.C \\
C_i &= C[f_i] \\
C_i' &= C'[f_i]
\end{aligned}
$$
$$\text{where } f_i = (a_i/a, \overline{a_{i+1}}/b, \overline{b_i}/d, b_{i+1}/c).$$

Then a ring of three cyclers is $C^3 = (C_1|C_2|C_3)\backslash\{a_1, a_2, a_3, b_1, b_2, b_3\}$. To permit interaction with the outside world we use a more complex process with "probes" a, a' (we also include rates) :

$$
\begin{aligned}
C &= (a, \mu).C' + (a', \lambda).C'' \\
C' &= (b, \mu).C'' + (b', \lambda).C \\
C'' &= (c, \mu).C + (c', \lambda).C'
\end{aligned}
$$

---

[2] all addition is mod n where n is the number of nodes in the ring

To form the ring of 3 cyclers, as in [10]:

$$C_i = C[f_i]$$
$$C_i' = C'[f_i]$$
$$C_i'' = C''[f_i]$$

where $f_i = \{((a_i, \mu_i)/(a, \mu), (b_{i+1}, \mu_i)/(b, \mu),$
$$(\overline{b_i}, \mu_i)/(c, \mu)), (a_i', \lambda_i)/(a', \lambda),$$
$$(b_{i+1}', \lambda_i)/(b', \lambda), (\overline{b_i'}, \lambda_i)/(c', \lambda)\}$$

and finally $C^3 = (C_1|C_2''|C_3'')\backslash\{b_1, b_2, b_3, b_1', b_2', b_3'\}$

2. Alternatively, we can simply define

$$C_i = (x_i, \lambda_i).C_{i+1} + (y_i, \mu_i).C_{i+2}$$
$$Ring = C_1 + C_2 + C_3$$

We obtain the state transition graph, Figure 7. Since this contains cycles—

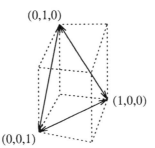

Figure 7: Graph for a 1 customer case

actually only one of length greater than two—we apply Kolmogorov's criterion to obtain the reversibility condition $\lambda_1\lambda_2\lambda_3 = \mu_1\mu_2\mu_3$.

6 Algorithm

We now present an algorithm to deduce the Markov process associated with an MPA system. Initially, all we are presented with is the source for the MPA system and the definition for the state function F. From this we would like to systematically derive our Markov graph.

In our terminology, we have the set V, the relation E, and the functions L and F. From these we would like to derive V_M, E_M and L_M. Towards this goal, we can define two new functions E_f and E_{Mf} of type $Process \rightarrow \{Process\}$ and $state \rightarrow \{state\}$ respectively. These correspond to the relations E and E_M i.e. $E_f P = \{Q \mid EPQ\}$ and similarly $E_{Mf} P = \{Q \mid E_M PQ\}$

We next define T, which is similar to E_f but only relates processes with defined states:

$$TP = \{Q \mid Q \in E_f P \wedge FQ \neq \perp$$
$$\vee \quad Q \in E_f^* P \wedge \forall Z[E_f^{-1*} Z = P \wedge E_f^{*-1} Q = Z \Rightarrow FZ = \perp]\}$$

Then we can define E_{Mf} by $E_{Mf} R = \{S \mid FP = R \wedge FQ = S \wedge Q \in TP\}$ since by the definition of T, we are guaranteed the constraint on paths. Thus we can define the function E_{Mf} if we can find the set $\{S \mid FP = R \wedge FQ = S \wedge Q \in TP\}$ for a given R. It can be seen that this can be accomplished by the function $(\text{map } (F)) \circ (\text{fold } \cup) \circ (\text{map } (T)) \circ F^{-1}$ which when presented with a state derives all processes that map onto that state by application of F, finds all processes with defined states that the process can immediately evolve into and returns the states associated with those processes. Thus the function basically finds the states out of a given Markov state. We can use the function as the definition for E_{Mf} and as a basis for an algorithm for its construction. The construction for the function L_M is similar. The only problem with this definition is that for most cases, F^{-1} is not easy or even possible to find. In this case we can only piecewise build up E_{Mf} by examining processes, P, such that $FP = R$. We are not guaranteed completeness, i.e. to find all of the state transitions out of a given Markov state since we cannot know if we have examined all processes that map onto a given state for their contribution to the graph. Only pathological systems seem to possess the property that for any two processes, P and Q, then if $FP = FQ$, we do not get identical state transition information from both, this means that usually only examining one of those processes will yield as much information as examining all of them.

The algorithm to piecewise build up a Markov graph when we cannot find F^{-1} is presented as a functional program in the style of Miranda [3]. It works by simply searching the derivation tree of a given process to find processes with defined states, thus showing which states can be reached in one Markovian transition from the state defined by the initial process.

6.1 Example : The M/M/1 queue

We might define a single server queue at the "implementation level" as a server and buffer process working in tandem:

$$C = (\text{arrival}, \top).(C^\frown C) + (\text{service}, \top).D$$
$$D = (\text{arrival}, \top).C$$
$$S = \overline{(\text{service}, \top)}.(\text{departure}, \mu).S$$
$$Q = (D|S)\backslash\{s\}$$

A C process can be thought of as being a single customer. On an arrival event it becomes chained to another C process, thus reflecting another arrival. Thus the buffer of customers waiting to be served can be viewed as the chained list of C processes. A D process can be thought of as the space left after a customer has been served. On an arrival event it becomes a C process. An S process can be thought of as being the server part of the queue. It accepts service events from the buffer and sends departures to the environment.

---

[3] Miranda is a trademark of Research Software Ltd.

Again, being interested in queue length, we can thus define the state function as:

$$
\begin{aligned}
F_1(C^\frown Rest) &= 1 + (F_1 Rest) \\
F_1 C &= 1 \\
F_1 D &= 0 \\
F_2 S &= 0 \\
F_2((d,\mu).S) &= 1 \\
F((P_1|P_2)\backslash L) &= F_1 P_1 + F_2 P_2
\end{aligned}
$$

Here we have a more complex derivation tree thus we will present it as a group of labelled transitions as opposed to a data-structure. let A^n represent $A|A|\cdots$ n times starting with process Q at the root we have the labelled transition :

$Q \xrightarrow{(\tau,\lambda)} (C|S)\backslash\{s\}$

this gives us transitions $Q = (0,[(1,\lambda)])$ Thus we now have state 0 and its transition out to state 1 at rate λ as part of the Markov graph.

We can analyse the new process further to build up the rest of the Markov transition graph. The next level in the derivation tree has just the node for the process $(C|S)\backslash\{s\}$, with labelled transitions :

1. $(C|S)\backslash\{s\} \xrightarrow{(\tau,\lambda)} ((C^2)|S)\backslash\{s\}$

2. $(C|S)\backslash\{s\} \xrightarrow{(\tau,\top)} (D|(departure,\mu).S)\backslash\{s\}$

giving us the transition : $(1,[(2,\lambda),(1,\top)])$ we now also have the transition from state 1 to state 2 at rate λ. We also get an instantaneous transition from state 1 back to itself (a customer moving from the buffer into service). This does not contribute to the Markov transition graph, since we are looking at the underlying *continuous time* Markov chain.

Proceeding to examine the next level of the graph :

1. $(C^2|S)\backslash\{s\} \xrightarrow{(\tau,\lambda)} ((C^3)|S)\backslash\{s\}$

2. $(C^2|S)\backslash\{s\} \xrightarrow{(\tau,\top)} ((C^\frown D)|(departure,\mu).S)\backslash\{s\}$

3. $(D|(departure,\mu).S)\backslash\{s\} \xrightarrow{(\tau,\mu)} (D|S)\backslash\{s\}$

giving us the transitions :

1. $(2,[(3,\lambda),(2,\top)])$

2. $(1,[(0,\mu)])$

We could proceed in this manner to constructively build up the remaining Markov graph to any desired size (since it has an infinite number of nodes). However, after analysis to this point we can 'see' how the queue process is developing and thus choose to examine it in a 'general configuration' in order to give us an inductive definition for the remaining Markov graph. Indeed, theorem proving techniques exist that mechanise this process.

1. $((C^n \frown D^j)|S)\backslash\{s\} \xrightarrow{(\tau,\top)} ((C^{n-1} \frown D^{j+1})|(\text{departure}, \mu).S)\backslash\{s\}$

2. $((C^n \frown D_j)|S)\backslash\{s\} \xrightarrow{(\tau,\lambda)} ((C^{n+1} \frown D_j)|S)\backslash\{s\}$

3. $((C^n \frown D^j)|(\text{departure}, \mu).S)\backslash\{s\} \xrightarrow{(\tau,\mu)} ((C^n \frown D^j)|S)\backslash\{s\}$

examining these we now get the following transitions defining the Markov graph:

1. gives us no useful information, i.e. $(n, n, \top)$

2. gives us $(n, n+1, \lambda)$

3. gives us $(n+1, n, \mu)$

Thus we find that we have the same underlying Markov process as for the more simply defined queue in section 4.1. Its reversibility is immediate.

7 Conclusion

Product-form stochastic models provide an efficient means for performance modelling and so they are much sought after, either as an exact representation or an approximation for physical systems. Markovian process algebra has good expressive power but in general leads to inefficient solutions based upon the direct solution of the underlying Markov chain. We have begun to develop a methodology for detecting and synthesising product form Markov models by considering their simplest subset, reversible processes.

Although only a small fraction of product form processes are reversible, the compositionality of process algebras offers the prospect of mechanically identifying a large proportion of them. Moreover, by developing ways of syntactically combining processes to preserve their *product form* property, a far greater proportion of such processes will become identifiable. This idea has been applied to *quasi-reversible* processes, which define a much more substantial subset of product form models, in [4]. Our work on reversibility will provide an important pool of base cases upon which to build a significant collection of processes with easily computable product form solutions. Noting that neither reversibility nor quasi-reversibility implies the other property, all that will be necessary will be to determine when any process we recognise as reversible is also quasi-reversible. This is a relatively simple task; e.g. the birth-death process is quasi-reversible by virtue of its Poisson arrival process.

References

[1] Harrison P G and Patel N M. *Performance modelling of Communication Networks and Computer Architectures.* Addison-Wesley, 1993.

[2] Harrison P G and Strulo B. *Process Algebra for Discrete Event Simulation.* Workshop on Formalisms, Principles and state-of-the-art, Erlangen, September 1993.

[3] Grant-Duff Z and Harrison P G. *Homomorphisms for parallel program synthesis.* Research Report, Department of Computing, Imperial College, 1995.

[4] Harrison P G and Hillston J. *Exploiting quasi-reversible structures to find product form solutions in Markovian process algebra models.* Proceedings of the 3rd workshop on Process Algebras and Performance Modelling, June 1995.

[5] Hillston J. *A Compositional approach to Performance Modelling* PhD thesis, University of Edinbrough, 1994.

[6] Hillston J. *The Nature of Synchronisation* Proceedings of the 2nd workshop on Process Algebras and Performance modelling, August 1994.

[7] Kelly F. P.*Reversibilty and Stochastic Networks* John Wiley and Sons, 1978.

[8] Koukoulidis V. N. and Comeau M. A. *Graph-Theoretic Characterisation and Modeling of Product-Form Queueing Networks* Preliminary report, Concordia University, Canada.

[9] Lamport L. LaTeX. Addison-Wesley, 1986.

[10] Milner R. *Communication and Concurrency.* Prentice Hall, 1989.

[11] Strulo B. *Process Algebra for Discrete Event Simulation* PhD thesis, Imperial College, 1993.

Access Networks

Wavelength Assignment Between the Central Nodes of the COST 239 European Optical Network

L. G. Tan

M. C. Sinclair

Dept. of Electronic Systems Engineering, University of Essex,

Wivenhoe Park, Colchester, Essex C04 3SQ.

Tel: 01206-872477; Fax: 01206-872900; email: mcs@essex.ac.uk

Abstract

Finding an optimised route and wavelength allocation plan is an NP-complete problem in WDM networks. A genetic algorithm to optimise the allocation plan on the eleven central nodes of COST 239 European Optical Network is presented. Use was made of real traffic data where possible, and several route and wavelength allocation plans were produced for the network. Both Wavelength Path and Virtual Wavelength Path routing schemes were considered in the route and wavelength assignment process.

1 The Background

The work described in this paper was undertaken as part of COST Action 239. The COST framework (European Cooperation in the field of Scientific and Technical Research) is run by the European Commission, aiming to promote pre-competitive R&D cooperation between industry, universities and national research centres. COST 239, 'Ultra-high Capacity Optical Transmission Networks', is studying the feasibility of a transparent optical overlay network capable of carrying all the international traffic between the main centres of Europe.

At the start of the project, twenty such centres were identified, each acting as the gateway for all (or half) of the European international traffic for their country.

The initial network design and analysis of a proposed European Optical Network (EON) based on those nodes was presented in [1, 2]. The initial topology design was carried out by making what were presumed to be reasonable assumptions about the possible traffic distribution, given the node populations, and incorporating suitable structures to enhance reliability. The topological design was then analysed, using existing software [3, 4], and incorporating the PD (population-distance) model for node-to-node traffic, two-shortest-link-disjoint-path routing and a simple model for link availabilities. This resulted in comparative link capacities, which were scaled to realistic levels for the immediate, medium and long term. Subsequently, an improved model for European international telephony traffic, the PFD (population-factor-distance) model, has been developed [5].

A recent development in COST Action 239, is the proposed partitioning of the network into an 11-node central network and three subnetworks [6], all utilising Wavelength Division Multiplexing (WDM). Fibre non-linearities constrain the number of wavelengths that can be used over particular distances. These effects become significant when transmission takes place at large capacity, on multiple wavelengths and over long distances. Consequently, the allocation of routes and wavelengths in such a network and the resulting network wavelength requirement are topics under current study.

2 The Problem

The problem this paper seeks to address is that of producing a route and wavelength allocation plan for the 11-node central network of the proposed partitioned COST 239 EON (see Figure 1). The capacity requirements for the 11-node EON, given in Table 1, have been generated from the real traffic data (scaled to long-term levels, and completed using the PFD model). As only 2.5 and 10 Gbit/s channels are considered here, as proposed in [6], a single channel is allocated to satisfy each multiple of 2.5 Gbit/s or 10 Gbit/s in the capacity requirement. This results in the optical channel allocation matrix given in Table 2. A total of 75 channels are needed in each direction to provide the required capacity for the $25 \times 2 = 50$ fibre network.

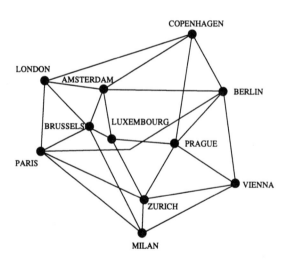

Figure 1: The 11-node COST 239 EON

A single route, chosen from the k-shortest paths, is used between node pairs. A hand allocation of routes and wavelengths on the network has shown that, allowing for single-fibre failure, a minimum network requirement of eight

Capacity Matrix												
Node	0	1	2	3	4	5	6	7	8	9	10	
0		10	2.5	10	10	10	2.5	2.5	2.5	2.5	2.5	Copenhagen
1	10		30	30	40	20	20	10	30	2.5	20	Berlin
2	2.5	30		10	10	10	10	2.5	10	2.5	2.5	Vienna
3	10	30	10		20	10	10	2.5	20	2.5	10	Milan
4	10	40	10	20		30	20	2.5	20	2.5	20	Paris
5	10	20	10	10	30		20	2.5	10	2.5	10	London
6	2.5	20	10	10	20	20		2.5	10	2.5	10	Amsterdam
7	2.5	10	2.5	2.5	2.5	2.5	2.5		2.5	2.5	2.5	Prague
8	2.5	30	10	20	20	10	10	2.5		2.5	10	Zurich
9	2.5	2.5	2.5	2.5	2.5	2.5	2.5	2.5	2.5		2.5	Luxembourg
10	2.5	20	2.5	10	20	10	10	2.5	10	2.5		Brussels

Table 1: Capacity Matrix of 11-node COST 239 EON

wavelengths could probably be achieved using Wavelength Path (WP) routing [6]. In this paper, the same WP scheme was considered, although the requirement for single-fibre-failure re-routing was dropped for simplicity. In addition, the effectiveness of a Virtual Wavelength Path (VWP) routing scheme [7] was examined in the route and wavelength assignment process.

The size of the problem space, for a given allocation strategy, is the number of different route and wavelength allocation plans possible. With the eleven central nodes, this space ranges in size from 2^{330} to 2^{900} for the different allocation strategies considered.

3 Genetic Algorithm

To search the vast problem space efficiently, a genetic algorithm (GA) was employed [8, 9], as has previously been used for topological optimisation of the full COST 239 EON [10]. The algorithm is based on mechanics of natural selection and natural genetics [8]. Instead of searching directly in problem space, the GA decouples the problem into *genotypes* and *phenotypes*. Genotypes are strings which can be decoded into phenotypes and phenotypes are solutions in the problem space.

A general sketch of a simple GA procedure appears in Figure 2. At first, an initial population of genotypes must be chosen. A randomised approach was used here in generating the initial population, ensuring a large variety of genotypes existed and hence maximising the chances of the GA exploring the greatest possible range of phenotypes (*i.e.* route and wavelength allocation plans).

Then, an iterative process starts. During iteration t (a generation), the GA maintains a population of genotypes, $P(t) = \{x_1^t, \ldots, x_n^t\}$. Each genotype

Optical Channel Allocation Matrix												
Node	0	1	2	3	4	5	6	7	8	9	10	
0		1	1	1	1	1	1	1	1	1	1	Copenhagen
1	1		3	3	4	2	2	1	3	1	2	Berlin
2	1	3		1	1	1	1	1	1	1	1	Vienna
3	1	3	1		2	1	1	1	2	1	1	Milan
4	1	4	1	2		3	2	1	2	1	2	Paris
5	1	2	1	1	3		2	1	1	1	1	London
6	1	2	1	1	2	2		1	1	1	1	Amsterdam
7	1	1	1	1	1	1	1		1	1	1	Prague
8	1	3	1	2	2	1	1	1		1	1	Zurich
9	1	1	1	1	1	1	1	1	1		1	Luxembourg
10	1	2	1	1	2	1	1	1	1	1		Brussels

Table 2: Required Channels for The 11-node COST 239 EON

x_i^t is then decoded into a phenotype and evaluated to reveal some measure of its *fitness*; in this case, an assessment of the network wavelength requirement of the decoded route and wavelength allocation plan. This determines if one genotype is better than another, and is used to decide which genotypes will survive and hence reproduce. Reproduction is done by stochastically choosing genotypes that are more fit to become parents of a new population.

This is followed by applying genetic operators such as *crossover* and *mutation*. Crossover exchanges corresponding genetic materials from two parent genotypes, allowing beneficial genes on different parents to be combined in their offspring; while mutation simply toggles, at very low probability the genes in a new genotype. In fact, the driving force behind GAs is the reproduction of individuals in proportion to fitness together with these genetic operators. Variations are introduced into the offspring using these genetic operators, after which the new genotypes replace existing ones and the next generation $(t + 1)$ begins.

A GA proceeds in this way, creating a population of strings, copying them with some bias toward the best, mating and swapping genes, and mutating an occasional gene for good measure. Each population will be evaluated and its members reproduced to form a new population. The iterative process stops when a termination condition is met. This could be when a required maximum number of generations is reached, when the population has *converged*, or when an optimal (or good enough) solution is discovered.

4 Allocation Plan Optimisation

The aim of the allocation was to obtain a route and wavelength for each channel through the network, in such a way as to minimise the number of wavelengths

```
procedure Ga
  begin
    t = 0;
    initialise P(t);
    evaluate genotypes in P(t);
    while termination condition not satisfied do
    begin
      t = t + 1;
      select P(t) from  P(t-1);
      recombine genotypes in  P(t);
      evaluate genotypes in P(t);
    end
end
```

Figure 2: A Genetic Algorithm

required overall. GENESIS v5.0 was used [9], with the strategies encoded as a simple binary string. Each string represents a matrix $\Omega[M, d]$, where M is the number of routes required between node pairs in the network and d the forward or backward directions of a connection path between each node pair. For a given node pair, each direction of a connection path could be represented by n bits, to allow a choice from the available k-shortest paths (*i.e.* $k = 2^n$). Hence, with $M = 55$, $d = 2$ and $n = 3$ (*i.e.* $k = 8$), a string of length $[55 \times 2] \times 3 = 330$ bits could be used to represent a route allocation plan.

With this approach, the GA is responsible for selecting an appropriate route from the k-shortest paths for each connection, while the wavelength assignment procedure is done heuristically. Thus, the GA indirectly optimises the route and wavelength allocation plan by choosing the path between each node pair in the network. The choice of routes has been limited to $k = 8$ shortest paths (*i.e.* $n = 3$) so as to limit the size of the string. This choice of alternative paths is sufficient to route the traffic in the network, as it is unlikely that a near optimal route and wavelength allocation would use very long routes, which will eventually lead to an increase in the number of wavelengths required.

4.1   Allocation Strategies

Bidirectional links (using two fibres on each link) have been considered for the 11-node COST 239 EON to average out the traffic flows. This gives a total of 55 ($11 \times 10/2$) connections and 25 links (and 50 fibres) for the network. Four route and wavelength allocation strategies have been developed, each an extension of the one before [11]:

- **Strategy 1** : a single channel is used to carry all the demand between a node pair *i.e.* unlimited capacity is assumed. String length was thus $[55 \times 2] \times 3 = 330$ bits.

- **Strategy 2** : the required number of channels is allocated between node pairs, but all such channels follow the same route through the network, and hence have the same string length as Strategy 1.

- **Strategy 3** : an extension of Strategy 2, except each of the channels may follow a separate route, giving a string length of $[75 \times 2] \times 3 = 450$ bits.

- **Strategy 4** : for Strategies 1 to 3, the GA was only responsible for allocating the routes each channel would use through the network—the allocation of the actual wavelength used was done heuristically. For Strategy 4, the GA allocated both routes and wavelengths (λ_1 up to λ_8). This resulted in a $[75 \times 2] \times (3 + 3) = 900$-bit string. A suitable penalty factor was used, where necessary, to reduce the number of wavelength convertors required in the network.

5 Results

Ten runs (with different seeds) were carried out on each strategy, giving ten possible allocation plans. When several plans with the same network wavelength requirement were found, those with fewer links requiring the maximum number of wavelengths were preferred. Remaining ties were resolved by selecting networks with fewer wavelengths on longer links.

Figure 3 shows the best allocation plan for each strategy with the WP scheme. A comparison of the best runs of Strategy 1 and Strategy 2 shows that the performance of the GA was not degraded even when more channels are required, as only one additional wavelength is needed per fibre when the number of channels required increases from 55 to 75. Also, separate routing of channels in Strategy 3 resulted in a smaller number of links requiring the highest number of wavelengths compared to Strategy 2.

The performance of GAs for the VWP scheme is rather similar to that for the WP scheme. The best allocation plans for Strategy 1 to Strategy 3 with the VWP scheme are shown in Figure 4. Also, Figure 5 illustrates the variations of the number of wavelengths assigned on each link of the network at selected trials from generation 0 to generation 800 in the best GA run for Strategy 3, *i.e.* the best strategy overall for the VWP scheme. There is a clear progression from (a), with far more wavelengths assigned per link, and consequently, higher cost, to (d). Of particular interest in Figure 4, for Strategy 3, is the number of wavelengths assigned on the Paris-Berlin link (1110km), as only 10 wavelengths have been assigned on this link and not the maximum of 12 wavelengths found. Hence, this solution has effectively reduced the non-linear effect.

Using Strategy 4, as shown in Figure 6, the same network wavelength requirement of six was obtained. However, in contrast to the 218 wavelength conversions required using a null penalty factor (and hence a VWP scheme), shown in part (a) of the Figure, the GAs have selected wavelength number, as well as the available routes, to reduce the total number of conversions to 51 using a fixed penalty factor in (b), and to 50 using a varying penalty factor in (c).

GAs have successfully optimised the network wavelength requirement in this scheme and, with the help of a penalty factor, reduced the number of wavelength conversions. However, the algorithms failed to produce a pure WP-routed plan, although the incorporation of a penalty factor has provided certain amount of local fine tuning. The wavelength conversions needed in this scheme will thus impose extra cost on the network.

5.1 Performance Comparison with ISHC

To check on the suitability of GAs for the problem, an attempt was also made to obtain route and wavelength allocation using *iterative stochastic hill climbing* (ISHC). However, a far poorer performance resulted—with the same number of trials as a GA run, ISHC could only manage to get down to a network wavelength requirement of 16 for the VWP scheme and 17 for the WP scheme.

6 Conclusions and Further Work

The results obtained from the VWP and WP schemes have shown that wavelength translation (in VWP scheme) is not necessary on the 11-node COST 239 EON, as the difference in wavelength requirements between them is very limited, which is in contrast with the general feeling in the literature. It has to be noted, however, that the difference may be larger if several other factors (*e.g.* management complexity, fault tolerence, restoration speeds, *etc.*) were considered

As shown in Strategy 4, a hybrid solution where wavelengths are only translated on specific nodes gives no significant drawback (in terms of network wavelength requirement) and thus could be used as an alternative to a pure WP scheme if network flexibility is required.

The usefulness and effectiveness of genetic algorithms in solving the route and wavelength assignment problem has been demonstrated. The best route and wavelength allocation plan found compares favourably with that obtained by hand-based heuristics [6], although no allowance was given here for single-fibre failure. Both network cost and susceptibility to non-linear effects were reduced compared to the heuristic solution. In addition, GAs have proven to be more effective than another robust optimisation technique, ISHC, in solving the problem.

The work presented in this paper could be extended by:

- Developing more rigorous cost models; for example, reliability issues could be incorporated in the work by considering single-fibre-failure re-routing in the allocation strategies.

- Providing finer granularity in the available alternative paths (k) with the use of different string encoding method, rather than only k-paths in powers of 2.

- Using a heuristic that assigns the wavelengths in terms of descending order of path length (and number of hops), as wavelength blocking is more pronounced in longer paths. This should result in an improved wavelength allocation plan.

- Investigating whether improved GAs search would result from variations on the simplistic objective functions or problem encodings used.

- Developing minimum-cost allocations (rather than simply trying to minimise the number of wavelengths needed).

7 Acknowledgements

The research described in this paper was undertaken by the first author (supervised by the second) as part of an M.Sc. in Telecommunications and Information Systems at the Dept. of Electronics Systems Engineering, University of Essex. The first author is supported by the kind generosity of his family.

References

[1] Sinclair MC, O'Mahony MJ. COST 239: Initial network design and analysis. Proc. 36th RACE Concertation meeting, Brussels, 1993; 89–94

[2] O'Mahony M, Sinclair MC, Mikac B. Ultra-high capacity optical transmission network: European research project COST 239. Information, Telecommunications, Automata Journal 1993; 12:33–45

[3] Sinclair MC. Single-moment analysis of unreliable trunk networks: two methods for the network grade-of-service. Proc. 8th UK Teletraffic Symposium, Beeston, 1991; 8/1–8/11

[4] Sinclair MC. Single-moment analysis of unreliable trunk networks employing k-shortest-path routing. Proc. IEE Colloquium on Resilience in Optical Networks, Digest No. 1992/189, London, 1992; 3/1–3/6

[5] Sinclair MC. Improved model for European international telephony traffic. Electronics Letters 1994; 30:1468–1470.

[6] Hjelme DR, Røyset A. Network partitioning. Proc. COST 239 Meeting, Colchester, 1994; 109–117

[7] Wauters N, Demeester P. Influence of Wavelength Translation in Optical Frequency Multiplexed Networks. IEEE/LEOS 1994 Summer Topical Meeting on Optical Networks and Their Enabling Technologies, Lake Tahoe, 1994; 23–24

[8] Goldberg DE. Genetic Algorithms in Search, Optimization and Machine Learning. Addison-Wesley, 1989

[9] Grefenstette JJ. A User's Guide to GENESIS Version 5.0. 1990

[10] Sinclair MC. Minimum Cost Topology Optimisation of the COST 239 European Optical Network. Proc. Intl. Conf. on Artificial Neural Networks and Genetic Algorithms, Alès, France, 1995; 26–29

[11] Tan LG. Wavelength Assignment for The COST 239 European Optical Network. M.Sc. Project Report, ESE Dept., University of Essex, 1995

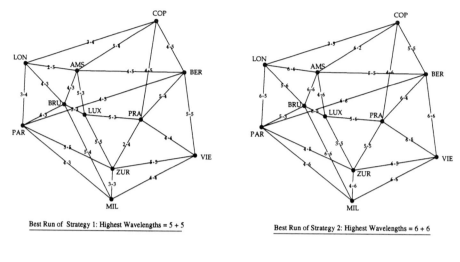

Best Run of Strategy 1: Highest Wavelengths = 5 + 5

Best Run of Strategy 2: Highest Wavelengths = 6 + 6

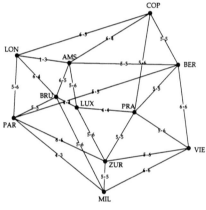

Best Run of Strategy 3 : Highest Wavelengths = 6 + 6

Figure 3: Best Runs of Strategy 1, 2 and 3 (WP Scheme)

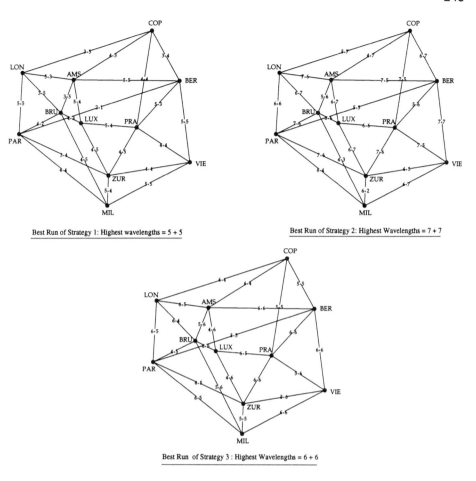

Best Run of Strategy 1: Highest wavelengths = 5 + 5

Best Run of Strategy 2: Highest Wavelengths = 7 + 7

Best Run of Strategy 3 : Highest Wavelengths = 6 + 6

Figure 4: Best Runs of Strategy 1, 2 and 3 (VWP Scheme)

246

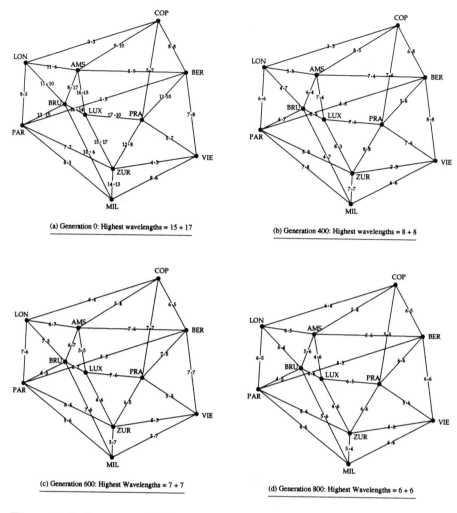

Figure 5: Performance of GAs in The Best Run of Strategy 3 (VWP Scheme)

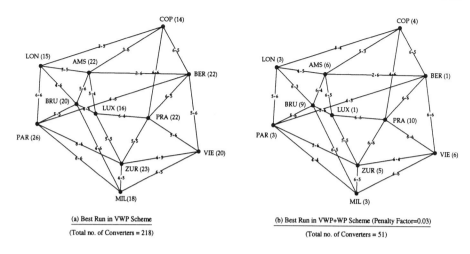

(a) Best Run in VWP Scheme

(Total no. of Converters = 218)

(b) Best Run in VWP+WP Scheme (Penalty Factor=0.03)

(Total no. of Converters = 51)

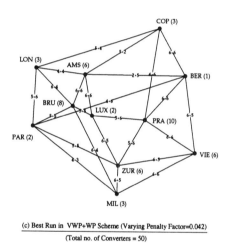

(c) Best Run in VWP+WP Scheme (Varying Penalty Factor=0.042)

(Total no. of Converters = 50)

Figure 6: The Best Run of Strategy 4 (# converters shown in brackets)

Algorithms for Wavelength Assignment in All-Optical Networks

L. Wuttisittikulkij and M.J. O'Mahony

Centre for Network Research
Department of Electronic Systems Engineering
University of Essex, Colchester CO4 3SQ
United Kingdom

Abstract

This paper describes algorithms for solving the wavelength assignment problem in multi-wavelength transparent all-optical networks. In such networks, each connection is assigned a unique wavelength along the physical route and it is assumed that no wavelength conversion is allowed. The objective is to assign wavelengths to all connections in such a way that the number of wavelengths required, M, is minimised. Two techniques are considered in this paper. The first is based on the use of genetic algorithms (GAs) and the second is a simple heuristic algorithm. It was found that the former can be used to achieve an optimal value of M with a considerable reduction of computational time requirement compared to the extensive search approach. Hence this technique is appropriate in the case where the number of wavelengths available is strictly limited and an optimal value of M is highly desired. The heuristic algorithm does not perform as well as the genetic algorithm; although in some cases it performs exceptionally well. However, its computational time requirement is almost negligible; and results can be obtained swiftly. Therefore, it could be used to obtain a good starting value before an extensive optimisation is carried out. It also assigns wavelengths on a single connection basis which is independent of the previous connections assigned. This makes it particularly attractive in the situation where one or only a small number of new connections are to be established without having to modify the existing connections.

1 Introduction

Multi-wavelength transparent optical networks are a topic of increasing research in Europe, the USA and Japan. Such networks are capable of providing high aggregate capacities and, for example, could overlay an existing transport network to meet future demands for capacity as well as enhancing flexibility and reliability. One example of this type of network is that of Multi-Wavelength Transport Network (MWTN) RACE project R2028 [1]. In that project, a realisation of MWTN nodes has been achieved with optical technology, for instance optical space switches, optical amplifiers and optical filters. The system was successfully demonstrated. In such a network, signals can be transmitted from a source node to a destination passing through nodes where the signal is optically routed and switched without regeneration in the electronic domain. In operation the fibre supports a number of signals at different wavelengths and routing is achieved by wavelength selection and space switches; this technique is known as wavelength routing [2].

In these networks the total number of wavelengths multiplexed in one fibre is constrained for practical reasons, for example fibre non-linearity limits the number of wavelengths x distance product. Therefore, it is important to utilise the wavelength resource in an efficient manner. There exist a few number of references on how the wavelength resources may be efficiently assigned according to the traffic demand. For example, [3] derived an upper bound of carried traffic given a number of wavelengths and [4] proved that the problem is NP-complete which means that any algorithm that correctly solves this problem will require in worst case an exponential amount of time and proposed a heuristic approach however no great detail was given. This paper presents and compares two alternative algorithms for implementing wavelength assignment within an optical network.

2 Problem Description

The requirements may be stated as follows. For a given network of N nodes and L bi-directional links with essentially two fibres in each link, a number of traffic connections C are to be established in the network. Each connection is assumed to be full duplex and would be assigned a unique wavelength channel along the physical path in both directions. The objective is to assign wavelength resources to all connections in such a way that the number of wavelength channels required, M, is minimal. In general, network resource allocation for an arbitrary connection involves two processes of selections: physical path and wavelength number. There will usually more than one path exists between a node pair, we assumed that K shortest paths are provided to choose from. If an appropriate physical path and wavelength number are properly selected for each connection, an efficient use of network bandwidth would result. One straight forward approach to find such proper path and wavelength number for all traffic is to try all the possibilities and the best outcome is selected. Apparently, there are in a total of K^C possibilities for only the physical path selections. Whereas the choices of wavelength numbers for each connection could be up to M (M is not known in advance), usually smaller than M depending on the status of network at the point when the connection is being established, i.e, the amount of free bandwidth remaining. Therefore, the total possibilities are $(MK)^C$ in the worse case. If all possibilities were considered, the optimal value of M would be achieved. However, as C gets large, it gets computationally more difficult to consider all combinations. For example, an arbitrary network structure of 18 nodes with every node fully interconnected (traffic demand) produces 1.7×10^{184} possibilities for only the choices of physical paths with K = 16 (a typical good value for small sizes of networks). Clearly, this is not practically feasible. Thus techniques that require less complexity are desirable. We proposed two such techniques here.

In all the investigations, the process of wavelength selection is achieved by a heuristic scheme described as follows: suppose that a particular physical path has been selected for a connection between an arbitrary node pair. In the remaining network resources, there usually be more than one free wavelength along the entire selected path. The preferred choice of wavelength number is the lowest one. Intuitively, this preference does not conflict with the objective defined above. This means that only the physical path then remains to be selected. For the physical path selection, two techniques based on the genetic algorithm (GA) and a heuristic approach are investigated. The basic principle of the GA is that it selectively and randomly considers a relatively small number of these combinations and can give an optimal result. On the other hand the heuristic approach considered here requires only a few iterations and yet gives a result close to the optimum under uniform traffic. In both techniques considered below, both the performance and computational time are assessed.

3   The GA Based Approach

In the Genetic Algorithm (GA), the problem parameters are coded as finite-length (binary) strings. For the application considered here, each connection is coded by m bits known as a gene. For a total number of connections of C, the string length needed is mC bits. The value of a gene associated with each connection determines the path number for the wavelength assignment of the connection. For a gene of m bits, up to 2^m different paths could be distinguished. When the GA generates a bit string, connections are iteratively assigned a free wavelength channel using the selected path defined by their corresponding genes. After all connections are established, the number of wavelengths M can be determined and returned to the GA, this is called the fitness. Then the GA performs its standard operations involving reproduction, crossover, and mutation. The detailed operations of the GA should refer to [5, 6]. This process would carry on until the optimal M is obtained. The optimisation is achieved by selecting an appropriate path for each connection. Note that the implementation of the GA was quickly accomplished by using a software package called GENESIS version 5.0, see [7].

4   The Heuristic Algorithm

The algorithm under study operates as follows. To establish a connection between a node pair, say nodes 1 and 2, a number of K shortest paths between them are considered. In each path, a free wavelength is sought from the remaining network resources and allocated to it. Only one of these paths would be selected to establish the connection and the path selection is based on the following procedure. The algorithm chooses the path that requires the lowest wavelength number, hence encouraging reuse of the same wavelength. If more than one such path exists, the shortest path amongst them is chosen so that the network bandwidth is used most effectively. If there are more than one path

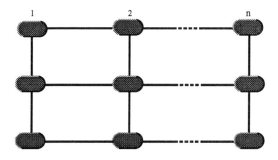

Figure 1: A structure of 3xn grid network

of the same length, which need the same wavelength, one of them is randomly chosen. This procedure reduces the total number of path selections from K^C to one.

5 Simulation Specifications

The performance of the algorithms was evaluated using 3xn grid network structures as examples, this is illustrated in Figure 1. The network contains nodes with the degree of 2-4 (the number of incident links on a node) which may represent a small section of a typical transport network. The structure could be particularly suitable to the transport network in countries like, UK and Japan where the geographical area is long and narrow. This structure has also been heavily used by the people in NTT for similar investigations [8]. In all cases, a traffic demand is assumed that gives full interconnection amongst all nodes.

For a given set of traffic connections to be established, one interesting problem encountered was the order of establishing connections. This situation may be described by means of an example. Suppose that two connections namely between nodes 1 and 3 and nodes 1 and 2 are to be set up in a network of 4 nodes, see Figures 2. These connections can be established with two differing orders. In the first, the connection between nodes 1 and 3 is established first and then followed by another connection. A possible outcome of the wavelength allocation is shown in Figure 2(a); two wavelengths are needed. In the second case, the order of connections established is reversed and the wavelength allocation requires only one wavelength, see Figure 2(b). As we can see, choosing a wrong path for the first connection as seen in the first case may lead to an increased number of wavelengths being required. Therefore, it is important to understand how the order of connection may affect the final results of the wavelength assignment. However, it is not possible to try all differing orders since there are in total of $C!$ combinations which is generally even greater than the problem space of K^C. Therefore, it was considered appropriate to examine only a few useful orders. Three particular orders of establishing connections

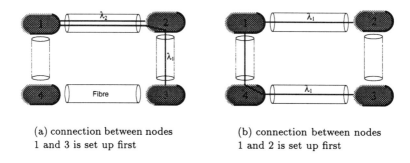

(a) connection between nodes
1 and 3 is set up first

(b) connection between nodes
1 and 2 is set up first

Figure 2: An example of wavelength allocation with different orders

were selected for the investigations. In the first, all connections between node 1 and other nodes are first established, and then followed by all connections between node 2 and other nodes and so on. In the second approach, all connections are established according to their respective shortest path lengths. A connection with the longest path length is established first, and then followed by those requiring shorter path lengths. This approach is expected to perform well for the following reason: a connection requiring a longer path is more difficult to establish than that of the shorter one. In the third method, connections are randomly established, which may represent a typical order. Both algorithms were applied to all of these three cases on 12, 15 and 18-node grid networks.

6 Results and Discussions

6.1 Number of Wavelengths

Figures 3 and 4 show the number of wavelengths required (M) against K achieved by the heuristic and GA approaches respectively. It is apparent that the GA approach always gives as good or better results in terms of M than the heuristic approach in all cases. In addition, for the same value of K the GA technique was able to find the same value of M under the three different traffic orders, whereas the heuristic approach performed well in some orders and not in the others.

The results also show that in both techniques K is an important factor in determining M. In all networks under the three differing connection orders, a dramatic reduction of M was achieved by increasing the value of K. For instance, in the 3x6 grid network M is reduced from 45 (with K=1) to 28 (with K=6) with the heuristic approach. However, a further increase of K does not result in an improved value of M. Despite such a similarity in this aspect, the effect of K on M for these two techniques are rather different.

In the GA approach, when K is set to a larger value, the size of the search space is effectively greater, thus requiring a longer computational time. Such an incremental change increases the chance that the optimal point is covered in the search space, which could lead to an improved result as shown in the simulation results when K was set from 2 to 4. However, when K was increased to 8 or 16, no further improvement of M was observed. This is probably because the optimal point may have been covered with K = 4. Therefore it implies that the increment of K may only increase the computational time.

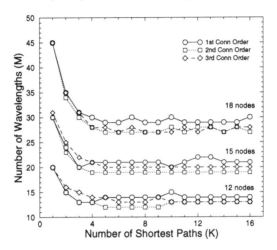

Figure 3: The number of wavelengths achieved with the heuristic approach

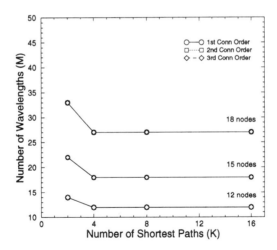

Figure 4: The number of wavelengths achieved with the GA

6.2 Average Distance

Unlike the heuristic approach the GA approach finds a number of different path assignments that result in the same minimal M. However, not all of them

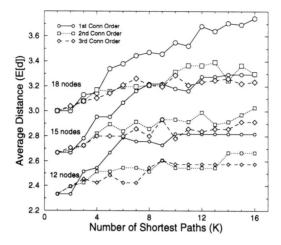

Figure 5: The average distance of a connection established achieved with the heuristic approach

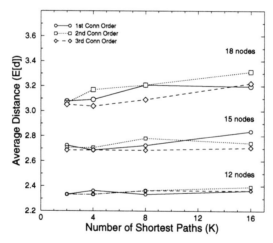

Figure 6: The average distance of a connection established achieved with the GA

require the same amount of bandwidth (network capacity) and the bandwidth requirements could be very different. Such features can be measured in term of the average distance of a connection established, E[d]. A smaller value of E[d] means that each connection was on average established on a shorter path length, hence utilising the bandwidth more efficiently and leaving more network capacity for future traffic. In this investigation, a total of 1000 different path assignments which gave the same optimal M were collected and their corresponding E[d] evaluated. The smallest value of E[d] was chosen and illustrated in Figure 6. Comparing this result to that of the heuristic algorithm in Figure 5, the GA approach always find a better or as good path assignment than the heuristic approach for the networks of 12 and 15 nodes. In addition, the E[d] found by the GA seems to be less dependent on the parameter K. However, in the network of 18 nodes with the second connection order the E[d] achieved with the GA is slightly greater than that of the heuristic approach. One explanation of this observation is that in the 18-node network the problem space is greater than the others hence it is quite possible that the 1000 results obtained may merely cover a small area of the entire problem space. This suggests that more results of the same M should be evaluated and this can be done by increasing the number of iterations or carrying out a new simulation with a different random seed. These approaches were taken and slightly better values of E[d] were achieved. Thus, with the GA, conducting more simulations increases the chance of finding the optimal result.

6.3   Computational Time Requirement

In term of the computational time requirement, the GA approach requires much fewer iterations than the full search technique. For example, in the case of 12-node grid network approximately in the order of 10^5 iterations were needed before the optimal value of M converged, see Figure 7(a), in comparison to the full search technique which requires in the order of 10^{79} iterations. Figure 7(a) shows that the GA initially selects a path for each connection on a random basis as evident from the value of M achieved in the first iteration is far greater than the optimum, i.e., from the initial 34 wavelengths to the optimal 12 wavelengths. After a few ten thousand iterations, the GA begins to find improved outcomes of wavelength assignment as more iterations are carried out, this is seen in the same Figure the values of M drops dramatically. When the values of M approach the optimum, the process starts to slow down until the optimal value of M is reached. This is a typical characteristics of the GA to this problem, and explains why the greater value of K the longer the computational time. For the cases of 15 and 18-node networks, the same tendencies occur, and are illustrated in Figures 7(b) and 7(c) respectively. These Figures show that the computational time is increased in a larger network with greater traffic demand.

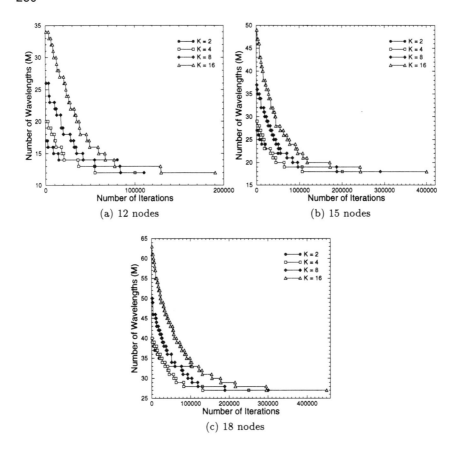

(a) 12 nodes

(b) 15 nodes

(c) 18 nodes

Figure 7: The computational time of the GA with the third order

7 Conclusions

The heuristic algorithm does not perform consistently well. Its performance depends on the connection orders and parameter K. Such characteristics may not be acceptable in the case where the number of wavelength channels is strictly limited and maximum efficiency of network bandwidth utilisation is required. However, its computational time requirement is almost negligible; and results could be obtained swiftly. Therefore, this technique could be used to provide a starting point for more extensive optimisation procedures. In addition the speed advantage makes it practically possible to find a result close to the optimal in a large network with a large number of connections within a short period of time. Moreover, it performs the wavelength assignment on a single connection basis which is independent of the previously assigned connections; no rearrangement of the previous connections is needed. This makes the algorithm particularly attractive in the situation where only one or a small number of new connections are to be established without having to modify existing connections. This scenario may be seen in practice when the traffic demand slowly increases with time and later optimisation may require considerable change to existing connections.

The GA based technique gives an excellent performance and is apparently always as good or better than the heuristic approach. It yields the same result for the three different connection orders. The computational time requirement is also greatly reduced in comparison with the full search technique (K^C iterations). Hence this technique is appropriate where a better value of M is required than that given by the heuristic approach.

Acknowledgement

This work was partially sponsored by the European Community under the MWTN RACE project R2028. The authors wish to thank M.C. Sinclair for his inspiration on the use of GAs and J.J. Grefenstette for making GENESIS version 5 publicly available.

References

[1] G.R. Hill, P.J. Chidgey, F. Kaufhold, and et al. A transport network layer based on optical network elements. *IEEE Journal of Lightwave Technology*, 11(5/6):667–678, May/June 1993.

[2] G. Hill. Wavelength routing approach to optical communication networks. In *Proc. INFOCOM'88*, pages 354–362, 1988.

[3] R. Ramaswami and K.N. Sivarajan. Optimal routing and wavelength assignment in all-optical networks. In *Proc. INFOCOM'94*, Toronto, Canada, June 1994.

[4] I. Chlamtac, A. Ganz, and G. Karmi. Lightpath communications: An approach to high bandwidth optical WAN's. *IEEE Transactions on Communications*, 40(7):1171–1182, July 1992.

[5] L. Davis. *Handbook of Genetic Algorithms*. Van Nostrand Reinhold, 1991.

[6] D.E. Goldberg. *Genetic Algorithms in Search, Optimization and Machine Learning*. Addison-Wesley, 1989.

[7] J.J. Grefenstette. *A User's Guide to GENESIS Version 5.0*, 1990.

[8] Y. Hamazumi, N. Nagatsu, S. Okamoto, and K-I. Sato. Number of wavelengths required for constructing optical path network considering failure restoration. *Trans. on IEICE Japan*, J77-B-I,(5):275–284, 1994.

Wave Propagation in Biaxial Planar Waveguides using Equivalent Circuit in Laplace Space

A.C. Boucouvalas
Bournemouth University
School of Electronics,
Talbot Campus,
Fern Barrow,
Poole,
Dorset BH12 5BB
U.K.
e-mail tboucouv@bournemouth.ac.uk

Abstract:
By applying Laplace transformations to Maxwell's equations, transmission line equations are derived for guided wave propagation in planar waveguides. For the first time an intuitive equivalent T-circuit model is derived for inhomogeneous planar waveguides of materials having **biaxial** dielectric permittivity tensor, which may have **conductivity**, and **attenuation** along the propagation direction. The materials may also have **arbitrary refractive index profile along any of the three axes**. TE and TM mode equivalent circuits for dielectric layers are derived and the mode propagation constants are calculated from the resonances of the cascade of the elementary T-circuits.

The method can be extended, so as complete waveguide characterisation would be possible, allowing dispersion, cut-off frequencies and mode field plots to be calculated incorporating the influence of material conductivity and attenuation.

Keywords: Waveguides, Electromagnetic waves, Biaxial Crystals Integrated Optics,

1. Introduction:

Fourier transform techniques when applied to Maxwell's equations, allow the study of wave propagation in planar and cylindrical waveguides [1,2]. This leads to the concept of modelling electromagnetic phenomena as equivalent circuits offering strong electrical engineering intuitive undertones. The approach has also been applied to many other areas of electromagnetics including antenna theory [3], scattering phenomena [4] and recently in the field of Quantum Mechanics [5,6].

The concept of this work is based on modelling waveguides as simple electric circuits, such that the waveguide modal properties translate and can be derived from the resonant frequencies of the equivalent circuits. This is common sense in electrical engineering, and standard circuit theory can subsequently be applied to the circuits.

In this paper, a model is developped for wave propagation in planar inhomogeneous inisotropic biaxial materials. The model is general and accounts for the material's wave attenuation along the propagation direction, and includes the material conductivity. This model should be useful therefore to modelling waveguide layers realistically.

The need to incorporate wave attenuation and material conductivity in this model, as contrasted to previous work, lead us to applying Laplace transforms, instead of Fourier transforms.

Numerical results are presented for the variation of the normalised propagation constant of the fundamental TE and TM modes with attenuation and conductivity.

2. Maxwell's Equations in Laplace Space:

A homogeneous planar waveguiding layer with the corresponding reference coordinates is shown in Figure 1. Propagation os assumed along the z-axis. The y-invariance in this propagation problem leads to $\dfrac{\partial}{\partial y} = 0$ for the relevant electromagnetic fields.

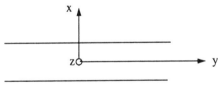

Figure 1: A thin waveguiding element with reference co-ordinates.

A uniform thin waveguiding layer can be described by its permittivity tensor, as follows:

$$\underline{\varepsilon} = \varepsilon_0 \begin{pmatrix} \varepsilon_1 & 0 & 0 \\ 0 & \varepsilon_2 & 0 \\ 0 & 0 & \varepsilon_3 \end{pmatrix}$$

The tensor nondiagonal elements are zero, and the tensor major axes aligned with the coordinate directions of Figure 1. The permeability of the material is assumed to be $\mu = \mu_0$, and σ, is the material conductivity.

The electric and magnetic fields E,H inside the layer are functions of (x,y,z) and time (t). The field components can in general be expressed by:

$$A(x,y,z,t) = \text{Re}[A(x,y,z)e^{j\omega t}].$$

Hence the Laplace transform of the fields along the propagation direction is given by:

$$\overset{0}{A}(z) = \int_0^\infty A(z)e^{-\gamma z}dz, \quad \text{where } \gamma = \alpha - j\beta. \text{ and } \alpha \text{ is the attenuation coefficient,}$$

and β is the mode propagation constant.

Maxwell's vector wave equations for time dependent fields are given by:

$$\left.\begin{array}{l} \nabla \times \underline{E} = -\dot{\underline{B}} = -\mu \dot{\underline{H}} \\[2mm] \nabla \times \underline{H} = \varepsilon \dot{\underline{E}} + \underline{J} \end{array}\right\} \quad \ldots\ldots\ldots\ldots\ldots\ldots(1)$$

In Laplace space, the components of equation 1 reduce to:

$$\left.\begin{array}{ll} -\gamma \overset{0}{E}_y = -j\omega\mu \overset{0}{H}_x & -\gamma \overset{0}{H}_y = (j\omega\varepsilon_1 + \sigma)\overset{0}{E}_x \\[3mm] -\dfrac{\partial \overset{0}{E_z}}{\partial x} + \gamma \overset{0}{E}_x = -j\omega\mu \overset{0}{H}_y & \gamma \overset{0}{H}_x - \dfrac{\partial \overset{0}{H_z}}{\partial x} = (j\omega\varepsilon_2 + \sigma)\overset{0}{E}_y \\[3mm] \dfrac{\partial \overset{0}{E_y}}{\partial x} = -j\omega\mu \overset{0}{H}_z & \dfrac{\partial \overset{0}{H_y}}{\partial x} = (j\omega\varepsilon_3 + \sigma)\overset{0}{E}_z \end{array}\right\} \quad ..(2)$$

The boundary conditions state that the field components across waveguide boundaries, are continuous, hence:

a) the tangential components of E_y, E_z, H_y, H_z are continous, and

b) the normal components of the vectors $\dot{B}$, and $J + \varepsilon \dot{E}$ are continuous.

In Laplace space this means that the corresponding transformed fields are also continuous across charge or current free bandaries.

3. Equivalent Transmission Lines:

3.1 TM Modes:

Let us define new variables as follows:

$$\overset{0}{V}_E = -j\gamma \overset{0}{E}_z$$

$$\overset{0}{I}_E = j\gamma \overset{0}{H}_y = (\omega\varepsilon_1 - j\sigma) \overset{0}{E}_x$$

Using equations 2, the following transmission line equations can be derived:

$$\left. \begin{array}{l} \dfrac{\partial \overset{0}{V}_E}{\partial x} = \dfrac{-\gamma_E^2 \overset{0}{I}_E}{j(\omega\varepsilon_1 - j\sigma)} \\[3mm] \dfrac{\partial \overset{0}{I}_E}{\partial x} = -j(\omega\varepsilon_3 - j\sigma)\overset{0}{V}_E \end{array} \right\} \quad \dots\dots\dots\dots\dots(3)$$

where $\gamma_E^2 = \beta^2 - \alpha^2 - \omega^2\mu\varepsilon_1 + j(2\alpha\beta + \omega\mu\sigma)$. Equations (3) represent a transmission line model for an infinitecimally thin waveguiding layer, transversly to the propagation direction.

The characteristic impedance of the transmission line is:

$$\overset{0}{Z}_E = \dfrac{\gamma_E}{j[\omega^2\varepsilon_1\varepsilon_3 + \sigma^2 - j\omega\sigma(\varepsilon_3 + \varepsilon_1)]^{\frac{1}{2}}} \quad \dots\dots\dots\dots\dots(4)$$

The planar layer of thickness d, can be represented by a T- equivalent circuit as shown in Figure 2.

The series and parallel elements of the circuit are given by:

$$\overset{0}{Z}_{S,E} = \overset{0}{Z}_E . \tanh(\gamma_E . d / 2),$$

$$\overset{0}{Z}_{P,E} = \overset{0}{Z}_E / \sinh(\gamma_E . d)$$

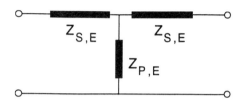

Figure 2: Equivalent T-circuits of TM modes of planar waveguide layer, of thickness d.

If the the layer is homogeneous, and of infinite thickness, such as a semi-infinite cladding layer, then it may be represented by its characteristic impedance $\overset{0}{Z}_E$, equation (4).

3.2. TE modes:

Similarly, for TE modes, let us define:

$$\overset{0}{V}_M = j\gamma \overset{0}{H}_z$$

$$\overset{0}{I}_M = -\omega\mu_0 \overset{0}{H}_x = j\gamma \overset{0}{E}_y$$

after some algebra the following transmission line equations can be derived representing an infinitesimally thin waveguiding layer:

$$\left. \begin{aligned} \frac{\partial \overset{0}{V}_M}{\partial x} &= \frac{-\overset{0}{I}_M \gamma_M^2}{j\omega\mu_0} \\[2em] \frac{\partial \overset{0}{I}_M}{\partial x} &= -j\omega\mu_0 \overset{0}{V}_M \end{aligned} \right\} \quad \dots\dots\dots\dots\dots\dots(5)$$

with $\gamma_M^2 = \beta^2 - \omega^2\mu_0\varepsilon_2 - \alpha^2 + j(\omega\mu_0\sigma + 2\alpha\beta)$.

The characteristic impedance $\overset{0}{Z}_M$ of equations (5) is given by:

$$\overset{0}{Z}_M = \frac{\gamma_M}{j\omega\mu_0} \qquad \dots\dots\dots\dots\dots\dots\dots\dots\dots\dots\dots(6).$$

In a similar manner to TM modes the transmission line can be represented by an equivalent T-circuit as shown in Fig. 3, with series and parallel elements given by:

$$\overset{0}{Z}_{S,M} = \overset{0}{Z}_M \tanh(\gamma_M d / 2)$$

$$\overset{0}{Z}_{P,M} = \overset{0}{Z}_M / \sinh(\gamma_M d)$$

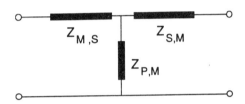

Figure 3: Equivalent T-circuit of TE modes of planar waveguide layer of thickness d.

4. Transverse Resonance of T-circuits:

An inhomogeneous waveguide can be represented as a set of thin homogeneous layers. This configuration can be represented as a sequence of equivalent circuits in tandem, as shown in Figure 4. The sequence is terminated by the characteristic impedance $\overset{0}{Z}_T$, and $\overset{0}{Z}_B$ of the top and base (semi-infinite) layers..

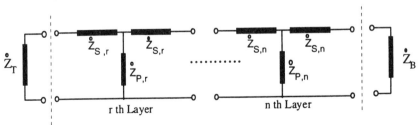

Figure 4: Cascade of equivalent circuits connected in tandem, representing inhomogeneous layers.

The propagation constant β, can be derived from the resonance frequencies of the above cascade. The impedance of the nth layer in terms of the impedance of the

(n-1)th layer, the series and parallel Impedances of the T-circuits, may be expressed as follows:

$$\overset{0}{Z_n} = \frac{(\overset{0}{Z}_{n-1} + \overset{0}{Z}_{S,n})\overset{0}{Z}_{P,n}}{\overset{0}{Z}_{n-1} + \overset{0}{Z}_{S,n} + \overset{0}{Z}_{P,n}} + \overset{0}{Z}_{S,n} \qquad \text{for n=1,2,3,4,.........n} \qquad .....(7)$$

This recurrence relationship, allows us to calculate the total impedances up to a boundary (such as a well defined cladding layer) from either side of the boundary. Resonance occurs when at the arbitrary boundary, the impedance from one side, $\overset{0}{Z}_1$, and the other side, $\overset{0}{Z}_2$, are related as follows:

$$\text{Im}(\overset{0}{Z}_1 + \overset{0}{Z}_2) = 0 \qquad .... (8)$$

Using a root searching technique, varying β, the resonances can be located.

5. Cut-off frequencies of TE and TM modes.

Considering for simplicity a symmetrical waveguiding structure with substrate equal to superstrate refractive indicies, and the guiding layer being of a material with conductivity and attenuation such as: $n_1 k_0 > \beta > n_2 k_0$. The cut-of frequencies of the modes are obtained when $\beta = n_2 k_0$. With this condition, the propagation factors of equations (3) and (5) become:

$$\gamma_{EC}^2 = n_2^2 k_c^2 - \alpha^2 - \omega^2 \mu \varepsilon_1 + j(2\alpha n_2 k_2 + \omega \mu \sigma)$$
and
$$\gamma_{MC}^2 = n_2^2 k_c^2 - \alpha^2 - \omega^2 \mu \varepsilon_2 + j(2\alpha n_2 k_2 + \omega \mu \sigma)$$
respectively, for the TM and TE modes.

In this case, ω becomes $\omega_c = k_c c$, where $c = \dfrac{1}{\sqrt{\mu_0 \varepsilon_0}}$, $k_c = 2\pi / \lambda_c$, , and

λ_c is the cut-off wavelength.

The cut-off wavelengths can be calculated in the same way as calculating the propagation constants. It is clear that both attenuation and conductivity affect the

6. Mode field plots:

The field plots of the propagation modes ban be plotted using this approach having first determined the propagation constant of the mode.
The proceedure can be illustrated with the aid of the following figure:

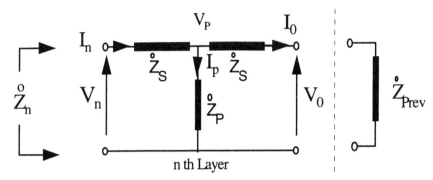

Figure 5: Voltages and currents through a representative transmission line.

From Figure 5, we can write for the voltage V_n , the "new" voltage (to the left of the transmission line) in terms of the Voltage at the node P , V_p , and the currents through node P as follows:

$$V_n = V_p + I_n \overset{0}{Z}_S$$

and

$$V_p = I_n \overset{0}{Z}$$

$$= I_n \frac{\overset{0}{Z}_P \left[\overset{0}{Z}_S + \overset{0}{Z}_{Prev} \right]}{\overset{0}{Z}_P + \overset{0}{Z}_S + \overset{0}{Z}_{Prev}}$$

with $\overset{0}{Z}_{Prev} = \dfrac{V_o}{I_o}$ and $I_n = I_o \left[1 + \dfrac{\overset{0}{Z}_S}{\overset{0}{Z}_P} \right]$.

The above recerrence equations can be used to compute the field plots, starting from infinity, (a few core radii, when the field has decayed to zero), and successively calculating the field up to the core centre axis.
V_E and V_M can be computed with respect to the x axis of the waveguide.
E_x and E_y and hence be directly calculated and plotted.

7. Conclusions:

A simple transverse transmission line technique has been developped to competely characterise planar waveguide layers of arbitrary refractive index profile, and at the same time allowing the conductivity and attenuation of the material to be incorporated in the analysis.

Complete characterisation is possible, allowing the calculation of cut-off frequencies as well as mode field plots, under the influence of attenuation and material conductivity.

8. References:

1. C.D. Papageorgiou and A.C. Boucouvalas: Computation of mode propagation constants and cut-off frequencies of optical planar layers using the resonance technique:OPTICA ACTA 1984 Vol. 31, no. 5, PP.555-562.

2. C. D. Papageorgiou and A.C. Boucouvalas "Propagation Constants of cylindrical dielectric waveguides with arbitrary refractive index profile using the resonance technique". IEE Electron. Lett., vol. 18, pp.786-788, 1982.

3. C.D. Papageorgiou and JD Kanellopoulos : "Equivalent circuits in Fourier space for the study of electromagnetic fields". J. Phys. A: Math. Gen. 15 , 1982, pp2569-2580.

4. C.D. Papageorgiou, A.D. Raptis and T.E. Simos:"A method for computing phase shifts for scattering" Journal of Computational and Applied Mathematics, 29, 1990, pp.61-67.

5. C.D. Papageorgiou and A.D. Raptis: "A method for the solution of the Schroedinger equation". Computer Physics Communications 43, 1987, pp325-328.

6. C. D. Papageorgiou, A.D. Raptis, T.E. Simos: "An algorithm for the solution of the Eigenvalue Schroedinger equation".Journal of Computational Physics, vol. 88, no.2, June 1990, pp.477-483.

Distributed Systems

An Architecture for the Measurement and Modelling of Service Provision in Distributed Systems

Clive M King, Ian C Pyle,
Centre for Intelligent Systems,
Department of Computer Science,
University of Wales,
Aberystwyth, SY23 3BD
cmk@aber.ac.uk icp@aber.ac.uk
+44 1970 622427

July 31, 1995

Abstract

This paper presents work in progress on the quantification and modelling of software in a distributed system. We have used the term "coordination layer" to denote the layer of system software that coordinates the communication and provision of services to an application programs within a distributed system.

Our model describes the interface between the layers of software and hardware present in a distributed systems. We discuss the level of abstraction provided by the system software to support distribution and present a technique for instrumentation of the coordination layer.

We present a metric for describing the coordination layer and suggest hypothesis concerning design tradeoffs within the coordination layer of a distributed system.

The work presented is aimed at providing a foundation for the study of the engineering principles and tradeoffs in the design of distributed systems software.

1 Introduction

This paper presents work in progress on the quantification and modelling of software in a distributed system. We use the term "coordination layer" to denote the layer of system software that coordinates the communication and provision of services to application programs in a distributed system.

Our model describes the interfaces between the layers of software and hardware present in a distributed systems. We discuss the level of abstraction provided by the system software to support distribution and present a technique for the instrumentation of the coordination layer. We concentrate on the quality of the services provided by the system software to an application. Functionality, speed of execution and reliability of machine instructions is not of direct interest in this study.

We propose that the level of abstraction provided by the coordination layer can be quantified. This is achieved by relating the number of coordination layer interface units required to service a request from the application and the number of kernel services used by the coordination layer to service the requests made by the application. We suggest some metrics for the measurement of the richness of the coordination layer based on the measurement of the use of kernel and coordination layer services and present hypothesis about the design tradeoffs made within the coordination layer.

The work presented in this paper is aimed at providing a foundation for the study of the engineering principles used and tradeoffs made in the design of system software for distributed systems.

Section two presents a layered model of a distributed systems that describes the flows of services between layers. The coordination layer is then placed in the context of related work. In section three we present a technique based on interposition for measuring the use of the layer that provides services to the application. Section four describes the implementation of interposition and associated tool support. Section five discusses the merits and limitations of using interposition to monitor the use of the coordination layer. Section six presents metrics for describing the coordination layer and hypothesis concerning design tradeoffs within the coordination layer. Finally, section seven presents conclusions and further work.

2 A Layered Model of Service Provision in a Distributed System

Each layer of software in a computer system takes requests for services from the layer above, requests and receives services from the layer below and returns services to the layer above. The attributes of the services exchanged can be described using functionality and Quality of Service.

Quality of Service(QoS) is defined by the QOSMIC Consortium [18] as "the user perceivable attributes that makes a service what it is. It is expressed in user-understandable language and manifests itself as a number of parameters, all of which have subjective or objective values." The relevant aspects of QoS depend on the nature of a service required by the client. The functionality of a service describes the type of service provided for a purpose. A clear distinction must be drawn between functionality and QoS. Functionality makes no reference to the quality parameters such as performance, security, predictability or reliability.

Figure 1 shows a distributed system viewed as a set of nodes that have hardware, kernel and coordination system software components. The coordination layer components use the services of the underlying kernel and provide services to the application. Applications may run on one or more nodes and communicate with other application components via the coordination layer. This model can be applied to the structure of most modern distributed system software.

The focus of our work is on the coordination of events and communication within distributed systems software. The functionality, performance and reliability of the

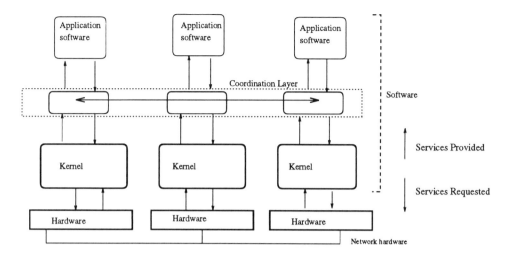

Figure 1: A conceptual view of a Distributed System

underlying hardware greatly affects the form of the system software [8]. Therefore, a model of a distributed system must consider the services requested from and provided by the hardware components. The modelling of performance, reliability and functionality of a hardware system must take the attributes of individual hardware components and their configuration into account. A review of current literature [6, 8] has not revealed a suitable method for describing the relevant aspects of performance, reliability and functionality of the hardware components of a node suitable for use in describing the services used by the kernel.

The kernel is the software layer responsible for the management of resources within a single node, including scheduling, I/O and memory management. Kernel functionality is described in the system call interface documentation such as POSIX [9]. However, little information is provided on the QoS provided by the kernel in most modern operating systems. King [11] describes a technique for measuring the QoS provided by the kernel in response to requests for service by monitoring system calls made and relating the results to experiments that measure QoS for each system call.

The coordination layer is the software layer that orchestrates the direction of requests for services from an application and directs the service requests to the appropriate nodes in a distributed system; that in turn service the request and communicates to the application via the coordination layer [1]. The coordination layer is responsible for handling inter-process communication and requests for system services from applications running on both local and remote nodes. Elements of the coordination layer can take a range of forms, examples include the RPC library [4], socket library [12], CORBA message broker [22], Ada distributed systems annex [7], Arjuna [20], ISIS [3] and the Domain Naming Service [16].

The model presented only addresses synchronous subroutine calls. Asynchronous interprocess communication such as signals or interrupts generated by hardware devices are not considered. Clark [5] proposes an alternative view of the flow of control within the operating system that considers asynchronous interprocess communications.

The coordination layer has some similarity with the often used term "middleware". Schreiber [19] describes middleware as the distributed application infrastructure that exists between an application and the operating system and network services on a single node. The coordination layer described in this paper includes the middleware describes by Schreiber, but also the elements that Bernstein [2] considers to be part of the operating system. Bernstein describes a range of non-performance attributes by which applications, middleware and the operating systems can be measured. Bernstein also notes that over time a migration of services between middleware and operating system can occur in both directions. We content that in addition to a change over time, the location of services will be influenced by design tradeoffs intended to make the system suitable for use in a particular environment.

The non-functional specification of an application should specify the QoS that the coordination layer is required to provide to fulfill the specification. With a knowledge of the services requested from the coordination layer and the Quality of Service requirements of the application derived from the non-functional specification, the functionality and Quality of Service requirements of the coordination layer can be determined and compared to current provision.

The richness of the services provided by the coordination layer is related to the level of abstraction and control provided to the application and the richness of the services provided by the kernel. If the coordination layer is deficient in providing a service required by the applications programmers, then the required services must be constructed from those presented by the current coordination layer or by making the compilation system responsible for the distribution of an application as in the Ada compilation system [7].

3 Interposition and Measurement of the Coordination Layer

Dynamic linking refers to the portions of the link editing process that produce executable programs that can share objects such as library routines at run-time. It is a fundamental element of program compilation. This allows the separation of the interface used during compilation and the library implementation used at run-time. Dynamic linking is not limited to UNIX, being a feature of most modern operating systems and is essential for using interposition at run-time.

Interpositioning involves a well known aspect of C and the dynamic linker. It enables the supplanting of a library function with a user written function of the same name [13]. The user written function that replaced the library call can be used to instrument and record the use of services by the coordination layer [21]. Interposition can be applied to any operating system that uses dynamic linking as a base such as System V UNIX or Windows NT [17].

Many program errors in existing applications have resulted from a lack of awareness of the effects of overloading program function names with those of existing library functions [13]. However, with care, interposition can be applied to replace the library function with alternative services or service providers.

A complementary technique for the measurement of system calls made by the coordination layer to the kernel is described in [11].

4 Implementation and Tool Support

The repetitive nature of the code required to implement the instrumentation of library calls through interposition suggests that appropriate tool support should be used to reduce scope for programmer error and improve productivity. Tool support has been developed to automate the generation of C function stubs from existing libraries and header files.

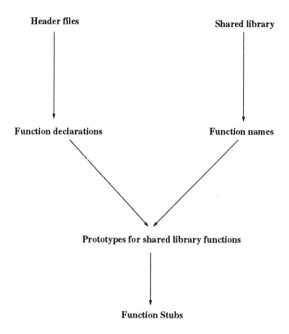

Figure 2: Stages in Function Stub Generation

Figure 2 shows the four stages in the generation of function stubs under UNIX.

- Extract the names of all global, defined functions from the symbol list of the appropriate shared library.

- Search all related header files to derive the set of all function declarations. The header file provide the definition of the function interface.

- Match library calls with appropriate function prototype.

- Parse the function declaration and generate an appropriate function stub.

The processing of libraries and header files is achieved by using simple UNIX shell scripts. The stub generator written using Lex and Yacc, required a parser capable of dealing with the full range of the C type system [10]. It is important to maintain type declarations to avoid compilation problems. The function wrappers are then compiled to produce a shared library [21].

The program is run with the wrapper library preloaded [21] and the functions in the wrapper library are used in preference to the default shared library. Environment

```
#ifdef memcmp
#undef memcmp
#endif

int memcmp(void * arg0, void * arg1, size_t arg2)
{
    static int (* fptr)() = 0;
    char buffer[BUFSIZE];

    fptr = (int (*) ())dlsym(RTLD_NEXT, "memcmp");
    if (fptr == NULL) {
        (void)fprintf(stderr,"dlopen: %s\n",dlerror());
        return(0);
      }

    log_instance("Called function :- memcmp\n",buffer);

    return((*fptr)( arg0, arg1, arg2));
}
```

Figure 3: Example Function Stub

variables are used to force the linker to load the wrapper library after the program text, but before the system library. Figure 3 shows an example of a function stub that has been automatically generated. The call to dlsym causes the run-time linker to use the next symbol to match the current function, thereby using the copy from the system shared library from then point onwards within the scope of the wrapper function, instead of the wrapper function.

The function log_instance records the invocation of the library function. Its contents and interface may be changed to suit current requirements such as recording function arguments or timestamps.

This technique and associated tool support can be used for any shared library that also has header files containing ANSI C prototypes of function declarations.

To date tool support only addresses Solaris and ANSI C. However, having proved the concept, we believe tool support could be easily constructed for other compiled languages and operating systems with facilities for dynamic linking.

5 Evaluation

The method presented has a significant performance overhead. Initial experiments indicate that an overhead of around 40% in program execution can be expected. However, we wish to establish more exact figures though further experimentation. We also consider that this figure can be significantly reduced by code profiling and suitable compiler options. Any measurement of program execution requires care, when timing a program. For example, interposition must only be applied to the timed program and not the program doing the timing in order to avoid substantial distortion of the measurements. A modified version of the shell command time [14] enables the preloading of shared libraries for the target program only.

As with any methods of intrusive program examination [15]. An overhead may affect the execution of time critical programs such as those with short timeouts or alarms thereby giving different results to non-instrumented programs.

Analysis of the results of an execution must discard the library and system calls required to load shared libraries.

6   Metrics for the Coordination Layer

We propose that the "thickness" of the coordination layer is a good indication of the level of abstraction and design tradeoffs made

$$\delta = \frac{Number\ of\ System\ Calls\ made}{Number\ of\ Library\ Call\ made} \tag{1}$$

In Equation 1, δ is a ratio. Where δ is equal to 1, this would imply format conversion. If δ had a value less than 1 it would infer a more complex interface or a service such as caching data. A value of δ of greater than one would infer a simplification of the interface and an increase in the level of abstraction.

For example, a call to malloc under Solaris 2.x using GCC 2.7.0 invokes the system call brk twice irrespective of the arguments given to malloc, δ would have a value of 2. This ratio is supported by source code inspection. However, manual source code inspection takes much longer and is more error prone.

Different arguments to the same library function may produce a different ratio depending on the implementation of the library. This implies a requirement to test across the range of possible argument values and combinations.

Using a measure of the ratio of library calls made by an application to the system calls made by the library calls, we can infer design decisions made in the design of the library routines called.

A review of a range of distributed systems suggest that transparency and control, reliability and performance are traded against each other within the coordination layer. We plan to substantiate these observations and propose the following hypothesis.

Transparency is directly proportional to the thickness of the coordination layer.

Control is inversely proportional to the thickness of the coordination layer.

Reliability is directly proportional to the thickness of the coordination layer.

Performance is inversely proportional to the thickness of the coordination layer.

If the hypothesis are substantiated, it implies that performance can not be considered in isolation and must be taken as one of a number of parameters to consider when building a model or implementation of a computer system.

7 Further Work and Conclusions

Based on our model of a distributed system, we have presented a technique for measuring attributes of the coordination layer and metrics that may allow the comparison of alternative designs. The work presented in both this paper and [11] is directed at modelling the design decisions made by architects of distributed systems. We aim to build an executable model of a distributed system, using measurements taken from real program execution as source data. This will enable the investigation of tradeoffs within each of the layers in the model.

Tool support for the generation of function stubs to be interposed in a running program for UNIX and C has been completed. Current work is focused measuring the thickness of the coordination layer across a range of distributed systems and the selection of appropriate techniques for modelling design decisions in distributed systems.

Further work on modelling a node in a distributed system requires a method of describing the functionality, performance and reliability of the hardware components. Further research opportunities exist in applying similar techniques that to those described to instrument the application/coordination layer boundary and the coordination layer/kernel boundary and apply it to the kernel/hardware boundary perhaps by instrumenting a device driver. This would provide more complete information to incorporate into models of a computer system.

The technique of interposition could be used for a range of other purposes such as providing additional functionality, fault-tolerance or error recovery to existing well defined and tested libraries and applications without change to the library or application.

References

[1] R M Alder. Distributed coordination models for client-server computing. *Computer*, 28(4):14–22, April 1995.

[2] P Bernstein. Middleware - an architecture for distributed system services. Technical Report CRL 93/6, Digital Equipment Corporation, Cambridge Research Lab, One Kendall Square, Cambridge, MA 02139, March 1993.

[3] K Birman. The process group approach to reliable distributed computing. *Communications of the ACM*, 36(12):37–53, December 1993.

[4] A D Birrell and B J Nelson. Implementing remote procedure calls. *ACM Computing Survey*, 2(1):238–41, 1984.

[5] D D Clark. The structuing of systems using upcalls. In *Tenth ACM Symposium on Operating Systems Principles*, 1515 Broadway, New York, 10036, 1-4 December 1985. Washington, ACM.

[6] J R English, L Yan, and T Landers. A modified bathtub curve with latent failures. In *Proc. Ann. Reliability & Maintainability Symp.*, January 1995.

[7] A Gargaro. Distributed programming in Ada 9x. In C Loftus, editor, *ADA Yearbook 1994*, pages 99–113. IOS Press, 1994.

[8] J L Hennessy and D A Patterson. *Computer Architecture: a Quantitative Approach*. Morgan Kaufmann, 1st edition, 1990.

[9] IEEE. POSIX 1003.1-1990: Standard portable operating system interface for computer environments. Reference number ISO/IEC 9945-1: 1990(E). Institute of Electrical and Electronics Engineers, NY, 1990.

[10] B W Kernighan and D M Ritchie. *The C programmming Language, Second Edition*. Prentice-Hall, 1992.

[11] C M King and I C Pyle. Reporting quality of service and functionality requests from operating systems. In R Poley, J Hillston, and P King, editors, *10th UK Performance Engineering Workshop*, pages 79 – 87. UKPEW, September 1994. Edinburgh University.

[12] S J Leffler, M K McKusick, M J Karels, and J S Quarterman. *The design of the 4.3 BSD UNIX operating system*. Addison-Wesley, 1988.

[13] P Linden. *Expert C programming, Deep C Secrets*. Prentice Hall, Englewood Cliffs, NJ 07632, 1994.

[14] D J MacKenzie. *GNU Time Manual Page*. Free Software Foundation, Inc., 675 Mass Ave, Cambridge, MA 02139,, 1.6 edition, October 1993.

[15] C E McDowell and D P Helmbold. Debugging concurrent programs. *ACM Computing Survey*, 21(4):594–622, December 1989.

[16] P Mockapetris. Domain Names - concepts and facilities. Request for Comments 1034, ISI, November 1987.

[17] C Petzold. *Programming Windows 3.1*. Microsoft Press, Redmond, Washington, 3rd edition, 1992.

[18] QOSMIC Consortium. QOSMIC - Deliverable D1.3C QOS and Performance Relationship. RACE Deliverable 82/KT/LM/DS/B/013/B1, QOSMIC Consortium, July 1992.

[19] R Schreiber. Middleware demystified. *Datamation*, 41(6):41–45, April 1995.

[20] S K Shrivastava, G N Dixon, M C Little, G D Parrington, F A Hedayati, and S M Wheater. The design and implementation of Arjuna. Technical report, Computing Laboratory, University of Newcastle upon Tyne, UK, March 1989.

[21] SunSoft. *SunOS 5.3 Library and Linkers Manual*. SunSoft, 2550 Garcia Avenue, Mountain View, CA 94043, USA, November 1993.

[22] S Vinoski. Distributed Object Computing with CORBA. *C++ Report*, 5(6), August 1993.

A Two-Tier Load Balancing Scheme for Distributed Systems

Christopher Edwards and Madjid Merabti
Liverpool John Moores University
E-mail: M.Merabti@livjm.ac.uk

Abstract

Workstations within a distributed system often contain powerful processors, although on average, the processing demands of the user are far less than the capacity of the workstation they are using. Load balancing uses the ability to remotely execute processes to allow for the spreading of the computational load over the workstations within a distributed system. This paper presents a summary of the issues and characteristics that must be considered when developing a load balancing algorithm. It looks at the development of a load balancing mechanism using techniques from a number of important solutions within the area, with emphasis being placed on scalability, stability, fairness of service, efficiency and a move away from an "all or nothing" approach where a single load threshold mark is used, using a High-Low scheme [1]. The paper provides both a theoretical and performance based analysis of the algorithm in order to provide a fair evaluation. Results show that improvements in user response time are evident when idle / lightly loaded processor(s) are available in order to accept remote processes.

1. Introduction

Distributed systems provide many advantages over centralised systems, including greater **fault tolerance** (when problems occur with hardware or software), **cost, openness** (the ability to simply but effectively extend systems in different ways) and are often appealing because of the ever increasing diversity of user requirements. However, workstation-server based distributed systems in general have no underlying method of spreading the computational load evenly over the available workstations. In some cases, when running processor intensive jobs, the performance of a workstation may be severely reduced, with the user having to wait long periods for the completion of background jobs. As a result of this, some workstations may be heavily loaded, while others are lightly loaded, or even idle. Management schemes that remedy this situation are highly desirable [2].

Load balancing uses a number of the fundamental aspects of a distributed system in order to provide a management facility that allows the spreading out of the work load of each processor, by executing processes remotely. Many load management policy problems exist that can by summed up by asking, "*which processes should be moved where and when, to improve performance*" [3].

Differences between the two phrases **load balancing** and **load sharing** are often confused. "*The term load balancing has sometimes been used to imply the objective of equalising the loads of the hosts, whereas load sharing simply means a redistribution of the workload*" [4]. Throughout the remainder of this paper, the phrase **load balancing** will be used to relate to the subject area as a whole, without any real distinction between the two. It should also be noted that within this paper, the words **workstation, computer** and **node** are used interchangeably.

The emphasis of the work described in this paper is the development a load balancing prototype that specifically deals with the areas of :

- scalability - with the network being divided into a series of domains
- stability - using, for example, a distributed method of storing load values, two way negotiation between nodes and a two tier threshold policy meaning there is more control over the movement of processes
- fairness of service - with an attempt being made to provide each node on the network with a guaranteed certain level of performance
- efficiency - with processes only being moved when necessary, and when dealing with a system containing mainly lightly loaded nodes, processes only being migrated to local, neighbouring nodes

A two-tier load-balancing scheme that meets the above requirements and its performance are described in the rest of this paper.

2. Requirements of a Load Balancing Management Facility

When considering the load balancing of a system, the basic goals of a load balancing algorithm can be stated as [3]:

- to distribute the load of a system fairly, providing acceptable performance regardless of where the job is executed
- to provide transparency of this distribution to the user
- to achieve the best system performance
- to meet the required user performance expectations
- to tolerate failure and allow systems to be extended and / or re-configured

During the development of a load balancing mechanism, numerous characteristics will have a strong influence on its performance. Examples of these include [5]; the methods used for **load estimation**, the amount of **information exchange** to take place between machines (and the transmission patterns to be used) and the **receiver's knowledge** of migrating processes.

Design choices and decisions that must be made when developing an algorithm are often classified around a series of components or policies. Specific definitions of these policies often differ between sources, although a broad outline is as follows [3] [6] ;

The transfer policy defines the method used to decide that a process should be executed remotely. In most cases, this refers to considering whether a process is eligible to be load balanced, and how to determine whether a computer is overloaded enough to require remote execution of a process. The most common transfer policy is one which considers the local load of a computer against a certain local threshold value, without initially considering the loads on other machines. *The information policy* defines the load state update strategy to be used. This determines the method used for updating the load state of the different computers within the network, in terms of the kinds and amount of information to be made available, and the way such information is accessed. *The location policy* determines which computer to send a process too. It is important that the location policy works in harmony with the information policy, in that when locating a computer to take a process, the most up-to-date and relevant load information is used, leading to a fair allocation of processor resources. A simple location policy would use the least loaded computer, provided that the load on the computer was below a certain threshold level. *The acceptance policy* determines the amount of control a computer receiving a process has in deciding whether or not to accept the process. The two main examples are a single request protocol where a computer receiving a process is unable to refuse the process (a fast method that can lead to instability due to incorrect decisions) and a negotiation protocol where negotiation between the two computers means that the receiving processor is able to refuse the process (meaning poor decisions can be tolerated and reversed, although the procedure will take longer). *The decision making policy* is often used to define the point at which the processor will decide where to execute the process.

Other parameters that require consideration during the development of a load balancing mechanism include;

- system size - whether to create a scalable algorithm, capable of allowing any number of nodes within the network
- centralised/distributed algorithm - whether to have a centralised area at which to store the load information relating to the workstations on the network, or to store this information in a distributed fashion. A centralised server would allow better monitoring of statistics, but would also threaten scalability
- initiation of remote placement - whether to have a source initiated or a server initiated algorithm, i.e., should it be the overloaded computer (the source of the overload) that initiates the remote placement, or the underloaded node (the server) that executes the process

3. Related Work

A selection of load balancing algorithms were studied during the early stages of this work. A number of these algorithms that relate directly to the different aspects of the prototype implementation are briefly considered below :

The Pairing Algorithm [7] This is a **distributed** algorithm using **co-operation** between nodes that are near to each other on the network in order to improve performance locally. The main actions of the algorithm are as follows ;

A node on the network with a high load sends a query to a node located near it (i.e. a neighbour node), in order to attempt to find a stable node able to accept a migrating process. The node receiving the message may then perform one of three actions :

- send a rejection message
- form a **pair** with the node that has sent the message
- force the node requesting the partnership to wait until it is in a stable state and able to either form a pair or reject the request

If the node receiving the message rejected the request, the overloaded node must send a query to another local node. Once a pair of nodes have been established, the overloaded of the two nodes selects and migrates processes to the other processor. The decision as to which processes to move is made by considering the improvement to the response time of each process, if run on the remote machine. Processes are moved one at a time, until no more improvement can be made. If no processes can have their performance improved, the pair is broken.

This algorithm is a good example of a co-operation based algorithm, although, because consideration is only given for the migration of processes to local neighbour nodes, the mechanism can be seen to provide only a very local solution, with no consideration being given for cases when all of the nodes local to the overloaded node are in an overloaded state. The extent to which this problem exists depends primarily on the average activity on the distributed system (i.e. are neighbour nodes often likely to be busy), and how one defines how many local neighbour nodes a single node may have. The simplicity of the idea behind the mechanism means that it may be a relatively simply idea to implement, although because the nodes within the network make no attempt to hold details of the load on the nearest nodes, the solution means that the potential overhead for finding an underloaded node may be significant.

The Vector Algorithm [8] As with the first algorithm, the vector algorithm also uses a **distributed** method, with the main idea behind the algorithm being that each node on the network keeps track of the loads on a small number of processors (i.e. a **subset**) of the network. Three basic components are present on each of the nodes within the network :

- The processor load algorithm
  used to keep up to date a small **load vector** present on the node, maintaining load values of a fixed number of other nodes on the network
- The information exchange algorithm
  present in order to exchange the load information it holds with other members of the subset of nodes, using a randomly selected path
- The process migration algorithm
  used to make the decisions as to if and when to migrate processes. This component works by periodically forcing each computer to consider migrating processes, by first considering its own load, and then considering the load of the other nodes that it holds information about

The improvement in performance using this algorithm depends on the size of the load vector held by each node. For the algorithm to function on a global scale, each vector would need to contain load information on all of the nodes within the network.

Although again, this algorithm can be seen as a co-operation based algorithm, in order to provide sufficiently up-to-date load information, the information is exchanged purely using a randomly selected path. This in theory can provide problems, as because of the random nature of the technique, updates must be frequent enough for the information to be accurate, but not so frequent as to mean that updates are taking place unnecessarily often. Using an update technique that is more concerned with actual changes in load states may allow for the efficient provision of more timely, up-to-date information. The algorithm is known not to work efficiently when a large number of processes are created at any one time, and may in some cases provide situations where processes are moved unnecessarily.

The Flexible Load Sharing Algorithm (FLS) [9] The aim of this algorithm is to provide a load balancing technique that is both **flexible** and **scalable**. The developers of the algorithm state that in general, because of mechanisms used within load balancing algorithms such as broadcast, the scalability of an algorithm is limited by the large overheads that are produced when dealing with the networking demands of more than a small number of nodes. In order to develop a solution to this problem, their algorithm makes : *"local decisions based on non-local state information that has been made available locally"*

FLS provides scalability by dividing the system into sets of nodes, called **domains**. Load balancing can then be applied to individual domains, irrespective of how it is applied to others. Domain membership is **symmetrical** between overloaded and underloaded nodes, meaning an overloaded node always has a supply of underloaded nodes to which it can allocate processes. Each node maintains a **cache** of information which holds details of the load values of the nodes within its domain. When a process is executed :

- If the local node is overloaded, the node checks the cache in order to find an underloaded node able to take the process
- A **cache manager** exists in order to make sure that an overloaded node has underloaded nodes available in its domain
- The FLS treats the details within a nodes cache as being a set of "hints", forcing the node to communicate with its counterpart located during the cache check. This method stops the allocation of components that are unable to accept the process, even if the node had recently been in a position to accept the process
- After the communication between the two nodes, if the underloaded node is still in a position to accept the process, the process is executed remotely, else it is executed locally on the overloaded node

Because of the emphasis to ensure that processes are only migrated to nodes that are able to accept them, the negotiation protocol used considerably increases the communication overhead within the algorithm. Although ensuring the correct allocation of processes, two way negotiation is required for the movement of all processes. As a result of this, if a node is unable to accept a process, the general rule is to execute the process locally, rather than attempting to find another more suitable node, as the additional communications overhead incurred when under a large load would deem this procedure too costly.

Consideration of a number of the ideas / techniques used within the above (and other) algorithms has lead us to the following proposal for a load balancing mechanism.

4. A Two-Tier Load Balancing Scheme

4.1 Algorithm Characteristics and Design Choices

An important parameter in the design of load balancing algorithms is that of the load threshold value which defines the load value used to decide whether a process is executed locally or remotely. The most common method used is to have a single threshold value. If the load on the node is below this value, it is underloaded, else it is overloaded. During the design, it was decided that the High-Low scheme [1] should be used. The main idea behind the High-Low scheme is to have two threshold values, used to help guarantee that the user of a local node is given a certain level of performance. To decide if a process should be executed remotely because the local load is too high, the **high mark** is used. If the current load of a node is above the high mark, the process will be migrated to another node. To decide if a process from a remote node can be executed on the local node, the **low mark** is used. The local load must be below the low mark if the process is to be accepted, thus helping to ensure that the local node will always have a pre-defined level of performance.

Within the implementation, to help maintain both **scalability** and **stability**, each node will only consider a partial view of the system, called a **domain** [6] [9], therefore increasing the scope of the algorithm beyond that of a relatively small distributed system. Developing a **distributed** implementation [10] means it is easier to implement, with the algorithm being replicated across all of the nodes in the system, which also implies that the maximum number of nodes within the system need not be specified.

Should the algorithm be **source** or **server** initiated ? A source-initiated algorithm (where the overloaded node initiates the remote placement) would place more load (finding an underloaded node) on an already overloaded node, but would also ensure that a process is only remotely executed when the load on the node has reached a certain level. A server-initiated algorithm (where the underloaded node initiates the remote placement) would place most of the overhead on the underloaded node and would also be more stable, as an underloaded node could prevent overloading by controlling the amount of work it requests. However, the server may not be aware of remote occurrences that prompt remote execution. The decision made was to develop a source-initiated algorithm, ensuring that processes will only be moved when absolutely necessary.

The correct timing and information to be used when deciding whether or not to move a process must be carefully considered. These are defined by the *decision making / transfer policies* of the algorithm. With regard to when to decide where to execute a process, migrating any relevant processes before they have started has been chosen, as it is the less complicated solution, with research showing that it is unclear as to how much benefit there is in providing migration after process start up [6]. The information to be used by the *transfer policy* when determining whether a node is overloaded enough to require the remote execution of a process is a simple threshold measurement, i.e., when the load of the node passes a certain threshold point, attempts will be made to migrate the process. The threshold measurement used will be the **high mark** of the High-Low scheme detailed earlier.

The strategy used to update the various load state values (the *information policy*) consists of a two tier mechanism for distributing the load states:

- an underlying method advertising the load states on a periodic basis, to all of the nodes in the same domain (known as a **periodic update**).
- an event driven state update between neighbour nodes [3] (known as an **event driven update**), limited to between neighbour nodes to prevent a large amount of network traffic that may be produced if an event driven update was used to send the updated load information to every node in the domain

In both cases, the information sent includes both the current load and the **low mark** of the particular node, to allow receiving nodes to determine whether the sender node is underloaded.

How do you determine which node to send a process too ? It is important here for the details of the *location policy* to relate directly to the *information policy*, in order to get the most from the available load information. In order to do this, a two tier method of locating a free node is used :

- firstly, the data stored about a neighbour node is checked to see if this node is able to accept the process that requires migration
- if the local data stored on the neighbour node means that it is unable to take the process, the node will then use the data stored relating to the other nodes within the domain, to select a suitable node

How much control does a node receiving a process have in deciding whether or not to accept the process ? The *acceptance policy* for the implementation has again been designed to relate to the two tier mechanism detailed above, as follows :

- if the *location policy* means that the neighbour node is able to take the process, it is sent using a single request protocol, with the request and the work being sent together, the neighbour node being unable to refuse the process. This is unlikely to cause instability or flooding, as, because of the **event driven** update between the neighbour nodes, the load information is likely to be correct
- if the neighbour node is unable to take the process, the node will then use the information on the node with the lowest load as a hint only, carrying out negotiation with this node to see if it is able to take the process. This negotiation is necessary, as the information from the **periodic update** may be out of date [9]

4.2 Algorithm's Prototype Design

Each node within the network must hold two data files containing details of load values on other nodes, and a series of constant values. The *neighbour data file* holds information relating to the current load of the neighbour node. This data allows the neighbour node to be identified, the age of the data to be monitored, and the load values of the node to be accessed. The *domain data file* holds the same style of data, but in this case, there is data available for each of the nodes within a single domain. *Constant values* are required for the high and low marks of the High-Low scheme, the number of nodes in each domain, along with a number of values required to control the periodic update, including figures to define the time periods between the updates, and the load differences used to consider the needs for an update.

Each workstation is required to contain three software modules capable of concurrent execution. Brief details of these modules are as follows :

The *ExecuteProgram module* is used in order to provide scheduling of processes on a workstation. It is executed each time a command is entered, and is used to determine whether the command should be executed locally, on a neighbouring node or on a different node within the domain. The most up to date information available (the neighbour node information) is used primarily to find a remote location for the process. If the neighbour node is unable to accept the process, the **domain table** is used to ensure that the node with the lowest load values is contacted first, but also allows for further searches of the table, should the first node found be unable to take the process (checks on the ages of load values within the table are made before communication takes place, to ensure the node is alive). Once a potential node has been found to execute the process, this module carries out the required communication between the local node and the node that is likely to accept the process, allowing for problem conditions to occur. This module also sends **event driven update** messages to its neighbour node, if an executing process causes a large change to the local load value.

The *NodeManager module* is executed continuously on each node, acting as a server for all of the incoming messages from different nodes within a domain. The module waits for an incoming message to arrive, and calls the correct function, depending on the message type. The four main types of message to be received are :

- an **event driven update** message from a neighbour node sent when an event on the neighbour node increases its own load considerably (used to ensure that the local node contains the most up to date load information regarding the neighbour node)
- a **periodic update** message from a node within the domain (used to ensure that the domain file on the local node is kept as up to date as possible)
- a message from a neighbour node stating that the neighbour node has already executed a command on the local node
- a request from a node within the domain wishing to execute a process on the local node (requiring negotiation between the two nodes to ensure that the load on the local node is of a suitable level to be able to accept the process)

Event driven updates (involving communicate with neighbour nodes) are also performed by this module, as accepting jobs from remote nodes may increase the local load value considerably.

The *PeriodicUpdate module* executes continuously on each node within the network in order to regularly inform the other nodes within the domain of the load on the local node. The update only takes place if the load value has increased by a set amount since the last update took place, or if a set period of time has passed. The function has been developed in this manner to ensure that the updates do not take place too often, or when they are not required. The communication takes the form of a **periodic update** message, and is received by the NodeManager module executing on each of the nodes within the domain.

The implementation of the prototype took place on a DEC Athena UNIX-based distributed system, running on an ethernet network. Interprocess communication (message passing) was developed using UNIX socket primitives [11] such as **socket, bind, listen, accept, connect, getsockname** etc., using Internet (IP) addresses (AF_INET) to identify sockets, and streams (SOCK_STREAM) for two way, sequenced, reliable communication. The UNIX **uptime** command [12] was used in order to provide the load average metric for the implementation (the actual value used was the average over the past minute). The UNIX primitive **hostname** was used in order to provide the Internet addresses of the workstations within the system.

5. Prototype Evaluation

The algorithm developed provides *scalability*, as each node within the network holds information on the load values of only a limited number of other nodes. The use of **domains** in order to divide up the network means that extension of the system is a relatively simple task. The use of mechanisms that create large overheads or congestion when used on a distributed system containing more than several dozen nodes such as broadcast, have been avoided within the implementation.

In order to provide a *stable environment*, the algorithm must contain certain characteristics, with *stability* being lost if a single element of the algorithm has been designed without consideration for it. The following features of the developed algorithm help to provide overall stability :

- *fault tolerance* - because of the method used to distribute load values to other nodes, the algorithm is not reliant on a single central server to hold the load information. The algorithm is also able to cope with the loss of nodes on the network
- *control of remote execution* - the use of a two tier threshold policy means that processes are not remotely executed unnecessarily, and nodes are not forced to take processes when their load value is above a certain level

- the two way negotiation protocol between nodes within a domain means that remote nodes have the opportunity to reject processes when they are unable to accept them
- the single request protocol is used only between neighbour nodes, when the load information is likely to be correct and up to date

Although difficult to measure, the following points define a level of *efficiency* within the algorithm :

- a system with mainly lightly loaded nodes will in general only require the migration of processes to local, neighbour nodes
- processes will only be migrated if the local load is sufficiently high enough to require remote execution

The algorithm provides *fairness of service* in that it attempts to guarantee the user of a workstation a certain level of performance, by ensuring that remote jobs are not accepted unless the local load is low enough, and by ensuring that new local jobs are executed remotely, if, and only if, the local load is above a certain load level. The method used to divide the workstations up into fixed domains may not be seen as a fair technique to use, as a domain containing a number of constantly highly loaded machines would mean that lightly loaded nodes within that domain would suffer, and load balancing would be limited. However, the threshold system with two levels ensures that the performance of the lightly loaded machines can be controlled.

A number of the features discussed above highlight the *flexibility* provided by the algorithm. Points that highlight a lack of flexibility include :

- the inability to move processes once they have started execution
- the fact that the potential for remotely executing a process is limited to the number of nodes within the domain

5.1 Experimental Results

The system used in order to carry out the experimentation and evaluation of the algorithm consists of a network of workstations. The evaluation results were compiled by using a single domain of **four** workstations. It should be noted that the remaining workstations within the distributed system were idle during the experimentation period. A user on a workstation has been simulated by a script developed to repeatedly execute a C program which performs numerous calculations. As its result, the script returns the time it has taken to execute, representing the **user response** time.

The results that follow show a comparison in relation to differing **low mark** values used, in order to show the effect on both the response time and job movement depending on how much work a lightly loaded workstation is willing to undertake (a similar technique can be found within [1]). The **high mark** within the results is constant, representing the point at which it is believed that a workstation moves from having a normal load to being overloaded.

The objective of the algorithm is to improve performance, which simply defined means the reduction of the average response time. The script used in order to produce the following results has been measured to take approximately 110 seconds when run

288

on an idle workstation, and approximately 207 seconds when run on an overloaded machine. Therefore, the difference between these two figures provides the scale for which improvement to response times is available. The results have been compiled under two different circumstances. Firstly, with the domain containing x overloaded nodes and x idle nodes, and secondly, with the domain containing x overloaded nodes and x nodes with a light load. We begin by showing the effects of having idle node(s) within the domain.

The graphs below show the response time of the overloaded node(s) where no load balancing is used, and the average response time with the load balancing software. It should be noted that when the low mark on each workstation is set to zero, no remote processes will be accepted by any machines, therefore giving the same effect as having no load balancing present.

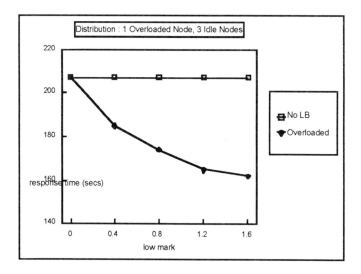

Figure 1 : Response Time of a Single Overloaded Node

Figure 1 represents a low system load, with a single workstation being overloaded, the remaining three workstations within the domain being idle. The curve on the graph represents a significant reduction in the response time as the low mark value increases, however, it should be noted that even with a small low mark (i.e. 0.4), and therefore only a small willingness to share, there is a substantial reduction in the response time. Results detailing the location of remotely executed processes show that as the low mark increases, so too does the number of jobs the neighbour node of the overloaded node is able to accept, until it reaches a state where it is able to accept almost all of the processes.

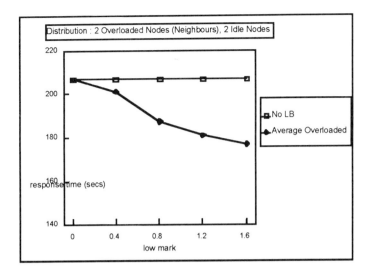

Figure 2 : Average Response Time of Two Neighbouring Overloaded Nodes

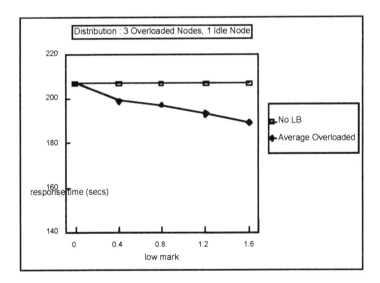

Figure 3 : Response Averages of Three Overloaded Nodes

Figure 2 represents a medium system load, with two of the four nodes being overloaded. As these results relate to the two overloaded nodes being neighbour nodes, they are unable to move any processes between each other, and therefore need to find other processors within their domain that are able to provide remote execution facilities.

290

Results recorded (although not graphed here) relating to two overloaded nodes that are not neighbours show a response time lower than that seen previously, even under a similar system load. Results detailing the location of remotely executed processes in this case show that most of the remote jobs are executed on neighbour nodes, thus reducing the amount of communication overhead, because of the single communication protocol used between neighbour nodes.

Figure 3 represents the system under a heavy load, with three of the four nodes being overloaded. Even when the system is overloaded, some reduction in the response time is possible. When considering the location of the remotely executed processes, there is only one node that is capable of accepting them, therefore the low mark value on this single node is critical to the amount of load balancing completed.

The above results show that allowing idle machines to participate to differing degrees in the execution of processes will lower the response time. However, the results only provide evidence of the benefits of the algorithm when having idle nodes able to accept the processes, and do not provide details relating to when a lightly loaded processor is not completely idle. The results gained in this area were less conclusive. In order to evaluate this, we ran the user script on the idle workstations as well as the overloaded machines, in order firstly to provide a degree of load on the idle workstations, and secondly to measure the effects to the response time on these now lightly loaded machines.

Results were compiled using various system load settings, a sample of which is as follows :

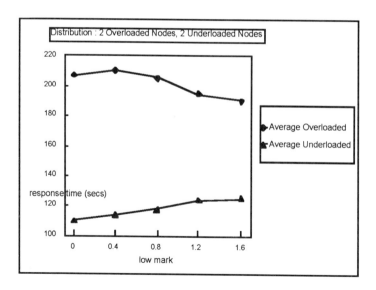

Figure 4 : Response Averages of Two Overloaded Nodes and Two Underloaded Nodes

Under the different system load settings, because there is a load being placed on the idle nodes by the execution of the user script, they are unable to accept processes when their low mark value is small. In some cases, because of the overheads of the load balancing software, the actual response times of the overloaded nodes increased. Benefits were only gained with increased low mark values (e.g. 1.2,1.6). The response time of the lightly loaded machines was not greatly effected by the added load of accepting a number of processes from other machines, especially when the low mark value was relatively small. This emphasises the fact that the High-Low scheme provides a good technique for controlling levels of load balancing participation.

Future work relating to experimentation with the algorithm in order to confirm the findings above may include the use of different settings within the load balancing software, and within the surrounding environment (the results provided above were gathered using a single set of settings).

6. Conclusions

Within this paper we have looked at the key characteristics of a load balancing management facility, including the basic goals and the key characteristics of an algorithm. We have discussed the need to use a series of components or policies in order to make all of the necessary design choices and decisions.

We have studied a number of existing load balancing algorithms highlighting both their positive elements and shortcomings in order to provide ideas for the prototype development. The load balancing scheme presented contains a number of important features. The use of the High-Low scheme [1] replaces the "all or nothing" techniques used within many algorithms, allowing for the control of the amount of participation an individual node can have in the load balancing process. The source initiated, distributed algorithm splits the network into domains of workstations in order for each node to only consider a partial view of the system. We have put forward the case for having a two tier mechanism when deciding where a process from an overloaded node should be executed. To tie in with this idea, a two tier method is also used for load updates, load information storage and the acceptance of processes, in order to make the movement of processes as efficient as possible.

The theoretical evaluation shows that the algorithm provides a certain level of scalability, stability, fairness of service, efficiency and flexibility (although a number of key areas that lack flexibility are highlighted).

The performance evaluation shows that remotely executing processes will reduce the overall response time, but as would be expected, the extent to which this is effective depends on the amount of idle resource available, and the amount this resource is willing to participate in the load balancing. Because of the use of a two-tier location policy, when wishing to remotely execute a process, the actual movement of processes can in some instances depend on the physical location within the network of the overloaded node. The low mark values provide an effective method of controlling the amount of participation a node may have within the load balancing scheme. Results when running the software within a domain where there are underloaded resources provided response times that were not as good as when running the software with idle resources. In some instances the low mark needed to be of a considerably high value for the load balancing scheme to provide any benefit at all. However, using the High-Low scheme meant that the effects on the more lightly loaded nodes when accepting processes in terms of increased response times could be controlled.

Further work will enable the fine tuning of the software in order to provide improved response times by determining correct values in areas such as load information distribution and effective domain sizes. On a broader scale, work to enhance the technique used for domain membership is seen as an area which may also provide large scale improvement.

7. References

1. Alonso R, Cova LL. Sharing Jobs Among Independently Owned Processors. Proc. 8th Int'l Conf. on Distributed Computing Systems, IEEE Computing Society Press, Los Alamitos, Calif., 1988; 282-287

2. Wilbur S, Merabti M. Towards Performance Management of Networks and Distributed Systems. IFIP 6.4a Conference on Networks Management. Berlin 2-3 July 1987

3. Goscinski A. Distributed Operating Systems - The Logical Design. Addison-Wesley Publishing Company, 1992

4. Zhou S. A Trace-Driven Simulation Study of Dynamic Load Balancing. IEEE Trans. Software Eng. Sept. 1988; 14:1327-1341

5. Fogg I. Distributed Scheduling Algorithms: A Survey. Department of Computer Science, University of Queensland, 1987

6. Kremien O, Kramer J. Methodical Analysis of Adaptive Load-Sharing Algorithms. IEEE Trans. Parallel and Distributed Systems, Nov. 1992; 3:747-760

7. Bryant R.M., Finkel R.A. A Stable Distributed Scheduling Algorithm. Proc. 2nd Int'l Conf. on Distributed Computing Systems, 1981; 314-323

8. Barak A, Shiloh A. A Distributed Load Balancing Policy for a Multicomputer. Software Practice and Experience, 1989; 15:901-913

9. Kremien O, Kramer J, Magee J. Scalable, Adaptive Load Sharing for Distributed Systems. IEEE Parallel and Distributed Technology, Aug. 1993; 1:62-70

10. Theimer M.M., Lantz K.A. Finding Idle Machines in a Workstation-Based Distributed System. IEEE Trans. Software Eng., Nov. 1989; 15:1444-1458

11. Frost J. BSD Sockets : A Quick and Dirty Primer. Software Tool and Die. madd@std.com, 1989

12. Sobell M. A Practical Guide to Unix, Second Edition, Benjamin / Cummings, 1991

Adaptive Distributed Simulation for Computationally Intensive Modelling

Kam Hong Shum

Computer Laboratory, University of Cambridge

Cambridge, United Kingdom

Abstract

The potential gain in speeding up the execution time of computationally intensive simulation in workstation clusters is tremendous. One of the challenges for exploring this gain is adaptation of the dynamic workload conditions and the heterogeneity of the workstation clusters. Static schemes of allocating simulation workload to distributed workstations are inadequate because the assigned workload may change over time and the workstation usage may vary as a result of the workload of other users. An adaptive distributed simulation should also take account of the differences in workstation architectures and networking technologies among workstation clusters. Hence, we propose an adaptive scheme that integrates a local approach and a global approach of workload distribution to achieve automatic adaptation at runtime.

In our experiments, the effectiveness of this scheme is demonstrated by the execution profile of a distributed simulation of a broadband communication network. We observe that the simulation exhibits a good speedup throughout its runtime under different conditions of resource availability in the workstation clusters.

1 Introduction

Advances in workstation and networking technology have facilitated the usage of workstation clusters for running many parallel and distributed applications. Among these applications, distributed simulation is one of the most important but computationally intensive applications. This paper explores the effectiveness of executing distributed simulation in workstation clusters. We address the problem of adapting to the dynamic workload environment and the heterogeneity of the workstation clusters. Static schemes of allocating simulation workload into distributed workstations are inadequate because the assigned workload may change over time such as the dynamic workload demand in the land-battle simulation [13]. Moreover, the workstation usage may vary as a result of the workload of other users. An adaptive distributed simulation should also take account of the differences in workstation architectures and networking technologies among workstation clusters.

Different approaches are adopted in recent research of adaptive execution of data-parallel applications [3], [6], [10], [15] for which the same algorithm is applied to a large data set. An adaptive version of master-worker parallelism is also implemented on top of Linda [2] by Piranha [1]. Unlike data-parallel applications, the parallelism of a simulation model is governed by the strategies [4], [12] to partition and co-ordinate distributed logical processes of the model. Distributed simulation is also different from the master-worker parallelism because communication between its logical processes is required during the simulation period. Thus, the design of adaptive functions for distributed simulation are related to the communication protocols between the logical processes. This paper focuses on the effective execution of distributed simulation applications based on conservation time window algorithms [9].

The remainder of the paper is organized as follows: Section 2 describes the simulation model of a broadband communication network. Section 3 proposes an adaptive scheme for workload distribution. Section 4 discusses the adaptability of the proposed scheme for workload distribution. Section 5 reports the results of experiments. Finally, section 6 presents the conclusions and future directions.

2 The Simulation Model

Simulation is regarded as an important performance modelling tool for communication networks. However, the simulation for a broadband communication network, such as an ATM network, is extremely time consuming due to an enormous number of packet-level operations for broadband switches and communication links involved in performance modelling. Parallel and distributed simulation is therefore used to speed up the time for execution. In [14] and [18], a parallel simulation of an ATM network was performed by dedicated multiprocessor machines.

The distributed simulation described in this paper is designed for evaluating the performance of the traffic flow control and call set-up algorithms for a broadband metropolitan area network (MAN) proposed in [7] and [8]. A generic model of a switching node in a broadband network mainly consists of an ATM switch, a packet loader, output buffers for the switch, and link buffers. The link buffers model the buffering effect of communication links in transmitting packets. The switching nodes and communication links of a broadband network can be decomposed spatially into two types of logical processes, namely *tasks* and *grains*. Tasks are the basic units of the distributed simulation program. Each task processes the events of multiplexing, demultiplexing, and buffering packets that are all related to a specific outgoing communication link. A group of tasks forms a grain if they are located in the same workstation. Tasks and grains are long-lived logical processes because they will be active throughout most of the simulation period.

Algorithms of *conservative time window* [9] are applied to process event messages of the tasks in distributed workstations. The time window is the time period when events can be processed safely without incurring causality errors. The outgoing events in a time window are packed into time-stamped event messages before they are sent to other tasks. If tasks are located at different grains, exchange of event messages between them is by explicit message passing using PVM [5], otherwise event messages are transferred by internal data movement. Basically, the size of a time window is determined by the propagation delay of the outgoing links, but the window size can be extended by the lookahead technique described in [11].

3 An Adaptive Scheme for Workload Distribution

The distributed simulation is written using a single program multiple data (SPMD) model meaning that all workstations run the same piece of program. The control of workload distribution is exercised at the application level. We propose an adaptive scheme that integrates a local approach and a global approach of workload distribution to achieve automatic adaptation at runtime. For the global workload distribution, grain migration and/or partitioning are invoked whenever any computation or communication unbalance is detected. The grain partitioning involves either splitting or merging the grains into different grain-sizes and the grain migration relocates grains to different workstations for execution. In conjunction with the global workload distribution, the workload of the partitioned grains can also be balanced locally to improve overall speedup. Both methods of workload distribution can also be carried out independently.

3.1 Local Workload Distribution

The basic operations of the distributed simulation conservative time window algorithm are described as follows: For every time window period, each task exchanges time-stamped event messages locally with its neighboring tasks according to the interconnection of the simulated network. A task is first synchronized with its preceding tasks for incoming event messages. Then any pending events in the event-lists for that time window period will be processed. Finally, its outgoing event messages will be sent to its succeeding tasks. Since the slowest grain performance will be the distributed simulation performance, the overall speedup of the simulation can be increased by sharing the workload of the slowest grain with its neighboring grains. A distributed protocol of *local workload distribution* (LWD) is proposed in this section. Figure 1 summarizes the simulation operations and the conditions of the LWD.

The protocol of the LWD will only be deployed if any one of the conditions shown in figure 1 is satisfied. These conditions are embedded in the operation of the simulation. A *sender-initiated* [17] load balancing approach is adopted in which the task in a heavily loaded workstation initiates the LWD. The load metric for comparing relative loading is the average task execution time t_{exec}. It is calculated by dividing the execution time of a time window from the time window period. It can be normalized by the relative task computing demand in the simulation. Both t_{exec} and a status vector called *Sharing_task status* are the information of tasks which will be exchanged between the neighboring

The Distributed simulation (Conservative Time Window) :

> **Do Parallel:** for each Grain,
>> **Do Parallel:** for each Task in the Grain,
>>> While simulation not completed,
>>>> Begin
>>>>> Wait and receive incoming time-stamped event messages from its preceding tasks,
>>>>> Process the events for the next time window period,
>>>>>> (Conditions for the local workload distribution are embedded),
>>>>> Send outgoing time-stamped event messages to succeeding tasks.
>>>> End

Conditions of the local workload distribution :

Condition 1
> If $\frac{t_{exec}}{t'_{exec}} \geq \gamma$ for more than η consecutive time windows,
>> Set its *Sharing_task status*.
>> Send its internal states to its takeover task.

Condition 2
> If the *Sharing_task status* of its succeeding tasks is set,
>> Send a duplicate copy of event message to the takeover task of its succeeding task.

Condition 3
> If it is waiting for any task for execution and,
>> If the internal states and incoming event messages are ready,
>>> Execute the virtual task.

Condition 4
> If it is a virtual task, and is intercepted by the original task,
>> Reset the *Sharing_task status* of the original task.

Condition 5
> If it is intercepted by its corresponding virtual task,
>> Send signals to its preceding tasks to deactivate all incoming event messages.
>> Deactivate itself.

Definition of Symbols :

t_{exec} is its normalized task execution time for a time window,
t'_{exec} is the normalized task execution time for a time window of its takeover task
γ, η are the constant great than or equal to one

Figure 1: Distributed simulation with the local workload distribution

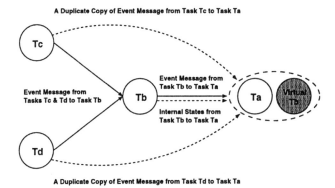

A Duplicate Copy of Event Message from Task Tc to Task Ta

Event Message from Tasks Tc & Td to Task Tb

Event Message from Task Tb to Task Ta

Internal States from Task Tb to Task Ta

A Duplicate Copy of Event Message from Task Td to Task Ta

a. The Virtual Task of Tb is invoked

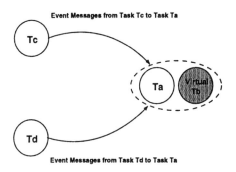

Event Messages from Task Tc to Task Ta

Event Messages from Task Td to Task Ta

b. The Virtual Task of Tb is taken over the original Tb

Figure 2: An Example of performing the LWD

grains. Each bit of the Sharing_task status vector represents the status of a task. If a task's average execution time t_{exec} is greater than or equal to γ times of the average task execution time of its takeover task t'_{exec} for more than η consecutive time windows, it will set its Sharing_task status and send its internal states to one of its succeeding tasks, called the *takeover task*. The values of γ and η can be fine-tuned to adjust the sensitivity of the LWD to the varying workload from the simulation and the other users. In receipt of the Sharing_task status, the task's preceding tasks will send a duplicate copy of event message to its takeover task. If the grain of the takeover task is idle and all the internal states and event messages are ready, it will execute a *virtual task* to take over the task. Once the virtual task is started, it will race with the original task for completion.

Figure 2 depicts a simple example of performing the LWD. For task T_b, it receives incoming event messages which come from tasks T_c and T_d (the

298

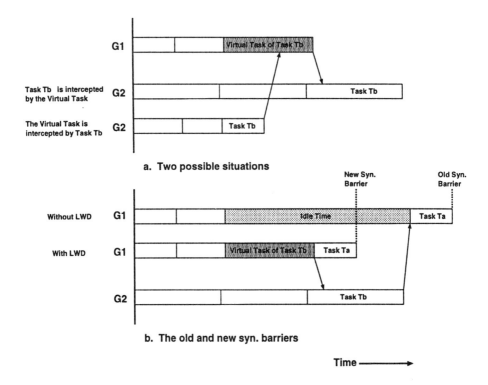

a. Two possible situations

b. The old and new syn. barriers

Figure 3: Timing Diagrams of the grains for the LWD

preceding tasks) and sends outgoing event messages to task T_a (the succeeding task). Tasks $T_a, T_b, T_c,$ and T_d are all located in different grains. If task T_b is placed in a heavily loaded workstation, task T_a will always wait for the incoming event messages to proceed. The transfer of the duplicate copies of event messages and the internal states of task T_b at the time when its virtual task is invoked in the grain of the task T_a is illustrated by figure 2a. If the LWD is deployed successfully, the original task T_b will be taken over by its virtual task (figure 2b).

Figure 3a shows two possible situations that may arise for the example as depicted in figure 2. Tasks T_a and T_b are located in grains $G1$ and $G2$ respectively. If the virtual task finishes before the original task, it will send an intercepting signal to deactivate the original task, otherwise the virtual task will be deactivated if it is intercepted by the event message from the original task. All the event messages from the preceding tasks will be stopped if a task is deactivated. And any outdated interception is discarded. Figure 3b shows the effect of the LWD in shortening the synchronization barrier of $G1$. If the LWD is not applied, there will be a long idle time for $G1$ to wait for the incoming event message from $G2$. Instead of waiting idle, $G1$ executes a virtual task of T_b to race with the original T_b. The successful completion of the virtual task shortens the idle time awaiting for the synchronization barrier.

3.2 Global Workload Distribution

Global workload distribution (GWD) mainly consists of grain partitioning and migration. For grain partitioning, several partitioning levels are predefined. The grains of the simulation can split or merge into appropriate partitioning levels in different situations. Furthermore, the grains can migrate to different workstations for execution. A special task called a *partitioner* is responsible for performing operations of the GWD which can be classified as :

- **Loading Information Collection**
  At every time interval t_{report}, each grain reports the average task execution time t_{exec} and the average communication and synchronization time t_{cs} of its tasks to the partitioner. For the workstations that do not have any residing tasks, the partitioner obtains their loading information from the broadcast information in their local network. Therefore, any available workstations and workload unbalance in computing or communication can be identified. The availability of a workstation is determined by the number of active processes and the amount of free memory in the workstation.

- **Decisions for Partitioning and Migration**
  Grain merging and grain splitting are the possible outcomes of partitioning. Partitioning is carried out whenever there is a need to improve execution performance. Grain merging is applied if there are not enough available workstations to maintain the present partitioning level and the unbalanced workload situation cannot be improved by the LWD alone. If more workstations are available, simulation performance can be improved by grain splitting. Migration occurs if any grain is mapped into a different workstation. After a partitioning level and the required workstations are chosen, the grains are mapped into the workstations and the partitioner will inform every task the time to invoke global synchronization for the GWD. Partitioning and migration can be carried out at the same time.

- **Partitioning and Migration Enforcement**
  To facilitate inter-task communication, every workstation, grain, and task has an identification (ID) which is the integer task identifier used for identifying processes in PVM. As shown in figure 4, the IDs of workstations, grains and tasks are stored in three different one-dimensional arrays. The workstation IDs are fixed at the time of program startup and the IDs of grains and tasks are updated whenever migration and partitioning have taken place. The one-to-one mapping of workstation IDs to grain IDs is altered if migration occurs, and the mapping of grain IDs to task IDs, which can be one-to-one or one-to-many mapping, is updated if partitioning is required. After the mapping, the partitioner will send each task a new copy of task IDs so that every task can relocate concurrently. Each task relocation requires the transfer of task internal states and the local event-list. During partitioning, all virtual tasks will relocate and convert

to the tasks of the grains in the new partitioning level. The events of tasks will not be processed until all of the tasks in a grain finish relocation.

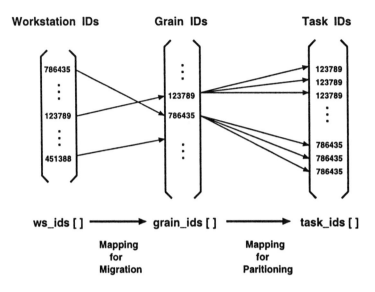

Figure 4: Mapping for Migration and Partitioning

4 The Adaptability of the Distributed Simulation

Both the GWD and the LWD are the key elements to support adaptive distributed simulation. The GWD helps to partition grains into a suitable partitioning level and relocate the grains into lightly loaded workstations. On the other hand, the LWD refines the size of the grains in each partitioning level to tackle local unbalanced workload situations. Their combined effects facilitate automatic grain-size tuning at runtime.

The decomposition of a switching node into grains and tasks affects the granularity of partitioning. The grains in each partitioning level can consist of different numbers of tasks. For example, a network of nine switching nodes connected bidirectionally in a mesh can be decomposed into two partitioning levels as shown in figure 5a and 5c.

Figures 5a and 5b depict the cases for the LWD when there are nine grains, if the workload of $G2$ and $G4$ are unbalanced, their tasks are shared by their neighboring grains, say the virtual tasks of $T3$, $T4$, and $T5$ of $G2$ are invoked by $G1$, $G3$, and $G5$ respectively. Similar situation occurs when there are three grains (figure 5c and 5d), if the workload of $G5$ is unbalanced, the virtual tasks of $T8$, $T11$, and $T15$ of $G5$ are invoked by $G2$ and the virtual tasks of $T10$, $T14$, and $T17$ of $G5$ are invoked by $G8$.

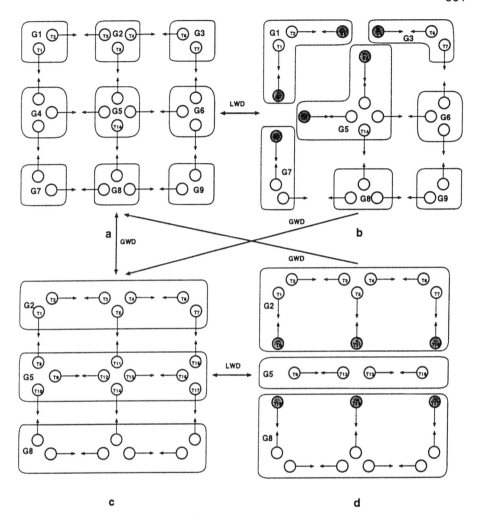

Figure 5: Effects of the GWD and the LWD

Grain merging and partitioning of the GWD will be activated to alter substantial unbalanced workload situations. Figure 5 also shows that the GWD can adjust the grain-size of the simulation. For example the 3-grains in figure 5c can be the result of the grain merging from the 9-grains in figure 5a or the 7-grains in figure 5b.

Each of the approaches of the GWD and the LWD does not itself provide a complete solution for adaptive distributed simulation. For the GWD, the predefined partitioning levels may not suit all the possible unbalanced workload situations and the overhead of global synchronization may grow as the number of tasks increases, thus the LWD is introduced to balance the workload of a grain with its neighboring grains more efficiently. However, depending on the connectivity of the simulated network, a task will only exchange event messages with its preceding or succeeding tasks. If the workload of a task is unbalanced, a virtual task can only be invoked by its takeover task. Thus, the LWD cannot be deployed if the workload of the takeover task is also unbalanced. The GWD can then be applied to alleviate this kind of problem. Since the GWD is more costly than the LWD, it should be deployed less frequently. These two workload distribution are therefore complementary to one another. Their limitations and strengths are summarized in table 1.

Table 1: Limitations and strengths of the GWD and the LWD

	Limitations	Strengths
GWD	• Limited partitioning • Overhead in global sync.	• Adapt to large load unbalance • Grain migration
LWD	• Limited load distribution	• Efficient local load sharing • Grain-size fine tuning

5 Experiments and Results

In this section, we present the experiments for performance testing under different conditions and evaluate the results of the experiments. Firstly, we evaluate the effect of the time window size on the speedup of the distributed simulation. Secondly, we investigate the effectiveness of grain-size determination when the simulation is executed in heterogeneous workstation clusters. Finally, the adaptability of the LWD and the GWD to external workload is examined.

The target model for simulation is an ATM MAN [7], [8] which has the same topology as the network shown in figure 5. The predefined partitioning levels of the network consist of a partitioning of 1, 3, 9, and 24 grains. The experiments are carried out on a wide range of workstations which consist of 50

DEC3100, 9 DEC Alpha, and 4 HP9000/700 workstations. All the workstations are connected by Ethernet in different subnets. The high performance DEC Alpha and HP9000/700 workstations are mainly used for research purposes. The CPU speeds of the DEC3100 workstations are relatively slow and they have less memory capacity. The experiments are carried out at midnight so as to minimize interference from other users and hence a controlled environment is ensured. External workloads are generated by computationally intensive loader processes.

In our simulation, the time window size is determined by the number of packets that can be stored in the communication link of the simulation network during packet transmission. In other words, the window size is proportional to the physical length of the links. Since, the bandwidth of each unidirectional communication link is 424Mbps and the packets are 53 byte ATM cells, the loading time for each packet is $1\mu s$. For example, if the length of the single-mode optical fibre communication link is 50km, then the size of the communication link buffer is 250 cells.

Tables 2 and 3 show the relative speedup of the distributed simulation without dynamic workload distribution when different sizes of the time window are used. The results show that the granularity has a direct influence on the performance of the simulation. If the length of the communication links is increased, that is, the computation to communication ratio of each task is increased, the speedup is improved.

The number of grains partitioned in each partitioning level also affects the relative speedup. As shown in figure 5, the number of tasks per grain in different partitioning levels are different. In the partitioning level that has 24 grains, there is only one task in each grain, so all the inter-task communication is done by message passing between workstations. The increasing rate of the speedup is far less than linear as the number of the grains increases. It is because inter-grain communication is more expensive than intra-grain communication. We also compare the relative speedup when the simulation runs on DEC3100 and DEC Alpha workstation clusters (tables 2 and 3). Although the computing performance of a DEC Alpha workstation is much higher than that of a DEC3100, the relatively low communication performance of Ethernet limits the gain in the computing performance.

Table 2: Effects of time window size on relative speedup (*DEC3100 Cluster*)

No. of Grains	No. of Tasks per Grain	Size of comm. link buffer				
		250	500	1000	1500	2000
1	24	1.00	1.21	1.30	1.32	1.38
3	7, 10	1.96	2.73	3.27	3.52	3.66
9	2, 3, 4	4.36	6.31	8.13	8.93	9.38
24	1	7.96	11.44	18.30	21.52	22.88

Table 3: Effects of time window size on relative speedup (*DEC Alpha Cluster*)

No. of Grains	No. of Tasks per Grain	Size of comm. link buffer				
		250	500	1000	1500	2000
1	24	1.00	1.14	1.23	1.24	1.29
3	7, 10	1.31	1.73	2.12	2.28	2.41
9	2, 3, 4	1.86	3.62	5.13	5.59	6.15

Table 4 compares the relative speedup of executing the simulation in heterogeneous workstation clusters with and without the LWD. In the DEC Alpha clusters, there is one DEC3000/500, four DEC3000/400, and four DEC3000/300. Their memory capacities also vary from machine to machine, so the simulation execution time of DEC Alpha workstations (Alpha) is 1.73 - 2.40 times faster than the HP9000/700 workstations (HP). For the combination of 5 Alphas and 4 HPs, all the eight tasks from the 4 HPs are taken over by the virtual tasks in the Alphas because of the differences in computing performances between Alphas and HPs. For the combination of 9 Alphas, two tasks are taken over by the virtual tasks from the slower Alphas to the faster Alphas.

Table 4: Adaptation for heterogeneous clusters

No. of Grains	Workstation Combinations	Without LWD	With LWD	
		Rel. Speedup	Rel. Speedup	Virt. Task No.
1	1 HP	1.00	-	-
	1 Alpha	1.73 - 2.40	-	-
3	2 Alphas, 1 HP	2.37	3.09	2
	1 Alpha, 2 HPs	3.13	3.05	0
	3 HPs	2.22	2.22	0
	3 Alphas	4.61	4.27	0
9	5 Alphas, 4 HPs	6.03	7.58	8
	9 Alphas	8.39	9.79	2

Figures 6 and 7 illustrate the adaptability of the distributed simulation in a DEC3100 cluster and a DEC Alpha cluster. These figures show the relative speedup with and without workload distribution when different numbers of workstations are loaded with loader processes. In the event when there is no workload distribution, the speedup drops sharply even when only one workstation is loaded. However, if the LWD and the GWD are deployed, the speedup will only be degraded according to the number of workstations that are loaded. The number of active grains are different when the number of available workstations varies. If the number of workstations is sufficient and they are available

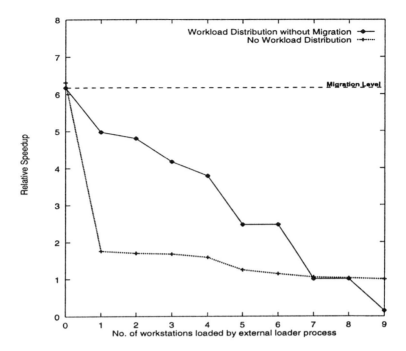

Figure 6: Workload Distribution in a DEC3100 Cluster

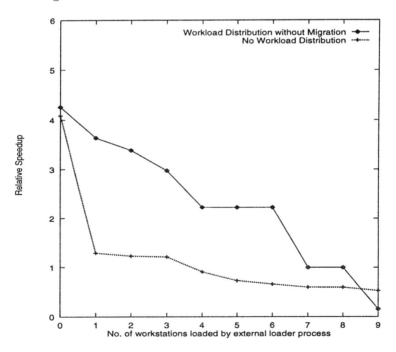

Figure 7: Workload Distribution in a DEC Alpha Cluster

when needed, grain migration can be carried out to maintain the maximum speedup to the migration level as shown in figure 6. There is no indication of migration level in figure 7 because migration can only be carried out within the nine workstations in the DEC Alpha cluster.

6 Conclusions and Future Directions

In this paper we integrate the approaches of local and global workload distribution for adaptive distributed simulation. We have tested the adaptability of a simulation of a broadband communication network under different unbalanced workload situations. The results of our experiments show that a good speedup can be maintained under different conditions of resource availability in the workstation clusters. Although the current implementation platform is workstation clusters, the same adaptive schemes of workload distribution can also be applied to other high performance distributed memory computing systems. Further studies are required to extend these schemes to different types of accelerated simulation algorithms. We plan to implement a scheduler based on the work in [16] to maximize resource utilization when multiple adaptive distributed programs are running in the same workstation clusters. Experiments will also be conducted in heterogeneous networks using both ATM and Ethernet technologies.

References

[1] N. Carriero, E. Freeman, D. Gelernter, and D. Kaminsky. Adaptive parallelism and piranha. *IEEE Computer*, January 1995.

[2] N. Carriero and D. Gelernter. Linda in context. *Communications of ACM*, 32(4), April 1989.

[3] G. Edjlali, G. Agrawal, A. Sussman, and J. Saltz. Data parallel programming in an adaptive environment. Technical report, CS-TR-3350, University of Maryland Institue for Advanced Computer Studies, 1995.

[4] R. Fujimoto. Parallel discrete event simulation. *Comm. of the ACM*, 33(10):30–53, October 1990.

[5] A. Giest, A. Beguelin, J. Dongarra, W. Jiang, R. Manchek, and V. Sunderam. *PVM 3.0 User's Guide and Reference Manual*, Feb. 1993.

[6] M. Kaddoura and S. Ranka. Runtime support for parallelization of data-parallel applications on adaptive and nonuniform computational environments. Technical report, School of Computer and Information Science, Syracuse University, 1995.

[7] S.-Y. R. Li. Algorithms for flow control and call set-up in multi-hop broadband ISDN. In *Proc. INFOCOM*, June 1990.

[8] S.-Y. R. Li and K. H. Shum. Distributed algorithms for flow control and call set-up in a broadband packet network. In *Proc. of the Twelfth Annual Conference on European Fibre Optic Communications and Networks*, pages 182–186, Heidelberg, June 1994.

[9] B. D. Lubachevsky. Efficient distributed event-driven simulations of multiple-loop networks. *Comm. ACM*, 32(1):111–123, January 1989.

[10] N. Nedeljkovic and M. J. Quinn. Data-parallel programming on a network of heterogeneous workstations. *Concurrency: Practice and Experience*, 5(4):257–268, June 1993.

[11] D. Nicol. Parallel discrete-event simulation of FCFS stochastic queuing network. *SIGPLAN Notice*, 23(9):124–137, September 1988.

[12] D. Nicol and R. Fujimoto. Parallel simulation today. *Annals of Operations Research*, 53:249–286, December 1994.

[13] D. Nicol and J. H. Saltz. Dynamic remapping of parallel computations with varying resource demands. *IEEE Trans. on Computer*, 37(9), September 1988.

[14] C. I. Phillips and L. G. Cuthbert. Concurrent discrete event-driven simulation tools. *IEEE J. Select. Areas in Comm.*, April 1991.

[15] R. Prouty, S. Otto, and J. Walpole. Adaptive execution of data parallel computations on networks of heterogeneous workstations. Technical report, CSE-94-012, Department of Computer Science and Engineering, Oregon Graduate Institute of Science & Technology, 1994.

[16] K. H. Shum and M. Bozyigit. A load distribution through competition for workstation clusters. In *Proc. of the Ninth International Symposium on Computer and Information Sciences*, pages 810–817, Antalya, November 1994.

[17] Y.-T. Wang and R. J. T. Morris. Load sharing in distributed systems. *IEEE Trans. Computers*, C-34, March 1985.

[18] T. Yokoi, N. Kishimoto, and Y. Fujii. ATM network performance evaluation using parrallel and distributed simulation technique. In *Proc. of the Teletraffic and Datatraffic, ITC-13*, pages 815–820, 1991.

Optimising Simple Statistical Calculations Using Distributed Memory Computers

Authors

D.M.Al-jumeily*, D.B.Clegg, D.C. Pountney, and P.Harris

*Author for correspondence, School of Computing and Mathematical Sciences,
Liverpool John Moores University, Byrom street, L3 3AF, England.

Abstract

In recent years there has been some interest in implementing statistical algorithms on parallel computers to process large data sets. We address the issue of optimally computing the arithmetic mean, on a distributed-memory parallel computing system. A key element of parallel algorithm design is speed-up modelling, whereby the likely improvement in processing speed-up is modelled mathematically. We present a comparison of such theoretical models with the performance of programs written in the occam concurrent language implemented on transputer arrays. Different parallel topologies and parallelisation strategies are considered. We conclude that for a simple computation such as the mean, performance models can eliminate the need to code all algorithm variations and can help to determine an optimal algorithm using characteristics of the hardware.

1. Introduction

In this article, algorithms for calculating the mean of large samples on arrays of distributed memory computers such as transputer networks are considered. Transputer technology [9] offers an inexpensive form of parallel computing well suited for undergraduate study programs. It has been asserted by Gill [4] "... that the advent of parallel programming may do something to revive the pioneering spirit in programming which seems to be generating into a rather dull and routine occupation."

It is the authors' experience that parallel computing using transputers is certainly more complex than sequential computing. It also offers a natural focus for attention to efficiency in coding, by comparing for example speed-up in parallel algorithms and also deducing the effective mega-flop rate for the execution of a block of coding. In addition as this paper illustrates there are many possible algorithms for solving a single task within a parallel environment. It is also evidently desirable to be able to effectively model algorithm performance and so eliminate the need to code all algorithm variations. With appropriate technical information relating to the speeds of arithmetic and communication between processors, it is shown that the performance of an algorithm on a transputer network can be realistically estimated.

2. Computational Methods for Calculating the Mean

The standard method for computing the mean of a set $\{x_i , i = 1,2,...,n\}$ is through the formula:

$$M_1 = \frac{1}{n}\sum_{i=1}^{n} x_i \tag{2.1}$$

where for convenience this computed estimate for the actual mean is called M_1. Neely [5] found the accumulated round off error in an implementation of this formula to be serious.

He also investigated the accuracy of the alternative methods through estimates M_2 and M_3.

$$M_2 = m_n, \text{ where } m_i = \frac{1}{i}((i-1)m_{i-1} + x_i) \ ; \ m_0 = 0 \ ; \ i = 1,2,3,......,n \tag{2.2}$$

$$M_3 = M_1 + \frac{1}{n}\sum_{i=1}^{n}(x_i - M_1) \tag{2.3}$$

Neely's experiments were performed on an IBM 7094, a machine with 27 bit mantissas. Of the three estimates, Neely found M_3, the two pass algorithm, to be the most robust and provided a considerable improvement on the accuracy of the estimate M_1. The estimate M_2 was demonstrated to be most unreliable. Youngs and Cramer [11] subsequently repeated these experiments and found similar results on machines with 32 and 36 bit mantissas. We report results for a 24 bit mantissa and IEEE [754-1985] arithmetic, implemented on trasputer.

3. Parallel Algorithms

Consider the parallelisation of the estimate M_1. It will be assumed that the data is initially on one processor. An approach is to sub-divide the data into a number of subsets and calculate the partial sums of each subset in parallel. The partial sums may then be gathered and the estimate M_1 then determined through the aggregation of the partial sums. Consider for example a linear chain network of p transputers and for convenience it is assumed that p divides n exactly and that m = n / p, an integer.

Figure 1: Linear Chain Network of p Transputer.

The stages of the algorithm are as follows:
- Stage 1 - The master processor which starts with n data items, exports (p-1) packets of m values into the network.

- Stage 2 - In parallel the master and all (p-1) slave processors sum their 'own' m data items.
- Stage 3 - The partial sums are gathered in from the network by the master processor.
- Stage 4 - The estimate M_1 is finally computed by the master processor.

The total time for the parallel algorithm is calculated by identifying the time for each of the stages. The component times are as follows:

- Stage 1 - The exporting of (p-1) packets each with m *real32* data items. The time required is:

$$(p-1)[T_s + mT_c] \tag{3.1}$$

  This can be achieved provided that each slave inputs and outputs data in parallel.

- Stage 2 - The parallel computation of all partial sums takes time:

$$mT_f \tag{3.2}$$

- Stage 3 - The time to gather in the data is:

$$(p-1)[T_s + T_c] \tag{3.3}$$

- Stage 4 - The time to compute the estimate M_1 on the master processor is:

$$(p+1)T_f \tag{3.4}$$

As described there will be parallel operations at stages 1,2 and 3. It is also possible at stage 4 for the master to receive data and in parallel simultaneously accumulate the values into a final sum. However the benefit is so marginal that this option is not considered. The interested reader may like to develop the corresponding result for themselves.

Thus the total time obtained by summing (3.1) through (3.4) is given by

$$T_p = (m + p + 1)T_f + (p-1)(m+1)T_c + 2(p-1)T_s \tag{3.5}$$

On the other hand on a single processor the total time for a sequential algorithm would be:

$$T_1 = (n+1)T_f \tag{3.6}$$

The effectiveness of the parallel algorithm is quantified by the ratio of the time on a single processor to the time on p processors of the same type, this is a definition of speed-up. Hence for this algorithm, the speed-up S_p is given by:

$$S_p = \frac{(n+1)T_f}{(m+p+1)T_f + (p-1)(m+1)T_c + 2(p-1)T_s} \tag{3.7}$$

As the packet size increases the effect of terms in T_s will diminish, and in the limit as n and hence also $m \to \infty$ we find:

$$S_p \to \frac{p}{1 + (p-1)T_c / T_f} \tag{3.8}$$

4. Algorithm Refinement

As we know, it is the communication costs which seriously influence the parallel algorithm. These are functions of the technology including the chosen network. For four processors a ternary tree network with one root and three branches would be preferable to a linear chain as the communication costs would be reduced. Hence consider the calculation of M_1 on a ternary tree. The four stages of the algorithm described in section 3 remain the same but the communication to the three branch processors can now be done in parallel and in one third of the time. If m = n/4 then the time can be described as:

$$T_s + mT_c \qquad (4.1)$$

The modified speed-up estimate is then given by:

$$S_p = \frac{(n+1)T_f}{(m+5)T_f + (m+1)T_c + 2T_s} \qquad (4.2)$$

with the corresponding limiting value equal to:

$$\frac{4}{1 + T_c / T_f} \qquad (4.3)$$

This compares with the limiting value $\dfrac{4}{1 + 3(T_c / T_f)}$ (Eq. (3.8) with p = 4) for

a linear chain of for transputers. A further improvement would be achieved by balancing the calculation over the tree and allowing the root processor to compute in parallel with communication to the branches.

To fully utilise the root processor the following balance equation must be satisfied:

Time to compute on the root = Time to communicate data to the branches
           plus the time to summate data locally on the branches
           plus the time for the root to receive the partial sums
           from the branches.

If the number of items summed by the root is $(1-\alpha)n$, where $(1-\alpha)n$ must be integer then the balance condition is:

$$(1-\alpha)nT_f = (T_s + \frac{\alpha nT_c}{3}) + \frac{\alpha nT_f}{3} + (T_s + T_c) \qquad (4.4)$$

From which for large n, we can write

$$\alpha = \frac{3T_f}{(4T_f + T_c)} \qquad (4.5)$$

Without the balance condition Eq. (4.4), the estimated speed-up is found to be

$$S_p = \frac{(n+1)T_f}{Max((1-\alpha)nT_f,(T_s+\frac{\alpha n}{3}T_c)+(\frac{\alpha n}{3}T_f)+(T_s+T_c))} \tag{4.6}$$

If balance is achieved as defined in Eq. (5.4), then

$$S_p = \frac{(n+1)T_f}{(1-\alpha)nT_f} \tag{4.7}$$

As the sample size increases ($n \to \infty$) and using Eq. (4.5) we find

$$S_p \to \frac{4+T_c/T_f}{1+T_c/T_f} = \frac{4+\rho}{1+\rho} \tag{4.8}$$

From which a speed-up of 2.125 is predicted for the worst case ($\rho = 1.67$). The improvement is welcomed but is still short of the target speed-up.

To achieve even better performance, the branch processors should also in parallel compute and communicate thus avoiding the bottleneck caused by the initial receipt of the full set of data. To achieve this, data is parcelled into a number of packets and the optimum packet size determined as described below.

Assume each branch of the tree is to receive y data items and these are to be split into q packets each packet containing y/q items. After receiving the first packet, the branch will in parallel receive the next packet and in parallel summate the data in the first packet. This process is repeated (q-1) times and finally the last received packet is summed.

The times for these elements are detailed below:

Receipt of first packet - $\qquad\qquad T_s + \frac{y}{q}T_c$

Parallel input and computation -

$$(q-1)\,Max(T_s+\frac{y}{q}T_c,\frac{y}{q}T_f)$$

(Note that it is the maximum of either the time for communication or the time for computation which is aggregated)

Sum last packet - $\qquad\qquad\qquad\quad \frac{y}{q}T_f$

Return local partial sum to the root - $\qquad T_s + T_c$

Now for the worst case (a T800 type transputer and link speeds of 10 mega bits per second),

$$T_s+\frac{y}{q}T_c > \frac{y}{q}T_f$$

where communication time exceeds computation time.

The total time is then given by:

$$f(q) = q(T_s + \frac{y}{q}T_c) + \frac{y}{q}T_f + (T_s + T_c) \tag{4.9}$$

This expression a function of the variable q will have a minimum value when

$$f'(q) = T_s - \frac{y}{q^2}T_f = 0 \tag{4.10}$$

The only practical stationary condition is consistent with a minimum when

$$q^2 = \frac{yT_f}{T_s} \tag{4.11}$$

Substituting this expression for q into Eq (5.9) and letting y increase and tend to infinity yields yT_c, a value which is readily justified when communication times are strongly dominant. Applying this to an algorithm in which the amount of calculation on the root is $(1-\alpha)n$ and y above is replaced by $\alpha n/3$ the balance equation in the limit simplifies to:

$$(1-\alpha)nT_f = \frac{\alpha n}{3}T_c \tag{4.12}$$

Solving for α and making appropriate substitutions yields the limiting speed-up:

$$\frac{3 + T_c/T_f}{T_c/T_f}; \quad \text{provided } T_c > T_f \tag{4.13}$$

In an unbalanced state, the estimated speed-up can in general shown to equal:

$$S_p = \frac{(n+1)T_f}{Max((1-\alpha)nT_f, (\frac{\alpha n}{3}(T_c + \frac{T_f}{q}) + (q+1)T_s))} \tag{4.14}$$

When computation time dominates i.e. $T_f > T_c$, the communication to the branches will be obtained 'free'. Hence the total time for the branches will be $(1-\alpha)nT_f$. Under such circumstances, each processor will compute the sum of one quarter of the data with $\alpha = \frac{3}{4}$. Hence the limiting speed-up will be 4.

Applied to all four combinations of transputer types and link speeds the following predicted values of limiting speed-up are applicable:

Transputer type	T_f	T_c	
		4.4×10^{-6}	2.2×10^{-6}
T414	12.2×10^{-6}	4.00	4.00
T800	2.3×10^{-6}	2.57	4.00

Table 1: Predicted Limiting Speed-up for Different Transputer Types.

5. Experiment Results

In this section we present the results for the algorithm

$$M_1 = \frac{1}{n}\sum_{i=1}^{n}x_i \quad \text{where} \quad i = 1,2,3,...,n$$

In our implementation we have made use of the following networks.

- Linear chain.
- Ternary tree.

And for the ternary tree the following cases have been investigated.

- Equal splitting of the data.
- Balanced algorithm with data sent to the branches in a single packet.
- Balanced algorithm with data sent to the branches in a sequence of small packets.

5.1 Linear Chain

Varying number of processors (p=2,3,4,5) have been used in the experiments. The speed-up has been measured and also theoretical estimates of the speed-up have been obtained using the crude value for the unit time to perform a single floating point operation $(T_f = 2.64 \times 10^{-6})$ and also the measured value $(T_f = 2.35 \times 10^{-6})$. Results are plotted below Figure 2 which show the speed-up decays as the number of processors is increased. This is due to the excessive communication costs. As might be expected the best theoretical estimate for the model is obtained using the measured value of T_f.

However the predicted performance using the unrefined value of T_f does realistically reflect the actual situation.

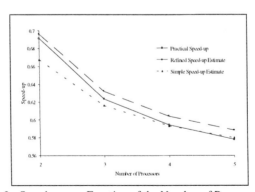

Figure 2 - Speed-up as a Function of the Number of Processors.

The variation of performance with sample sizes has also been investigated. The results below Figure 3 are for 5 processors and are typical of those obtained with an alternative number of processors.

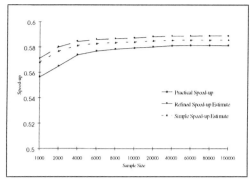

Figure 3: Speed-up for a Linear Chain With Varying the Sample Size.

These results show that as the sample sizes increases, there is a limiting speed-up which in this case is less than one.

5.1.1 Conclusions
- Theoretical and observed performances have shown good agreement.
- For current technology, the communication costs prohibit the algorithm being worthwhile.
- The elementary performance models provide a realistic estimation of speed-up.

5.2 Ternary Tree - Equal Splitting of the Data

In this experiment four processors have been used. Again the speed-up has been obtained by experiment and the theoretical speed-up has been obtained using both the crude and the measured values of T_f. The speed-up results are plotted below for different sample sizes, and these show that the model using the crude T_f is inferior to the model using the measured T_f. No obvious explanation has been found for the poor comparison for small sample sizes. Actual limiting performance is marginally better than estimated by the simple model, which indicates that the simple model is a useful indicator of performance.

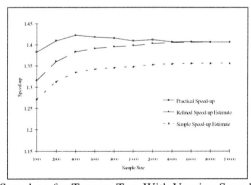

Figure 4: Speed-up for Ternary Tree With Varying Sample Size.

5.2.1 Conclusions

- Good agreement between the practical speed-up and the refined speed-up has been achieved with large sample sizes.
- The overall performance is considerably better than the speed-up in a linear chain with the same number of processors.
- The unexplained variation with low sample sizes is probably attributable to the over-estimation of communication costs.

5.3 Balanced Ternary Tree - Data Unequally Distributed

In this case data is unequally shared between the root and the branches. Initially we investigate the optimal fraction determined from the Eq. (4.4). For simplicity a fixed sample size (n=60000) has been used. To verify the fraction, speed-up results were obtained by varying the amount of the computation performed by the root processor. The results illustrated in Figure 5 show that the model behaves perfectly until fraction equal 0.47, and after that the model break down. This can be attributed to the root processor which is not behaving as expected. The reason for this is due to the parallel overlapping of communication and calculation which leads to a reduction in the performance of the root processor. In theory, the optimal fraction would be 0.49 compared to a measured value of 0.47, confirming that the root processor is under performing, and in this case the optimal overall performance will be achieved when the root handles slightly fewer data values than theory predicts.

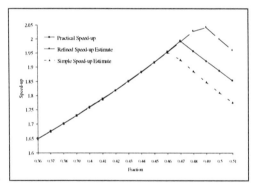

Figure 5 - Speed-up as a Function of the Optimal Fraction.

To refine the model requires separate specification of T_f on both the root and the branches. These values are simply determined by experiment, which have revealed that on each branch, T_f is unaffected. However on the root, experiments have shown that T_f diminishes as the amount of the computation by the root increases.

It is an elementary matter to refine the model Eq. (4.6), and the speed is given by:

$$S_p = \frac{(n+1)T_f}{Max((1-\alpha)nT_{f_r}, ((T_s + \frac{\alpha n}{3}T_c) + \frac{\alpha n}{3}T_{f_b} + (T_s + T_c)))} \quad (5.1)$$

where T_{f_r} is the unit time for the floating point operation on the root and T_{f_b} is the equivalent time on each of the branches.

Speed-up results have shown that 0.47 is the optimal fraction. This fraction has also predicted theoretically by the model Eq. (5.1) through the incorporation of measured variable values for T_{f_r} & T_{f_b}. This is confirmed in Figure 6.

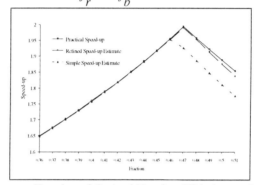

Figure 6: Speed-up as a Function of Optimal Fraction With Appropriate Unite Time for the Floating Point Operation on the Root.

5.3.1 Conclusions
- An effective increase in speed-up can be achieved by balancing the calculation on the root and branches.
- Specifically it is necessary to increase the work on the root by using the parallelism of the transputer.
- A very good agreement has been achieved between the theoretical and practical value of speed-up.

5.4 Balanced Ternary Tree - Branch Data Sent in a Sequence of Small Packets

In this implementation we consider sending the data from the root to the branches using a number of packets all of equal size. The optimal number of packets determined from the Eq. (4.10) is found to be 25. For simplicity a fixed sample size (n=60000) has been used. Figure 12 shows the results of the speed-up with varying the amount of calculations on the root, and sending the data to the branches using 25 packets all of equal length. Theory predicted an optimal fraction of 0.36 determined by the Eq. (4.11), and this is also verified by the experiment. The reader should note that different values for T_f can be used on

318

the root and the branches, and in this case a modified version of the speed-up
Eq. (4.13) is given by:

$$S_p = \frac{(n+1)T_f}{Max(n(1-\alpha)T_{f_r}, (\frac{\alpha n}{3}(T_c + \frac{T_{f_b}}{q}) + (q+1)T_s))}$$ (5.2)

where T_{f_r} is the unit time for the floating point operation on the root and T_{f_b} is
the equivalent on each of the branches. Results are indicated in the figure below.

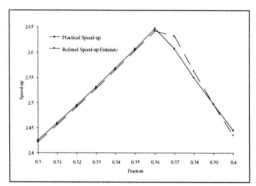

Figure 7: Variation of Speed-up With Root Fraction.

For a fixed fraction equal to 0.36, the speed-up has been investigated using a
varying number of packets and a fixed sample size (n=60000). For these values,
the optimal number of packets predicted by applying Eq. (4.11), is also verified.
Results show that as the number of packets approach 20, optimal performance is
achieved.

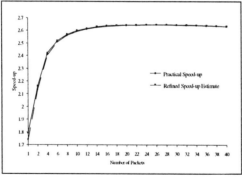

Figure 8: The Effect of Changing the Number of Packets on the Speed-up.

5.4.1 Conclusions

- A good speed-up can be achieved by optimising the amount of calculation on the root and sending the data to the branches in a number of small packets, which has to be determined.
- Very good agreement has been obtained between the theoretical and the measured speed-up.
- The actual performance is a 4.5 fold increase above the results for a simple linear chain with 4 processors on a T800 transputer network.

6. Summary

With the availability of the hardware, parallelism becomes a truly attractive and viable approach to the attainment of high performance computational speeds. One of the interesting questions is the optimal design of an algorithm on a distributed memory parallel computer.

The performance of a parallel algorithm is not fixed, but will vary with the current state of technology in relation to the relative time for communication and computation. Applicable conditions could render that either the algorithm is effective or it is simply not worthwhile.

The performance of different models has been considered in this paper. The model which would achieve the best parallel performance has been specified.

The results for calculating the mean using M_1 have been considered in terms of parallel performance, and it has been proven that the performance of the algorithm depends on the type of network and the manner of splitting the data between the processors.

The best performance for M_1, was achieved on a ternary tree network (4 processors), by balancing the data on the main processor, and sending the data to other processors as a sequence of small packets. The purpose of sending the data to the tree in small packets is to make all the processors perform arithmetic in parallel, whilst they are receiving data. In other words the best implementations require minimum communication overhead and best utilisation of the processors.

A high correlation of the performance model and actual results is possible using carefully refined times for computation.

Ideas described here are applicable to other methods such as M_2 and M_3 for calculating the mean, and indeed many other statistical applications. This work is continuing and we hope to report results for M_3, other statistical type calculations and also to address question of accuracy.

References

1. Atkin, P.(1988), Performance maximisation, Inmos technical Note 17.
2. Flynn, M.J.(1986), Parallel Architectures, Systems International.
3. Freeman, T.L., and Phillips, C.(1992), Parallel Numerical Algorithms, Prentice Hall.
4. Gill, S. (1958), Parallel Programming, Computer Journal, Vol. 1.
5. Neely, P.M.(1966), Comparison of Several Algorithms for Computation of Means, Standard Deviations and Correlation Coefficients, Comm ACM, Vol. 9.
6. Occam 2 Toolset User Manual (1990), INMOS Ltd, CSA Architects.
7. Pountain, D., and May, D.A.(1987), A Tutorial Introduction to Occam Programming, BSP Professional.
8. Transputer Development System.(1988), INMOS Ltd, Prentice Hall.
9. Transputer Reference Manual.(1988), INMOS Ltd, Prentice Hall.
10. Transputer Technology and Applications.(1992), Distance Learning Video, Liverpool John Moores University.
11. Youngs, A.E., and, Cramer, M.E.(1971), Some Results Relevant to Choice of Sum and Sum-of-Product Algorithms, Technometrics, Vol. 13, No. 3.

Simulation

Object Oriented Simulation Models for Performance Modelling

Rob Pooley

Department of Computer Science

Edinburgh University

Kings Buildings

Mayfield Road

Edinburgh EH9 3JZ UK

E-mail: rjp@dcs.ed.ac.uk

Abstract

This paper examines how to make use of object oriented design and programming approaches in discrete event simulation. It reviews the concepts of classes and objects in general terms. From these it outlines the structure of an object oriented simulation package and its construction in C++. This is compared with SIMULAs original implementation of such facilities. Using the C++ package performance models are constructed along process based object oriented lines and the ideas of component based, hierarchical modelling are examined.

1 Introduction

Much interest has been generated in recent years in the use of object oriented structuring in simulation modelling. This is perhaps unsurprising, as object oriented approaches were first defined in the SIMULA languages of the late 1960s, which evolved to address the difficulties of expressing simulation models in computer programming languages. With the widespread popularity of C++ as an object oriented programming language, recent attention has focussed on how to write discrete event simulation packages in this language. Unfortunately C++ lacks some of the features of SIMULA for this purpose.

In [6] the main features of a language suited to implementation of a simulation package in the process style were identified as:

1. encapsulation of data in structured form (objects) capable of replication from a template (classes);

2. ability to hold ordered lists of such structures;

3. ability to specialise and extend structures from a parent class (inheritance) and to hold different specialisations in the same list;

4. ability of structures to possess autonomous actions and to suspend and resume these at desired points (co-routines).

Points 1-3 allow lists of waiting processes to be constructed, where each process may differ in certain specific ways from others while retaining the ability to behave as a process. This supports the time ordered event list and various kinds of conditional waiting in a simulation. Various forms of automatic statistical collection and reporting can also be supported by this means. These features are essentially those required in object oriented languages.

Point 4 adds an efficient and flexible way of defining actions for processes and suspending and restarting these without explicit scheduling by a monitor. A process encapsulates the behaviour of a component as a sequence of actions (events) which may contain looping and conditional branching. Most object oriented languages have not included this feature, although the related Actors language has. Packages for threads and lightweight processes offer such facilities for some operating systems.

Taken together these features allow the building of efficient simulation packages with a minimum of intrusion at the level of model description. SIMULA [8] possesses these features. It has been used to build simulation packages such as CLASS SIMULATION [8] and DEMOS [1]. Its use in this way is explained in [3] and [5].

2 Object oriented design of package

2.1 Analysis

Using the design methodologies outlined in books such as [2], we can quickly find obvious objects for a simulation package. These are:

Processes, which will represent active components. They need to be capable of waiting for simulation time to reach a certain point and of waiting for conditions associated with passive objects to be true. As active objects they need to know their own pattern of behaviour and to be capable of resuming activity after waiting for time to pass or other conditions to become true.

Statistical objects, such as those keeping count or maintaining averages. These must be able to be updated and to report their measures. Ideally they should operate unobtrusively.

Condition objects, which are those tracking blocked processes and allowing interactions with these. In many models these will be the basis of important statistical information and so should share the abilities of statistical objects.

Data generating objects, such as random number generators and input data streams. These may usefully record statistical information as well as supplying appropriate input values.

2.2 Design

From this analysis we move towards a design, including more concrete objects needed to support a view of the eventual implementation. We begin by noting that queueing and list handling are likely to be widely used and so define the need for classes for handling lists. To exploit inheritance, we define a type of list with some elementary operations to add and remove items.

We also note the need for some sort of coroutine-like mechanism. As C++ and most object oriented programming languages do not support this directly, we have to put some careful thought into this and it is discussed separately below. We note for the moment the need for some coroutine scheduling and a scheduler.

Since Statistical, Conditional and Data objects all require reporting and recording facilities, we define a parent class for these. We also define a notion of lists to hold reportable objects in groups, to allow structured, automatic reporting. Since update and reporting functions are specific to the derived classes, we expect them to be virtual. All reportable objects must therefore be queueable.

The time dependent waiting will be handled by defining an event list. The operations Hold and Schedule, as well as the constructor for a process must operate on this. Considering our design as a layered model, we tentatively decide on the following structure:

```
        General Purpose Object Oriented Language
        ------------------------------------------
                List handling package
                ----------------------
    Event list, report lists, reportable objects, processes
    ----------------------------------------------------------
            Data input objects, data generating objects,
        statistical collection objects, conditional objects
        ----------------------------------------------------------
            User specialisations for application
            ------------------------------------
                        Models
```

3 Building core features in C++.

We move to the first simulation specific layer and consider processes and reporting.

3.1 Processes and the event list

As mentioned before, the implementation of waiting is the most problematic in C++. Prefixing the parent class of all processes with list item capabilities solves the problem of how to enqueue them, but their scheduling requires more complicated strategies. Thus the operations Schedule and Hold cannot be implemented directly, as in SIMULA, where objects can Detach themselves and Resume each other.

There are at least three strategies for suspension and resumption of processes. These are:

1. Create or use an existing library which supports co-routines (variously known as threads and lightweight processes). Stroustrup adopted this approach in the process.h package which accompanied the early UNIX C++ releases. So does the Newcastle University C++SIM package. In these it is possible to follow the approach used by SIMULA, but such

libraries are not portable and often run only on one version of UNIX. The intention here is to write only in C++, even if that means a certain loss of elegance.

2. Write processes as classes holding a list of event objects, to be executed in turn, followed by a wait in some list. Define the parent Event class to have a virtual function called, for instance, DO(). Execution of an event means evoking the Do() function. The scheduler which runs the simulation then examines the event list and calls Do() in turn for each Process held there. The Process Do() function calls the Do() function for the first event object in its list and removes this, possibly inserting it at the end of the list if the process loops. This approach is elegant, but means that writing and generating processes is quite complicated. It is also quite difficult to embed loops and conditional branches into the lists of events. For an implementation of event lists which could be used in this way, but explained in a non-simulation context, see [4].

3. Write processes as classic state machines, with a switch statement as the main control structure, but no loop outside this switch as in monolithic finite state automata. Each event is a case branch of this switch and must set the next event value in a data member before exiting. If the Process is placed into a queue before exit it will be rescheduled from there when its next event is enabled. This switch forms the body of a (virtual) function which must be written for any sub-class of Process and the event list and scheduler work as in version (2) above. Loops and conditional branching can be defined by appropriate setting of the state variable before exiting. This is not very elegant, but is totally mechanical to write from a process description. It seems especially suited to automatic generation from a graphical modelling tool. It is the approach used below.

3.2   The C++ package used

The header files for this part of the simulation show the major features which are used in simulation models. Processes are required to define their behaviour in the virtual function Body(). This must contain the state machine switch a a standard form, as we shall see. Two important public functions are Schedule(float), which re-inserts the Process in the event list with a given delay, and Cancel(), which removes the Process from any list in which it currently is waiting. This will be useful in moving the Process from a resource queue to the Event list, for instance.

The use of the other features is fairly self-explanatory when we come to examples.

Functions which are only to be used within Body() are declared in the protected section of the class, since they should not be accessible from other Processes. The other functions are mostly used to read protected data, such as the next state.

```
// Header of Processes and Event
List

class Proc: public mylink {
```

```
private:
   float spTime;
   char* spTitle;
   int spState, spWanted;
protected:
   Proc(char*,float);
   int NextState();
   void SetState(int);
   void Terminate();
   void Set_Want(int);
   int Get_Want();
public:
   float Ev_Time();
   char* Title();
   virtual void Body();
   void Schedule(float);
   Proc* Cancel();
   friend void Hold(float);
   friend void Sim_Run(float);
   friend void Resource::Release(int);
   friend void Resource::Acquire(int);
};

class seSimEnv {
private:
   float seTime;
   head seEv_List;
public:
   friend float Time();
   float SetTime(float);
   seSimEnv();
   friend void Hold(float);
   friend Proc* Current();
   friend Proc::Proc(char*,float);
   friend void Sim_Run(float);
   friend void oosAdd_in_Order(Proc*);
};

float Time();
Proc* Current();
void oosAdd_in_Order(Proc* P);
void Hold(float);
```

3.3 Other modelling primitives provided

As well as the basic scheduling of processes, the package built to illustrate the object oriented simulation approach in C++ contains statistical collection and automatic reporting. These are based on two classes, Stats, which is the parent of all reportable objects, and Report_List, which contains lists of such objects. By calling the function Report on a Report_List, the virtual function

`Report()` is called for all objects in that list. This allows reporting to be implemented a a manner specific to the form of ststistics being collected while exploiting a single form of `Report_List`.

By a trick involving virtual functions and a string identifying which group of objects this sub-class is to be associated with, various `Report_Lists` are automatically generated and used to provide structured reports. Users are free to add new reportable objects if they use this mechanism to do so. For full details, you are referred to the MODES User Guide [MODES 1995b].

As well as providing the basis of explicit statistical collection devices, many other objects have implicit ststistics associated with them and can usefully be prefixed by the parent class Stats. By assigning uniques return values to `My_Type_Name`, these can be separated in the reporting.

```
// Statistical collection devices

class Stats: public mylink {
private:
   static int
   scCurr_Stat;
   int My_Type;
   void Set_Report();
protected:
   static int Next_Stat(const char*);
   Stats(char*);
   char* My_Title;
   long Obs;
   virtual const char*My_Type_Name();
public:
   virtual void Report() = 0;
   int Your_Type();
   friend void Sim_Run(float);
};


class Report_List:public head {
protected:
   const char* My_Title;

public:
   const char* Title();
   void Report();
   Report_List(const char*);

};
```

The other major group of features are those concerned with waiting in states where a condition other than the elapsing of a fixed time interval is causing blocking. This is represented in the package used here by a Resource object, which forces Processes to wait until a desirec amounbt is available before continuing. This uses the Stats class to make such objects reportable.

```
// Now the Resource interactions
```

```
class Resource:public Float_Stats {
private:
   int srAmount, srMax, srCum_Sum;
   float srLast_Time;
   head My_List;
   void Update(float);
protected:
   const char*My_Type_Name(){return "Resources";}
public:
   void Acquire(int);
   void Release(int);
   void Report();
   Resource(char*,int);
};
```

Finally the package builds a family of random number generators using a base class called Random. Again these are all made sub-classes of Stats, so that they can provide reports on their use automatically.

```
// Now the random number
stuff
class Random: public Stats {
private:
   long rnSeed;
protected:
   float rnDraw();
   Random(char*,int);
   const char* My_Type_Name(){return "Random numbers";}
};

class R_Random: public Random {
protected:
   float rnMean, rnSum;
   R_Random(char*,int);
   const char* My_Type_Name(){return "Real Randoms";}
public:
   virtual float Sample() = 0;
};

class I_Random:public Random {
protected:
   int rnMean, rnSum;
   I_Random(char*,int);
   const char* My_Type_Name(){return "Integer Randoms";}
public:
   virtual
   int Sample() = 0;
};
// Here's a typical floating point random number
```

```
class Neg_Exp: public R_Random {
protected:
   const char*My_Type_Name(){return "Neg Exps";}
public:
   Neg_Exp(char*,int,float);
   void Report();
   float Sample();
};
```

The rest of the reporting features of the package are listed below.

```
class Int_Stats: public Stats {
protected:
   long scSum;
   Int_Stats(char*);
   virtual const char*My_Type_Name();
public:
   virtual void Update(int) = 0;
};

class Float_Stats: public Stats {
protected:
   float scSum;
   Float_Stats(char*);
   virtual const char* My_Type_Name();
public:
   virtual void Update(float) = 0;
};

// Here's a simple integer statistic

classCounter: public Int_Stats {
protected:
   const char*My_Type_Name(){return "Counters";}
public:
   Counter(char*);
   void Update(int);
   void Report();
};

// Here's a floating point statistic

class Clock:public Float_Stats {
private:
   float spSum;
protected:
   const char*My_Type_Name(){return "Clocks";}
public:
   Clock(char*);
   void Update(float);
   void Report();
```

```
};
```

4   Examples of the use of MODES

There follow two simple models using MODES in typical performance modelling
problems. These are not very large and some remaining problems in the use
of C++ as a vehicle for such models are considered in the final section of this
paper. They are, however, sufficient to show how models must be written using
this approach.

Each model is accompanied by its output, as a demonstration of the automatic tracing and report generation built into the package. As well as summary
statistics a trace of all events can be produced, and some of this is included to
show how the simulation proceeds.

4.1   M/M/1 queue

The first model shown is an M/M/1 queue built from a process and a resource
class.

```
#include "statslib.h"

 class Test_Proc: public Proc {
private:
   float Start_Time;
    int Reps;
public:
   Test_Proc(char*,float);
   void Body();
};

   Counter* Tries1,* Tries2;
   Clock* Elapsed;
   Neg_Exp* Delay1, * Delay2;
   Resource* Buffers;

   void Test_Proc::Body() {
      switch (NextState()) {
case 0:
      Start_Time = Time();
      Tries1->Update(1);
      Reps = 0;
      new Test_Proc("TP",Delay1->Sample());
      SetState(1);
      break;
case 1:
      Buffers->Acquire(1);
      SetState(2);
      break;
case 2:
```

```
        Hold(Delay2->Sample());
        SetState(3);
        break;
case 3:
        Buffers->Release(1);
        SetState(99);
        break;
case 99:
        Elapsed->Update(Time()-Start_Time); Tries2->Update(1);
        SetState(-1);
        break;
default:
        Error("attempt to reactivate terminated Proc ",Title());

        }
    }

    Test_Proc::Test_Proc(char* Title, float When): Proc(Title,When){}

    Test_Proc *Tp1, *Tp2;

void main() {
    float T;
    Tries1 = new Counter("First");
    Tries2 = new Counter("Second");
    Elapsed = new Clock("Duration of Procs");
    cout<<"Interarrival Time?\n";cin>>T;
    Delay1 = new Neg_Exp("Delay1",2345,T);
    cout<<"Service Time?\n";cin>>T;
    Delay2 = new Neg_Exp("Delay2",1234,T);
    Buffers = new Resource("Buffers",1);
    Tp1 = new Test_Proc("Proc1",0.0);
    cout<<"Simulation length?\n";cin>>T;
    Sim_Run(T);
}
```

4.2 The output of the M/M/1 queue model

```
Welcome to simple simulation in C++
Author Rob Pooley

Interarrival Time?
Service Time?
Proc1 created, with delay 0 at 0
Simulation length?
TP created, with delay 0.00684179 at 0
Proc1 acquired 1Buffers at 0
Proc1 holds for 1.27931 at 0
TP created, with delay 2.88148 at 0.00684179
TP blocked attempting acquire of 1Buffers at 0.00684179
```

```
Buffers released 1 at 1.27931
TP scheduled delay=0 at 1.27931
Proc1 terminates at 1.27931
Proc1 terminated at 1.27931
```

. . . .

```
TP blocked attempting acquire of 1Buffers at 190.295
TP created, with delay 2.1232 at 190.397
TP blocked attempting acquire of 1Buffers at 190.397
Buffers released 1 at 191.029
TP scheduled delay=0 at 191.029
TP terminates at 191.029
TP terminated at 191.029
TP holds for 16.2822 at 191.029
TP created, with delay 5.3285 at 192.52
TP blocked attempting acquire of 1Buffers at 192.52
TP created, with delay 24.4822 at 197.849
TP blocked attempting acquire of 1Buffers at 197.849
Simulation run ended at 200

Resources
Buffers: Observations 66 Avge 0.27154
Neg Exps
Delay1: Observations 54 Avge 4.11724
Delay2: Observations 50 Avge 3.10879
Clocks
Duration of Procs: Observations 49 Avge 6.00851
Counters
First: Observations 54 Avge 1
Second: Observations 49 Avge 1
```

4.3 A looping process

```
#include "statslib.h"

class Test_Proc: public Proc {
private:
   float Start_Time;
   int Reps;
public:
   Test_Proc(char*,float);
   void Body();
};

   Counter* Tries1,* Tries2;
   Clock* Elapsed;
   Neg_Exp* Delay1, * Delay2;
   Resource* Buffers;
```

```
   void Test_Proc::Body() {
    switch (NextState()) {
case 0:
   Start_Time = Time();
   Tries1->Update(1);
   Reps = 0;
   Hold(Delay1->Sample());
   SetState(1);
   break;
case 1:
   Buffers->Acquire(3);
   Hold(8.0);
   SetState(2);
   Reps++;
   break;
case 2:
   Buffers->Release(3);
   Hold(Delay2->Sample());
   if (Reps<=20) SetState(1); else SetState(99);
   break;
case 99:
   Elapsed->Update(Time()-Start_Time);
   Tries2->Update(1);
   SetState(-1);
   break;
default:
   Error("attempt to reactivate terminated Proc ",Title());
    }
}

Test_Proc::Test_Proc(char*
Title, float When): Proc(Title,When){}
Test_Proc *Tp1, *Tp2;

void main(){
   float T;
   Tries1 = new Counter("First");
   Tries2 = new
   Counter("Second");
   Elapsed = new Clock("Duration of Procs");
   Delay1 = new Neg_Exp("Delay1",2345,2);
   Delay2 = new Neg_Exp("Delay2",1234,5);
   Buffers = new Resource("Buffers",4);
   Tp1 = new Test_Proc("Proc1",0.0);
   Tp2 = new Test_Proc("Proc2",3.0);
   cout<<"Simulation length?\n";cin>>T;
   Sim_Run(T);
}
```

4.4 Output from the looping process model

```
Welcome to simple simulation in C++
Author Rob Pooley
Proc1 created, with delay 0 at 0
Proc2 created, with delay 3 at 0
Simulation length?
Proc1 holds for 0.00342089 at 0
Proc1 acquired 3Buffers at 0.00342089
Proc1 holds for 8 at 0.00342089
Proc2 holds for 1.44074 at 3
Proc2 blocked attempting acquire of 3Buffers at 4.44074
Proc1 holds for 8 at 8.00342
Buffers released 3 at 16.0034
Proc2 scheduled delay=0 at 16.0034
Proc1 holds for 2.13218 at 16.0034
Buffers released 3 at 16.0034
Proc2 holds for 10.2244 at 16.0034
Proc1 acquired 3Buffers at 18.1356
Proc1 holds for 8 at 18.1356
Buffers released 3 at 26.1356


. . . . . . . . . . . . . . . . . . . . . . . . . . . . . . . . . . . .


Proc1 holds for 8 at 375.668
Proc2 blocked attempting acquire of 3Buffers at 377.189
Proc1 holds for 8 at 383.668
Buffers released 3 at 391.668
Proc2 scheduled delay=0 at 391.668
Proc1 holds for 7.67787 at 391.668
Buffers released 3 at 391.668
Proc2 holds for 1.84161 at 391.668
Proc2 terminates at 393.509
Proc2 terminated at 393.509
Proc1 terminates at 399.346
Proc1 terminated at 399.346
Simulation run ended at 1000
 Resources
Buffers: Observations 65 Avge 1.56852
 Neg Exps
Delay1: Observations 2 Avge 1.44416
Delay2: Observations 42 Avge 6.75162
 Clocks
Duration of Procs: Observations 2 Avge 394.928
 Counters
First: Observations 2 Avge 1
Second: Observations 2 Avge 1
```

5   Component based and hierarchical modelling

The importance of object oriented simulation lies not only in its easy implementation of the process view, however. On a modelling level it represents a natural description of many systems, particularly computer and telecommunications networks. Work on exploiting this for network protocols [7] and interconnection networks [11, 12] shows the potential.

As well as allowing the component based structure of most computer based systems to be represented, an object based approach offers a natural hierarchy, through a compositional view of objects composed of sub-objects. This supports a natural re-use of components and may offer a way of exploiting more efficient hybrid simulation/numerical/analytic solution of models, with components being pre-solved and represented by aggregate behaviour tables embedded in hierarchical models.

As a further extension of this approach, it has been shown that process based components can also represent the functional behaviour of systems and be analysed by techniques used for process algebras [9]. Again, graphical representation of models can be used to generate forms suitable for behavioural analysis.

6   Conclusions and unresolved issues

From the work on MODES described here it is possible to claim that object oriented methods can be applied successfully to the building of discrete event simulation packages, even without the co-routine features of SIMULA. The resulting models are easy to write and can contain conditional branching and looping behaviour. However, it is not possible to write them in a normal high level language style. Instead a structure of states and successor states based on a switch must be used.

A problem not addressed here, but vital to large object based simulations, is that of memory management. While SIMULA provided garbage collection as a non-intrusive means of recovering memory freed when processes terminate, C++ requires that this be programmed explicitly, using destructors. The best solution to this problem remains an open issue.

One more issue remains to be tested. It was claimed in this paper that the process structure needed in these models would be suited to automatic generation from graphical editing tools. This has not yet been tested, but, given the success of tools for the DEMOS package the potential seems very real.

In general MODES is less complete than DEMOS. Even if completed to the same level it would be inherently less flexible and less elegant in many ways. However, given the ever-increasing popularity of C++ it seems likely that it could surpass DEMOS in use and so the effort expended in making it as well built as possible is probably wothwhile.

References

[1] Birtwistle, G.M., *Discrete Event Modelling on Simula*, Macmillan, 1979

[2] Booch, Grady, *Object Oriented Design with Applications*, The Benjamin/Cummings Publishing Company, Inc., 1991

[3] Franta W.R. 1978. *The Process View of Simulation*, North-Holland, Amsterdam

[4] Henderson, Peter, *Object Oriented Specification and Design with C++*, McGraw-Hill, 1993

[5] Mitrani, I. 1982. Simulation Techniques for Discrete Event Systems, Cambridge Computer Science Texts 14, Cambridge University Press

[6] Pooley, R.J. September 1986. "Languages for discrete event simulation", Proceedings of 13th SIMULA Users Conference, Calgary, pp 69-81, Association of SIMULA Users, Postbox 4403 Torshov, N-0402 Oslo 4, Norway

[7] Pooley R.J. and G.M. Birtwistle 1986. "Process based modelling of communications protocols", in S. Schoemaker Ed. Computer Networks and Simulation III, pp 81-101, North Holland, Amsterdam

[8] Pooley, R. J. *An Introduction to Programming in SIMULA*, Blackwell Scientific Publications, 1987

[9] Pooley, R. J. *Formalising the Description of Process Based Simulation Models*, PhD Thesis, University of Edinburgh, 1995

[10] Shannon, Robert E. *Systems Simulation the Art and Science*, Prentice-Hall International, 1975

[11] Wanke, C. *Very Large Crossbar Switches in Multistage Interconnection Networks*, M.Sc. Dissertation, May 1993, Dept of Computer Science, University of Edinburgh

[12] Wanke, C. "Object-Oriented Structuring Of A Discrete Event Simulation Model", in Pooley and Zobel Eds *Proceedings of UKSS '93*, Keswick, United Kingdom Simulation Society

Note on MODES availability

Device Driver Workload Modelling through an Abstract Machine

Gaius Mulley

Department of Computer Studies, University of Glamorgan
Treforest, Mid Glamorgan, CF37 1DL, UK

Abstract

This paper shows, by example, how different classes of hardware devices and their associated drivers can be accurately modelled through a system configuration language and simple model processor. A complex network interface requiring a combination of processor polling status registers, direct memory access has been modelled. Results show that the model is 90% accurate for block sizes of 2 bytes and 100% accurate for block sizes of 8192 bytes or more.

1. Introduction

In the life cycle of a realtime system, performance analysis is often undertaken after a commitment has been make to key components. Sometimes performance analysis and benchmarking are undertaken after a product has been completed and the opportunity to alter significant aspects of design are thus denied. To compound the problem of realtime system design it is not easy to extrapolate the final performance figures from individual component specifications. Furthermore the interrelated nature of devices, executive timeouts, DMA cycle stealing, interrupt overhead, application code, device driver code means that optimizing any one entity might not deliver the required performance improvement.

A realtime system based on a multitasking executive controlling interrupt driven devices can exact a significant performance limitation on a system. Previous research[1, 2] has shown that in addition to the simple processor demand, system performance is limited by the necessary context switch and synchronisation required to support asynchronous processes. Martin[3] has shown that in a network protocol engine 20% of the available processor time was consumed by the realtime executive. During small data packet size transfer 34% of the processor time was consumed by the timeout device driver. Conversely, during large packet transfers, 68% of the processor time was consumed by the DMA when moving packets between the memory and the network.

This paper reports on performance results gained while modelling a realtime system and how the associated workload was constructed. The systems of interest are constructed from multiple uniprocessors each running a realtime executive[4] to coordinate process activity. The multiprocessor systems are assumed to be MIMD and interprocessor communication is achieved through specialist hardware devices.

There have been many techniques presented to model computer systems[5, 6, 7, 8, 9, 10, 11]. The technique adopted in this research is discrete event simulation[12] as it exploits the correspondence between the implementation of next event simulation and the multitasking executive[13]. The event generation method is a model program operating on several abstract machines termed model processing elements (MPE). Previous research has shown that an accurate workload can be automatically generated for the software components[14, 15]. The software workloads are allocated to the SMP (simple model processor) within the MPE.

2. The Design Tool and the Model Processing Element

A realtime system design tool has been built and it consists of a number of components: next event simulator, Modula-2 workload generator, SMP object code manipulation tools (as, ld, nm, size, dis), debugger and SCL. The key component interaction is shown in figure 1.

The goal is to develop, debug and predict the performance of a realtime system before it is constructed. The simulation design tool allows different solutions to be explored together with their performance implications. The user expresses most of the software model in Modula-2 which is later transformed into a SMP workload description[14]. As the design progresses the Modula-2 model becomes closer to final production code[16]. Finally the model and production code are the same and it is possible to debug the production realtime system through the simulation design tool. The discrete event simulator has many advantages when debugging the realtime system:

(i)   ability to *suspend* time while the system is examined.

(ii) ability to *rewind* time and to single step in reverse.

(iii) ability to single step all code: interrupt service routines, device drivers and application code.

(iv) set break points on hardware and software events alike such as: start of DMA activity, timer interrupt, `iret` instruction

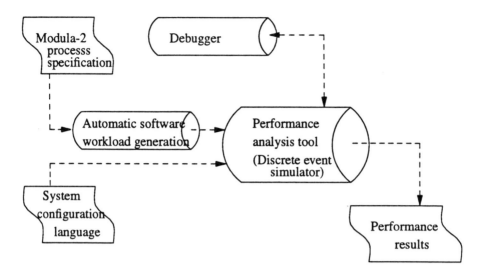

Figure 1: design tool components

Previous research has shown that accurate modelling of the all the constituent parts of the realtime system can predict results to within 4% of a production system[13]. The success of these results was due, in part, to the low level modelling of the executive behaviour. However the workload can be expressed in Modula-2 and compiler directives allow low level or high level process modelling.

One of the key entities of realtime systems is the device driver; the interaction between device control and the multitasking executive. Measurement and simulation of protocol engines[3] showed that the executive and device driver consumed up to 52% of the available Z80 processor time when transmitting basic block sizes of 2 bytes across the Cambridge Ring. A functionally equivalent but redesigned protocol engine showed that up to 70% of the available MC68000 processor time was consumed in the same two areas[17]. In modelling future systems it would seem prudent to pay particular attention to these two areas.

This paper will discuss, through examples, how the complex device driver workload can be constructed using the SCL, MPE and SMP instruction set. It reports on the accuracy of this workload model for a given range of block sizes.

3. Event Generation Through a Model Machine

Detailed modelling of a realtime system implies representing conditional behaviour of processes. Thus the process representation must allow for examination of conditional variables. Any manipulation of synchronisation primitives by a process will have a direct or indirect affect on the remaining processes. Furthermore device workload upon the processor can be closely modelled if the concept of interrupt mask register, DMA cycle stealing, processor cycles, interrupt priority and interrupt activation schema are present. In the simulation of a multiprocessor system there is one program script for each modelled processor, each of which is regarded as being executed in parallel. The interpretation of the program script is performed by a Model Processing Element (MPE). Each MPE contains a model of a processor and devices. It is intended that these models contain many of the important features found in traditional processors and devices.

3.1. Software Event Generation

It is desirable that software loads from a variety of different microprocessors can be mapped onto the processor within the MPE. The software workloads are allocated to the SMP (simple model processor) within the MPE. A detailed model of software can be automatically built to faithfully model functionality and execution cost[14, 15], alternatively, a higher level model can be produced which represents software in skeleton form (processes are briefly described in terms of fundamental interprocess communication primitives and an aggregate execution cost).

The simulator advances the current process one SMP instruction at a time. A simulation event is associated with the beginning of the execution of a SMP instruction which, will usually correspond to one or more instructions on a real processor. The events associated with the current instruction are placed onto the event queue and the next event simulation technique effectively interprets a process. A time of completion (or time of new event) must be calculated for each instruction. Figure 2 shows the relationship between the MPE, SMP and event generation.

The SMP instruction set can be categorised into executive calls wait, signal, initprocess, killprocess and initsemaphore; assignment, iteration and conditional primitives; device control such as in, out, iret, disable, enable; request for the processor pr.

The request for the processor, pr, represents an average load of a block of *real* machine code instructions whose functionality is not modelled in detail at the current refinement level.

3.2. Hardware Event Generation

The MPE must be configured with real system overheads. The MPE hardware characteristics such as processor speed, instruction speed and rate of device transmission are specified in the SCL. Within the SCL a basic unit of a device is termed a port. A port is specific to one MPE and its performance characteristics remain constant

342

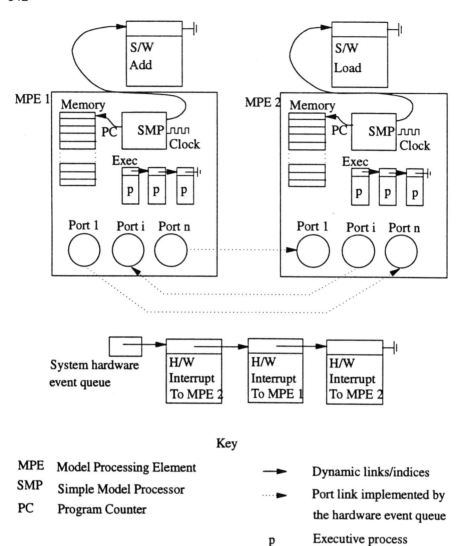

Key

MPE	Model Processing Element		→	Dynamic links/indices
SMP	Simple Model Processor		····►	Port link implemented by
PC	Program Counter			the hardware event queue
			p	Executive process

Figure 2: two MPEs communicating via ports and
the relationship to event generation

throughout a simulation experiment. This language describes a system in terms of a
collection of interconnected MPEs and each MPE is individually described in terms
of machine clock frequency, speed of processor, memory configuration and ports.

The description of a port determines whether it is a model character device, clock device or block DMA device, the amount of cycles DMA steals from the processor, duration of DMA burst, priority of DMA and the speed of operation. Ports can contain links to other ports and therefore more complex device models are created by combining these units. The EBNF description of the SCL is shown in figure 3.

The ports are accessible to the SMP by explicit input/output instructions and the processor communicates to a model device controller by sense status polling or a ready interrupt. The MPE port is a simple component, in which there are only a few processor accessible registers. None of the accessible registers have an initialization order. Erroneous device operation by the SMP invokes the simulation monitor which, at the software engineers request, may display all the current attributes of a port. Hence writing device drivers for simulation ports is easier than writing drivers for real world devices because the simulation ports are regular in operation, employ a simple set of control/status registers and the simulation provides substantial debugging. Together with debugging facilities the simulation tool offers detailed analysis of processor and DMA activity within a realtime system.

4. Modelling Devices Through the SCL

This section describes through two examples how a serial device can be modelled. Realtime systems are intricately linked with their devices and therefore reliable performance prediction can be achieved through close modelling of the hardware and software interface.

The serial device is chosen for several reasons: the device is frequently found in computer systems, the device may be modelled in detail or in an abstract form and the device requires a simple software driver. The hardware structure of the MPE and the characteristics of its components are derived directly from the design specifications, or, as in this case, manufacturers data sheets[18, 19]. A simple abstract example of a serial device could be regarded as an output port communicating to a terminal at a predefined baud rate. This example excludes modelling hardware and software handshake and the data transmission is unidirectional. The output port will issue an interrupt when the data register becomes empty. The terminal is assumed to accept all data characters without delay: the terminal, is regarded as an infinite sink. Figure 4 shows how this serial device is modelled in the SCL.

The code for this device driver is shown in figure 5. The model could be extended to include acknowledgements. Effectively one could imagine the infinite sink terminal being replaced by another machine which must consume any incoming character before another may be sent. The following assumptions are made:

(i)   The rate of transmission is 9.6k baud and therefore one character is transmitted every millisecond.

```
Configuration := " SYSTEM " " DEFINITION "
                   { MachineDefinition } " END "

MachineDefinition := " MACHINE " Cardinal
                        " CLOCK " " FREQUENCY " Cardinal
                        " MEMORY "
                        ( " PROTECTED " | " UNPROTECTED " )
                        [ ProcessorDef ] [ ClockDef ]
                        { PortDef } " END "

ProcessorDef := " PROCESSOR "
                   " CLOCK " " FREQUENCY " Cardinal " END "

ClockDef := " CLOCK " Cardinal { Cardinal } " END "

PortDef := " PORT " Cardinal
                BaseRateDef   LinksDef        PortModeDef
                PortTypeDef   [ DmaDef ]
           " END "

DmaDef        := DmaDelayDef   DurationDef
                 RateDef       CyclesDef      PriorityDef

BaseRateDef := " BASE " " RATE " Cardinal
LinksDef    := { " LINK " Cardinal " DELAY " Cardinal }
PortTypeDef := " PORT " " TYPE " ( " dma " | " nondma " )

PortModeDef := " PORT " " MODE "
                ( " read " | " write " | " neither " )

DmaDelayDef := " DMA " " DELAY " Cardinal
DurationDef := " DURATION " " OF " " DMA " Cardinal

RateDef       := " RATE " " OF " " DMA "
                 " TRANSFER " Cardinal

CyclesDef   := " CYCLES " " NEEDED " Cardinal
PriorityDef := " PRIORITY " Cardinal
```

Figure 3: EBNF description of the SCL

```
SYSTEM DEFINITION

MACHINE 1

    CLOCK FREQUENCY _
    MEMORY PROTECTED
    PROCESSOR CLOCK FREQUENCY _ END

    PORT Tx
        (* 1 msec transmission time per char *)
        BASE RATE 1000
        PORT MODE write
        PORT TYPE nondma
    END
END
```

Figure 4: serial device model in SCL

(ii)   The transmitter port issues a ready interrupt to the SMP in machine 1 when the SMP in machine 2 has read the current character from the receiver port.

(iii) The receiver port issues an interrupt to the SMP in machine 2 when a new character has arrived in the receiver data register.

In this model the performance of machine 1 is dependent upon the ability of machine 2 to consume incoming characters and device transmission speed. Figure 6 contains the necessary SCL code to model a serial device with hardware acknowledgement.

Machine 1 is configured with a transmitting port, the SMP writes a character to port Tx and initiates transmission by writing to the port control register. One millisecond later in simulation time the character arrives at port Rc on machine 2 and an interrupt is issued. The SMP on machine 2 services the interrupt by reading the received character and writing to the control register of port TxAck. Port TxAck is a model of hardware acknowledgement. When TxAck is initiated it immediately activates port RcAck which issues the ready interrupt to the SMP of machine 1. The ready interrupt on machine 1 indicates that the receiver has read the data character and so machine 1 is clear to send more data. The pseudo code for the device driver is shown in figure 7.

This section has shown how a RS232 device can be modelled in detail and with different characteristics. The RS232 device was modelled with and without hardware handshake. The SCL allows an initial abstract model of devices to be constructed which can later be refined into a very close model of the real device. It also

```
(*                                            isrTX:
   Write - char to TX device.                    saveRegs
*)                                                load _ReadyInt
                                                  signal
PROCEDURE Write (ch: CHAR) ;                      restoreRegs
BEGIN                                             iret
   (* Wait until device is ready
      to receive a character *)
   Wait(ReadyInt) ;
   (* put char in data register 1. *)
   OutPort(TX, DataReg1, ch) ;
   (* tell TX device to transmit. *)
   OutPort(TX, ControlReg, 001110B) ;
END Write ;


(*
   Init - initialization code that sets up the runtime
          characteristics of the device.
*)

PROCEDURE Init ;
BEGIN
   (* device to interrupt after each char departed *)
   OutPort(TX, ControlReg, 001010B) ;
   InitSem(ReadyInt, 1)
END Init ;
```

Figure 5: model device driver code

allows more complex devices to be built by combining SCL ports.

5. Modelling a Network Device

This section reports on how a complex device was modelled. A network protocol engine poses realtime performance constraints and thus makes a suitable case study. Additionally the design of a protocol engine is well documented[3] and the performance non intrusively monitored. These measurements can therefore be used to validate the device model.

Following the ISO 7 Layer model[20], network protocols are defined in terms of the exchange of messages between peer layers which use the services of an adjacent layer. Thus the purpose of each layer is to offer a set of services to the higher

```
SYSTEM DEFINITION              MACHINE 2

MACHINE 1                         CLOCK FREQUENCY _ END
                                  MEMORY PROTECTED
   CLOCK FREQUENCY _ END
   MEMORY PROTECTED               PORT Rc
                                     (* Interrupt when
   PORT Tx                            Rc reaches
      (* transmission                 completion *)
         time for 1                BASE RATE 0
         char = 1 msec *)          PORT MODE read
      BASE RATE 1000               PORT TYPE nondma
      (* Link to Rc             END
         1 msec after           PORT TxAck
         activation *)             BASE RATE 0
      LINK Rc DELAY 1000           LINK RcAck DELAY 0
      PORT MODE write              PORT MODE write
      PORT TYPE nondma             PORT TYPE nondma
   END                           END
   PORT RcAck                  END
      BASE RATE 0
      PORT MODE read           END
      PORT TYPE nondma
   END
END
```

Figure 6: SCL code representing a serial device with acknowledgement

layers, shielding superiors from the details of how the offered services are actually implemented. In a practical implementation of the protocol engine the messages from a given layer pass vertically down the protocol tower across the physical connection and up a similar tower to a corresponding peer layer. At the lowest level there is physical communication with another machine and only at this point are data really transferred.

The simulation model consisted of two protocol engines communicating across the Cambridge Ring network and each protocol engine modelled the minipacket and basicblock layer[21]. The Cambridge Ring was assumed to be error free and the device model was constructed after extensive analysis of the protocol engine design.

The implementation of the basicblock layer consisted of a DMA device interface controlled by four interrupt service routines. The variable length packet

```
(* Machine 1 *)                  (* Machine 2 *)
isrRcAck:                        isrRc:
   saveRegs                         saveRegs
   load _TxReadyInt                 load _RcReadyInt
   signal                           signal
   restoreRegs                      restoreRegs
   iret                             iret

PROCEDURE Write (ch: CHAR)        PROCEDURE Read (VAR ch: CHAR)
BEGIN                             BEGIN
   Wait(TxReadyInt) ;               Wait(RcReadyInt) ;
   OutPort(TX,                      (* Read character *)
           DataReg1, ch)            InPort(Rc, DataReg1, ch)
   (* Tell TX device to            (* Send flow control
      transmit. *)                     acknowledgement *)
   OutPort(TX,                      OutPort(TxAck,
           ControlReg,                      ControlReg,
           000100B) ;                       000100B) ;
END Write ;                       END Read ;

(*                               (*
   initialization code              initialization code
   for the device.               *)
*)
                                 PROCEDURE Init ;
PROCEDURE Init ;                 BEGIN
BEGIN                               OutPort(Rc,
   OutPort(RcAck,                           ControlReg,
           ControlReg,                      001010B) ;
           001010B) ;              InitSem(RcReadyInt, 0)
   InitSem(TxReadyInt, 1)        END Init ;
END Init ;
```

Figure 7: pseudo code for serial device driver with acknowledgment

presents constraints upon the software device driver. The interrupt service routine pseudo code in figure 8 shows that the receiver can only initiate DMA transfer once the length field has arrived. Whereas the transmitter must poll the minipacket status register and wait until all previous data has been accepted. The combination of these constraints means that a basic block will be transferred in four units. These units were encoded into the SCL as distinct ports which were individually configured to interface to the processor under interrupt and polling mechanisms.

The basicblock protocol essentially consists of interrupt service routines, which are invoked when the minipacket hardware needs servicing. The transmitter end of DMA interrupt occurs after DMA has transferred all the basicblock except the checksum onto the ring. However when the interrupt is issued the minipackets which have been sent may not be accepted by the receiver. Therefore it is necessary for the transmitter protocol engine to poll the status of the minipacket hardware before transmitting further data. The checksum sent interrupt occurs when the transmitter passes the checksum to the ring hardware.

The receiver software waits for a legal basicblock header to arrive, waits for the basicblock length minipacket to arrive and then configures the DMA to receive the rest of the basicblock. When the hardware minipacket layer detects a legal header packet the legal header interrupt is raised. The end of DMA interrupt occurs when the minipacket hardware layer has completed the transfer of all minipackets to form one basicblock.

5.1. Results

The network device model was verified against the theoretical basic block throughput by assuming an infinite processor speed. The results from this experiment are given in table 1 and they show that the device model is 90% accurate for block sizes of 2 bytes and 100% accurate for block sizes of 8192 bytes or more. The model only deviates from the theoretical limit when the block size is small and this is due to a small rounding error which is just noticeable with small transmission times. However, in all expected experiments with small basicblocks, the system will be processor saturated and so this error will have little effect on simulation results.

6. Conclusions

This paper has described how an abstract machine can be used to generate reliable scripts for a realtime next event simulator. It has shown how the SCL together with the abstract machine instruction set can model device driver hardware and software workloads.

The Cambridge Ring network interface is in many ways a complex device to model accurately, due to the possibility of basicblock fragmentation. However the flexibility of the abstract machine ports allowed this device to be successfully modelled in detail and the simulation results were encouraging.

In the future the SCL will be extended to include an memory management device and the design tool will be used to explore the performance of a specialist application server. The performance analysis tool will be used to investigate bottlenecks from the device hardware software interface up to the application layer. The

```
(* Transmitter *)                    (* receiver *)


PROCEDURE EndOfDMAISR                PROCEDURE HeaderArrivedISR
BEGIN                                BEGIN
    turn interrupts off                  turn interrupts off
    save registers                       save registers
    REPEAT                               retrieve message
    UNTIL all DMA'd                          header
          minipackets                    set ring hardware
          are accepted                   to receive the length
    write checksum                           minipacket
    minipacket to the                    start receiver timeout
    ring                                 REPEAT
    restore registers                    UNTIL length minipacket
    turn interrupts on                         has arrived
END EndOfDMAISR                          retrieve length
                                             minipacket
                                         set up DMA chip to
PROCEDURE ChecksumSentISR                perform the remaining
BEGIN                                    basicblock transfer ;
    turn interrupts off                  start DMA
    save registers                       restore registers
    REPEAT                               turn interrupts on
    UNTIL receiver has               END HeaderArrivedISR
          taken checksum
    cancel transmitter
       timeout                       PROCEDURE EndOfDMAISR
    set up DMA chip to               BEGIN
    transfer all next                    turn interrupts off
    basicblock except                    save registers
    the checksum ;                       set ring hardware
    start DMA                            to receive next
    start transmitter                    minipacket to build
       timeout                           the next basicblock
    restore registers                    cancel receiver
    turn interrupts on                      timeout ;
END ChecksumSentISR                      restore registers
                                         turn interrupts on
                                     END EndOfDMAISR
```

Figure 8: basicblock pseudo code

Block size	Analytical network limit	Simulated network limit
2	15875	14417
4	13229	12201
8	9922	9331
16	6615	6346
32	3969	3690
48	2835	2689
64	2205	2116
128	1167	1141
256	601	594
512	305	303
1024	154	153
2048	77	77
4096	39	38
8192	19	19

Table 1: Comparison of the theoretical network limit
and the simulation device model limit

next event simulator and associated utilities will be used during the next academic year within the realtime system teaching module.

References

1. Loader, R.J. and Martin, P.A., A High Performance Concentrator for the Cambridge Ring. In: Dallas I.N. & Spratt E.B. (ed) Ring Technology Local Area Networks, North - Holland, Amsterdam, 1984.

2. Brereton, O.P., Performance Figures for Message Passing over a Cambridge Ring, Software Practice and Experience 1982; 12(1):95-96.

3. Martin, P.A., The Exploitation of Cambridge Ring Performance Through the Design and Analysis of Intelligent Access Logic, PhD. Thesis. University of Reading, 1985.

4. Comer, D., Operating System Design: The XINU Approach, Prentice-Hall International, Englewood Cliffs, 1984.

5. Ferrari, D., Computer Systems Performance Evaluation, Prentice-Hall International, Englewood Cliffs, 1978.

6. Sherman, S., Baskett, F., and Browne, J.C., Trace driven modelling and analysis of CPU scheduling in a multiprogramming system, CACM 1972; 15(12):1063-1069.

7. Noe, J.D. and Nutt, G.J., Validation of a trace driven CDC 6400 simulation, AFIPS Spring Joint Computer Conference 1972; 40.

8. Nielsen, N.R., The simulation of time-sharing systems, CACM 1967; 10(7):397-412.

9. Nielsen, N.R., An analysis of some time-sharing techniques, Computer Systems 1971; 14(2):79-90.

10. MacDougall, M. H., Computer system simulation: an introduction, Computing Surveys 1970; 2(3):191-209.

11. Audsley, N.C., Burns, A., Richardson, M.F., and Wellings, A.J., STRESS: a Simulator for Hard Real-time Systems, Software Practice and Experience 1994; 24(6):543-564.

12. Fishman, G.S., Principles of Discrete Event Simulation, John Wiley and Sons, New York, 1978.

13. Mulley, G.P.C. and Loader, R.J., The application of a realtime systems simulation design tool to protocol engine design. In: Burkley D (ed) Procedings of the International Conference on Data Communication Technology, The National Institution for Higher Education, 1988, pp. 240-250.

14. Mulley, G.P.C, Automatic calculation of process workloads through a Modula-2 compiler, Department of Computer Studies, University of Glamorgan, Treforest, Mid Glamorgan, CF37 1DL. (to be published).

15. Park, C.Y. and Shaw, A.C., Experiments with a Program Timing Tool Based on Source-Level Timing Schema, IEEE Computer 1991; 24(5):48-57.

16. Wirth, N., Program development by stepwise refinement, CACM 1971; 14(4):221-227.

17. Mulley, G.P.C., A design tool for performance prediction of realtime systems, p. 126, PhD Thesis, University of Reading, 1989.

18. Motorola, MC68681 Dual Asynchronous Receiver/Transmitter (DUART), Motorola Semiconductors, East Kilbride, Scotland, 1985.

19. Semiconductor, National, NS16450/INS8250A/NS16C450/INS82C50A Asynchronous Communications Element, National Semiconductor Corporation U.S.A., 1985.

20. ISO, Reference Model of Open Systems Intercommunication, International Standards Organisation 1979; ISO/TC97/SG16/N227.

21. JNT, Cambridge Ring 82 Protocol Specifications, JNT, 1982.

Wide Area Networks

Reduction of Channel Idle Time in a Trunked Mobile Radio Network

S. R. Robson

National Band Three Ltd., Hedgerows Business Park, Springfield,
Essex, CM2 5PF. Tel. 0171-396-3374; Fax 0171-396-3377

M. C. Sinclair

Dept. of Electronic Systems Engineering, University of Essex,
Wivenhoe Park, Colchester, Essex, CO4 3SQ
Tel. 01206-872477; Fax 01206-872900; email: mcs@essex.ac.uk

1. Introduction

1.1 User Perception

Public Access Mobile Radio (PAMR) networks offer a wide choice of services
including selective or group voice and data calls, PABX interconnect and callback
for unattended radio units. A high grade-of-service is achieved through trunking,
coupled with the ability to queue. The user's perception of network performance is
usually obtained through highly visible parameters such as call set-up delay and
quality of speech. Inefficiencies in the management of the individual network
resources tend to remain hidden primarily because they impact on network capacity
rather than performance. Such a network will expand in accordance with growth
in its user base. To release more capacity we must reduce the time that valuable
resources remain idle once allocated. In this paper we will address the idle time of
the radio traffic channel resource.

1.2 Channel Resource Allocation

To establish a communications path between two distant mobile radio units
requires two radio channels, one at each end of the link. If neither party has to
queue for a radio channel, both will be allocated at the same time. If on the other
hand at least one party has to queue, there will inevitably be a differential queueing
delay introduced. That is, there will be two different queueing times with the
consequence that one channel will be allocated and remain idle until the other
becomes available. The amount of channel idle time generated will be seen to be

dependent on the traffic level, the number of radio channels in the resource pool at each end of the link and the radio channel resource queue type.

To obtain an exact expression for channel idle time requires complex analytical work, thus we will use an approximate method of analysis here. Also, using a purpose-built simulation tool we shall observe the level of channel idle time as a function of number of channels and traffic intensity and compare the performance using both the standard FCFS (first-come-first-served) queueing discipline and a queue-jumping discipline designed to reduce channel idle time. We shall consider the benefits to the user, in terms of grade-of-service, and to the network, as it affects channel utilization efficiency. For the purpose of the analysis given here the grade-of-service is defined as the percentage of users who queue for longer than 20 seconds.

2. Queueing for the Radio Channel Resource

2.1 Queueing Model

We will compare the two queueing systems - $M/M/c/\infty$ (or $M/M/c$ for short - i.e. c radio channels in the pool with unlimited queue size) and $M/M/c/2c$ (c radio channels in the pool and queue size limited to c) as used on a real network. Demands for the radio traffic channel resource by either a calling or called party (or both for a single site, i.e. *local* call, requiring only one resource) are assumed to have negative exponential inter-arrival times, in which case the probability that a new call arrival will have to queue is given by the Erlang C formula for the $M/M/c$ queue [1] as

$$P_{q\infty} = \frac{A^c}{A^c + c!\left(1 - \dfrac{A}{c}\right)\displaystyle\sum_{n=0}^{c-1}\dfrac{A^n}{n!}} \tag{1}$$

where, A = traffic intensity = λH , λ = mean call arrival rate
H = mean call duration (holding time)

For the $M/M/c/2c$ queue (a special case of the $M/M/c/k$ queueing system) the probability a call will have to queue is given as

$$P_{q2c} = \frac{A^c\left(1 - \dfrac{A^c}{c^c}\right)}{A^c\left(1 - \dfrac{A^{c+1}}{c^{c+1}}\right) + c!\left(1 - \dfrac{A}{c}\right)\displaystyle\sum_{n=0}^{c-1}\dfrac{A^n}{n!}} \tag{2}$$

(see, for example, [2]). These two queueing probability distributions differ as shown in Figures 1, 2 and 3.

3. Determination of Channel Idle Time

3.1 Calls Generating Channel Idle Time

Now consider calls made between two base stations X and Y in a large network, each having a separate pool of radio channels. We will assume that the probability of queueing at either base station is the same (equal to P_q), and that the traffic is uniformly distributed throughout the network such that, of all the calls that queue at both X and Y, half have a wait time which dominates at X and half at Y.

For calls between X and Y there are three categories where channel idle time is generated:

(a) Calls queued at Y but not at X generate idle time at X
(b) Calls queued at X but not at Y generate idle time at Y
(c) Calls queued at X and Y generate idle time either at X or Y, depending on which end has the non-dominant waiting time in each case

It could be argued that the idle time in the first two cases is inconsequential since it is generated at base stations where there is no queueing, hence no shortage of the radio channel resource when the call arrives. Whilst the allocated channel remains idle, however, a queue could develop if the pool of channels becomes depleted due to a sudden increase in the level of traffic, in which case all such channel idle time would add to the performance degradation of the other queued calls. We shall see, however, that it is only through the existence of the third category (c) that a reduction in channel idle time is possible.

At base station X the proportion of calls in category (a)

$$= P_q(1 - P_q)$$

and calls in category (c)

$$= \frac{P_q^2}{2}$$

giving a total probability of generating idle time at X

$$= P_q - \frac{P_q^2}{2} \tag{3}$$

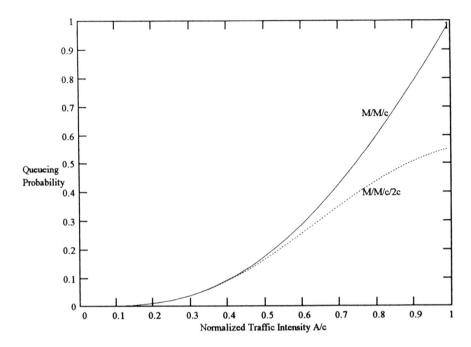

Figure 1: Probability of Queueing vs Loading: c=4

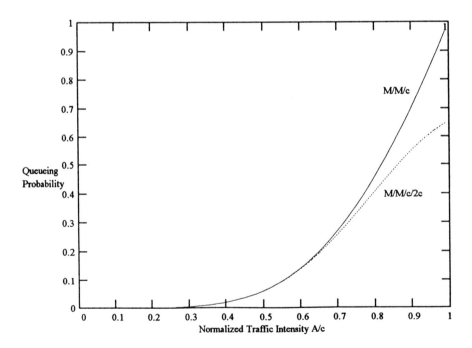

Figure 2: Probability of Queueing vs Loading: c=8

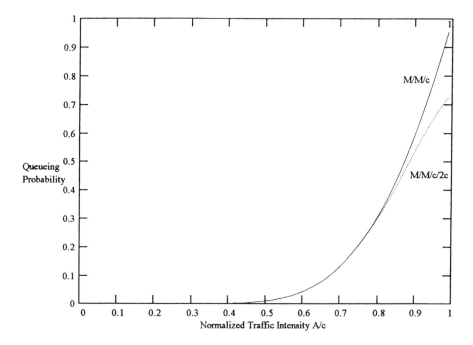

Figure 3: Probability of Queueing vs Loading: c=16

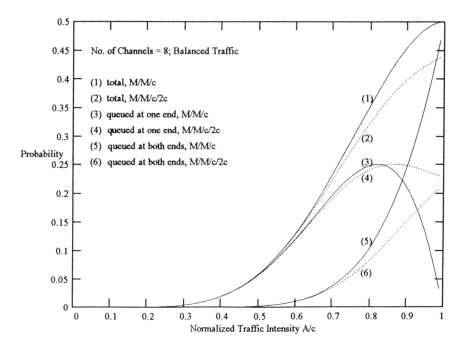

Figure 4: Probability of Generating Idle Time vs Loading: Balanced Traffic

As the level of normalized traffic approaches unity, $P_{q\infty}$ tends to 1.0 whereas P_{q2c} tends to some value between 0.55 and 0.75 (over the range $c = 4$ to 16 - as seen in Figures 1, 2 and 3). If we plot $P_q(1 - P_q)$, $P_q^2 / 2$ and $P_q - P_q^2 / 2$ against traffic intensity (Figure 4), we see calls in categories (a) and (b) falling off towards zero at high traffic levels only for the $M/M/c$ case. Those in category (c) reach 0.5 for the $M/M/c$ queue but a value well below that for the $M/M/c/2c$ case.

3.2 Queue Jumping

In [3] we considered bypassing the strict FCFS queue discipline by introducing queue-jumping as a means of reducing the unusable channel idle time. A party is allowed to jump the queue once the radio channel has been allocated at the other end of the call. We showed that the maximum achievable reduction in waiting time in a single queue jump was equal to the amount of queueing time that could be re-distributed amongst the displaced parties with non-dominant waiting times.

Clearly, for those calls queued only at one end, the single wait time will always be dominant. Any overall reduction in waiting time in the queue can only come about to the extent there are non-dominant wait times in the queues. If queued calls fall only in categories (a) and (b) but not (c) then no overall reduction in unusable channel idle time is possible. In practice, such a situation is unlikely under balanced loading conditions, but will become more likely when the load becomes unbalanced, i.e. some base stations with heavy queueing.

3.3 Waiting Time Distribution

For the $M/M/c$ queue the waiting time distribution is well documented and given in [2] as

$$W_{q\infty}(t) = 1 - P_{q\infty}e^{\frac{-(c-A)t}{H}} \qquad (4)$$

where $W_{q\infty}(t)$ is the probability the waiting time $\leq t$

For a given level of traffic the mean waiting time $\bar{t}_\infty$ for queued calls is found from

$$\bar{t}_\infty = \int_0^\infty e^{\frac{-(c-A)t}{H}} dt = \frac{H}{c - A} \qquad (5)$$

For the $M/M/c/2c$ queue the mean waiting time is also obtained from [2] as

$$\overline{t}_{2c} = \frac{A^c p_0 \dfrac{A}{c}}{\lambda_a c! \left(1 - \dfrac{A}{c}\right)^2} \left[1 - \left(\dfrac{A}{c}\right)^{c+1} - (c+1)\left(\dfrac{A}{c}\right)^c \left(1 - \dfrac{A}{c}\right)\right] \qquad (6)$$

where

$$p_0 = \left[\sum_{n=0}^{c} \frac{A^n}{n!} + \frac{A^c}{c!} \sum_{n=1}^{c} \left(\frac{A}{c}\right)^n\right]^{-1}$$

$$\lambda_a = \lambda(1 - p_{2c})$$

and

$$p_{2c} = \frac{A^c}{c!}\left(\frac{A}{c}\right)^c p_0$$

3.4 Derivation of Channel Idle Time

We can now estimate the channel idle time for both queue types. For a given level of traffic the mean channel idle time for base station X in the above example is the product of the probability a call will generate idle time and the mean differential queueing delay. Since we have assumed P_q is the same at base stations X and Y (i.e. same level of traffic and number of channels) we conclude that for the $M/M/c$ case the distribution of the waiting time in the queues at X and Y will be given by Eqn. (4) and both will have the same mean as given by Eqn. (5). The mean of the differential queueing delay for calls in all three categories will also be given by Eqn. (5). Thus the mean channel idle time per call, $\overline{t}_i$, will be given as

$$\overline{t}_i = \overline{t}_\infty \left(P_{q\infty} - \frac{P_{q\infty}^2}{2}\right) \qquad (7)$$

This is converted to fractional channel idle time and plotted against normalized traffic intensity in Figure 5. The expression in Eqn. (7) is only an approximation since the queues do not have complete independence.

By the same argument we can obtain an estimate of the channel idle time for the $M/M/c/2c$ queue although the assumption that the mean differential queueing delay is equal to the mean waiting time in the queue is less valid since the mean waiting time given in Eqn. (6) is not for an exponential distribution. Estimated channel idle time for the $M/M/c/2c$ case is plotted in Figure 6.

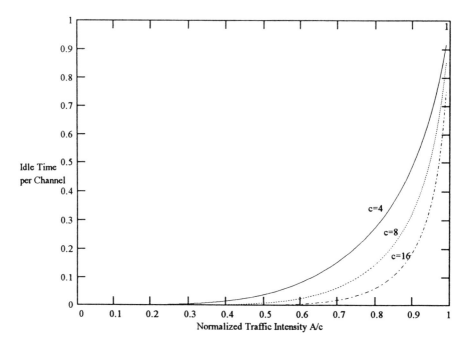

Figure 5: Calculated Idle Time per Channel vs. Traffic Intensity; M/M/c

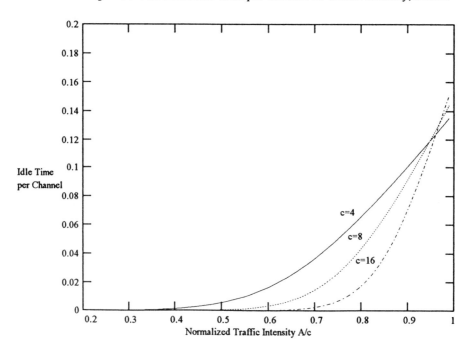

Figure 6: Calculated Idle Time per Channel vs. Traffic Intensity; M/M/c/2c

Although the plots appear to indicate that channel idle time increases as the number of channels decreases, it is important to realize that the normalized traffic intensity for $c = 16$ will be approximately double that for $c = 4$ at constant grade-of-service. Thus the idle time per channel turns out to be greater, the higher the value of c. Figures 5 and 6 also indicate that generated idle time is greater for the $M/M/c$ queueing system than for the $M/M/c/2c$, with channels becoming 100% idle as normalized traffic intensity approaches unity. This is due to the fact that the number in the queue grows rapidly and the queueing delay tends to infinity for the $M/M/c$ queue at this level of traffic (see Eqn (5)). For the $M/M/c/2c$ case the finite queue size limits the queueing delay but results in a growing proportion of blocked calls as the level of traffic increases.

3.5 Simulation Tool

A purpose built discrete event simulation tool, implemented in C++ by the second author, enabling a network to be modelled with adjustable parameters such as number of base stations, number of channels, traffic level, holding time distribution and queue type. A queue-jumping algorithm was introduced, which enabled a party to jump ahead of other queued parties as soon as a channel had been allocated at the other end of the call. This could be switched on and off as required.

As a means of program validation a number of simulation results for the $M/M/c$ and $M/M/c/2c$ queues were compared to analytical results and showed excellent agreement. A network of 8 base stations with identical parameters was subsequently modelled. Local to cross-network traffic ratio and mean call duration were set to correspond to those encountered on NB3 (National Band Three), a typical PAMR network. All base stations used the $M/G/c/2c$ queue type with traffic level varied from 0.9 to 1.2, where 1.0 signified a 2% grade-of-service for the corresponding (i.e. identically loaded) $M/M/c$ queue.

10 runs of 35,000 calls were made at different levels of balanced traffic for 4, 8 and 16 channel base stations with and without queue-jumping. The base station parameters and distribution of traffic were then modified to represent a section of the NB3 network. 10 runs of 100,000 calls were made at each level of unbalanced traffic over the range 0.95 to 1.20. (In this scenario some base stations would be operating at above-average levels of traffic and others at below-average levels.)

During the simulation runs, a number of statistics were calculated, enabling channel idle time, grade-of-service, queueing time CDF (cumulative distribution function) and differential queueing delay to be plotted as functions of traffic intensity (Figures 7 to 14). Figures 7 to 10 contain error bars corresponding to 90% confidence level.

3.6 Simulation Results

Figures 7 and 8 show channel idle time, measured in Erlangs, i.e. proportion of time each channel is allocated but idle. Over the traffic range covered, between 1% and 11% of the channel is idle and unusable in the balanced traffic case. It is clear from Figure 7 that provided the traffic intensity increases from a level that corresponds to a reference grade-of-service, idle time increases with number of channels. The effect of the queue-jumping algorithm is a very slight improvement at $c = 4$, but the channel idle time is almost halved when $c = 16$ at a traffic level of 1.20. This is because channel utilization is approaching 1.0 much more quickly when $c = 16$ than when $c = 4$. In a typical network the number of channels per base station would tend to be much higher than 4 in order to obtain trunking efficiency. Comparison of Figure 7 with Figure 6 shows that the estimated figures are the right order of magnitude. Figure 8 shows that in the case of typical network loading, channel idle time increases rapidly (to around 7.5%) and may be reduced substantially by the queue-jumping algorithm (by 37% at a traffic loading of 1.2).

Figure 9 reveals a sharp increase in grade-of-service for large values of c. This is again due to channel utilization approaching 1.0 more rapidly when $c = 16$. However, with queue-jumping the high grade-of-service at $c = 16$ and traffic loading = 1.1 is virtually halved. With only 4 channels the improvement appears small although it must be pointed out that the horizontal distance between the two plots gives an indication of how much extra capacity can be obtained whilst maintaining grade-of-service. As the plot itself becomes horizontal (as in the $c = 4$ case) so the horizontal separation between the "jump" and "no jump" plots can still be quite large. In Figure 10 the grade-of-service is rising rapidly. With queue-jumping it is reduced by as much as 50% at high traffic levels. To maintain a 5% grade-of-service, traffic could be increased by about 2.3%; at 10% grade-of-service a 4.3% increase is possible.

Figures 11 and 12 are plots of queueing delay CDF at different traffic levels. It is clear that the area between the curve and the upper horizontal axis, which represents the proportion of queued calls, is reduced by queue-jumping. However, the fact that each pair of curves intersect, indicates there is a small proportion of users who will have increased wait times.

The differential queueing delay plotted in Figures 13 and 14 shows a very consistent trend. In every case the spread is reduced by queue-jumping and the number of calls with zero differential delay (the peak value in each case) is increased. The extent of the redistribution is increasing with the traffic load and with number of channels as borne out by Figure 13 where all base stations have 16 channels.

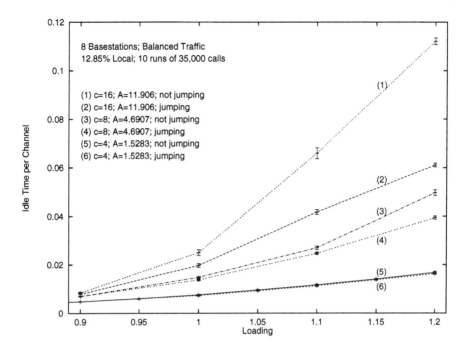

Figure 7: Idle Time per Channel vs. Loading: Balanced Traffic

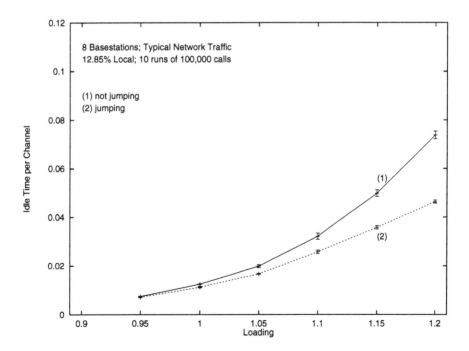

Figure 8: Idle Time per Channel vs. Loading: Typical Network Traffic

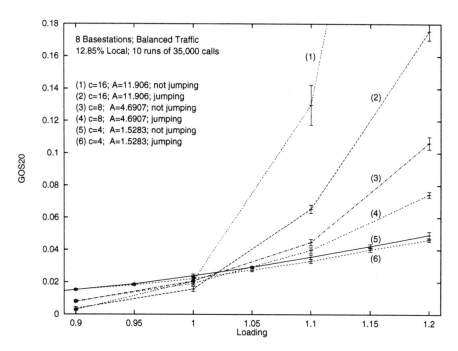

Figure 9: GOS20 vs. Loading: Balanced Traffic

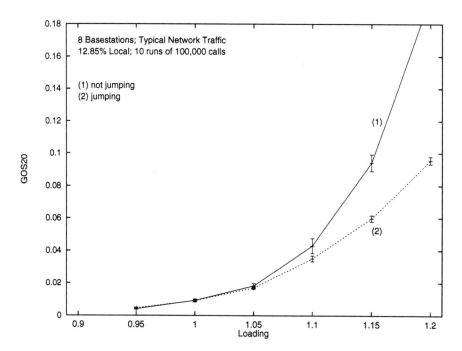

Figure 10: GOS20 vs. Loading: Typical Network Traffic

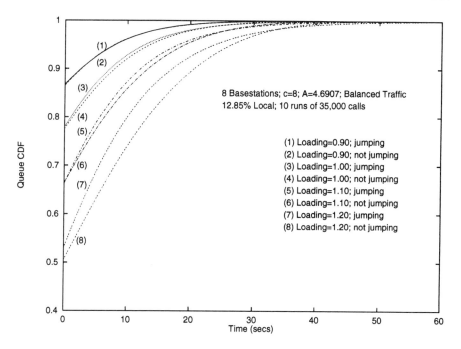

Figure 11: Queue CDF: Balanced Traffic

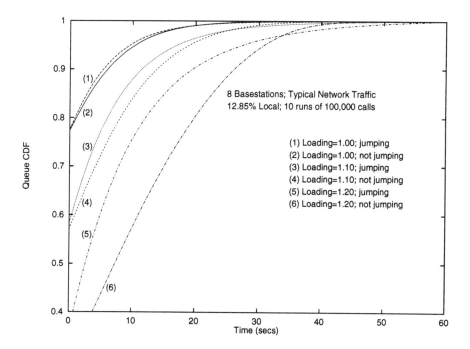

Figure 12: Queue CDF: Typical Network Traffic

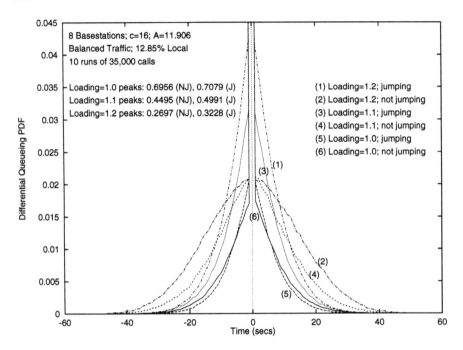

Figure 13: Differential Queueing PDF: Balanced Traffic

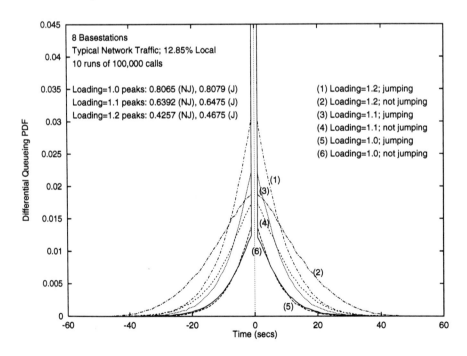

Figure 14: Differential Queueing PDF: Typical Network Traffic

4. Conclusions

We have seen that if the radio channels in a mobile radio network are allocated on a first-come-first-served basis that as much as 10% or more of the scarce radio channel resource can become unusable due to the differential queueing delay of queued calls. A reduction in channel idle time is possible only to the extent calls are queued at both ends. When $M/M/c$ queues are used a higher proportion of calls are queued at both ends than with $M/M/c/2c$ queues. However, due to the longer queueing delays in the case of the $M/M/c$ queue, the channel idle time, as calculated by an approximate method, is much greater than that generated by the $M/M/c/2c$ type and can approach 100% at very high traffic levels. Discrete event simulation has been used to model an 8-base station network with convincing evidence to support the operation of a simple queue-jumping algorithm which has been used to reduce the channel idle time by as much as 30-50%. Results indicate improved grade-of-service brought about by the reduction in differential queueing delay.

5. Further Work

There is scope for further comparison of the $M/G/c$ and $M/G/c/2c$ queue types using the simulation tool. The optimum threshold for queue jumping has yet to be determined and the effects of unbalanced traffic loads is still under investigation.

6. Acknowledgements

The research described in this paper was initially undertaken by the first author (supervised by the second) as part of an M.Sc. in Telecommunications and Information Systems at Essex University, sponsored by National Band Three Ltd.

7. References

1. Erlang A. K., Solution of Some Problems in the Theory of Probabilities of Significance in Automatic Telephone Exchanges, P.O Elect. Engineers J. 10 1917-1918, pp.189-197

2. Allen A.O., Probability, Statistics, and Queueing Theory, Academic Press, Orlando, Florida, 1978

3. Robson, S.R., & Sinclair, M.C., An Improved Channel Resource Queueing Algorithm for Trunked Mobile Radio, Proc. Twelfth UK Teletraffic Symposium, Old Windsor, Mar 95, pp.17/1-17/10

A Comparative Study of k-Shortest Path Algorithms

A. W. Brander

M. C. Sinclair

Dept. of Electronic Systems Engineering, University of Essex,
Wivenhoe Park, Colchester, Essex C04 3SQ.
Tel: 01206-872477; Fax: 01206-872900; email: mcs@essex.ac.uk

Abstract

Efficient management of networks requires that the shortest route from one point (node) to another is known; this is termed the shortest path. It is often necessary to be able to determine alternative routes through the network, in case any part of the shortest path is damaged or busy. The k-shortest paths represent an ordered list of the alternative routes available. Four algorithms were selected for more detailed study from over seventy papers written on this subject since the 1950's. These four were implemented in the 'C' programming language and, on the basis of the results, an assessment was made of their relative performance.

1   The Background

The shortest path through a network is the least cost route from a given node to another given node, and this path will usually be the preferred route between those two nodes. When the shortest path between two nodes is not available for some reason, it is necessary to determine the second shortest path. If this too is not available, a third path may be needed. The series of paths thus derived are known collectively as the k-shortest paths (KSP), and represent the first, second, third, ..., kth paths typically of least length from one node to another. In obtaining the KSPs, it is normally necessary to determine independently the shortest path ($k = 1$) between the two given nodes before computation of the remaining $k - 1$ shortest paths can be carried out.

Computation of shortest paths has been the subject of much research since the 1950s. It is important to realise that the term 'shortest' does not just apply to the distance between two nodes, but can involve any single component made up of one or more factors, including cost, safety or time, that put a weighting on the route. KSP algorithms are thus widely used in the fields of telecommunications, operations research, computer science and transportation science.

2   The Problem

The work presented here was driven by the desire to find a faster algorithm to calculate the KSPs between nodes in a network than that, Yen [1], used to date by the second author. It was realised that Lawler [2] had investigated this problem and improved on Yen [1], and this raised the question of whether there were even faster methods.

3 Definitions

In order to follow the solutions for the KSP problem, it is necessary to understand the basic ideas associated with many areas of modern mathematics, in particular graph theory. The following definitions explain some of the terms used in this paper.

The network is represented as a graph $G = (V, E)$ where V is a finite set of n *nodes* (or vertices) $V(G)$ and E is a finite set of m *edges* (*i.e.* links or arcs) $E(G)$ that connect the nodes. An edge is often represented as $e_p(i, j)$ with a certain weight, where i and j are the nodes at the endpoints of the edge e_p. A simple graph representation of a network is shown in Figure 1, where the nodes are shown as numbered discs and the edges as lines linking the nodes. A graph is said to be *connected* if there is a path between every pair of nodes in the network.

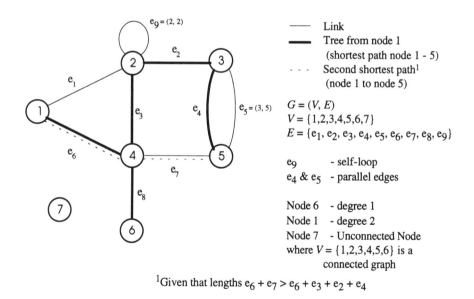

Given that lengths $e_6 + e_7 > e_6 + e_3 + e_2 + e_4$

Figure 1: Network drawn as a graph

The terms *path* and *route* may be defined as the sequence of nodes and edges that need to be alternately followed in order to move from a start node to an end node. An *undirected* network is one where all edges in the network are bi-directional, such that travel is possible along the edge in both directions, and generally $e_p(i, j) = e_p(j, i)$. It is possible that there may be two edges between some pairs of nodes and these edges are referred to as *parallel* (or multiple) edges. A *self-loop* is an edge that connects a node to itself.

The *hop count* is a count of the number of edges (or hops) in a path. An *elementary path* or *simple path* is a path containing no *loops*, such that each node is visited at most once and the maximum path hop count possible is given by the total number of nodes minus one (*i.e.* $n-1$). The weight associated with an edge may be positive or negative, but in general loops of overall negative length may not exist, as the tendency would be for the algorithms to go round the loop indefinitely, to minimise the total length of the path.

In order to calculate the KSPs it is necessary to force the algorithms to choose different routes through the network. This is conveniently done by *marking* the nodes and edges as not being allowed in further path calculations (between the two specified nodes). This operation is often referred to as *removal of a node* and *removal of an edge*, respectively. Removal of an edge e_p from a graph G is simply the operation that results in the subgraph consisting of all the nodes and all the edges in G except the removed edge e_p. Removal of a node i from a graph G is the operation that results in the subgraph consisting of all the nodes and edges in G except node i and those edges incident on node i. Edge marking is implemented by setting edges to have infinite length in some algorithms. This means that they will never be selected to form a path (or if they are, all valid paths have already been found); a path of infinite length is useless.

In many algorithms it is necessary to split a path into two sections in order to calculate other paths; these are known as the *root* and *spur* respectively. The root path is often taken from a section of a previously found [k-]shortest path and a new spur found by calculation.

Depending on the weights assigned to the edges in a network, it is possible that the graph of that network will not obey the rules of *Euclidean space*. Euclidean space requires that one side of a triangle is smaller than (or at the extreme, equal to) the sum of the other two sides. This means that the shortest route is also the most direct route. However the weights applied to the edges are not necessarily just concerned with the distance and therefore may not obey these rules. This is an important point when mentally visualising the expected operation of a KSP algorithm.

The *binary minimum heap* [3] is a convenient method for storing data that is entered in a random order and output in an ordered fashion, and it is used here to store a record of the paths awaiting selection as the next shortest. This heap has at most two (binary) children for each element and the top element (1) has the smallest (minimum) value, and thus here is the next shortest path. It is known that *Fibonacci heaps* [3] provide a more efficient means of data storage, but for simplicity the binary minimum heap has been retained.

4 Assumptions

In order to limit the search for appropriate KSP algorithms, it proved necessary to make some assumptions about the networks that the algorithms would be applied to. Given that our interest was primarily telecommunications networks, the following assumptions were made: networks are undirected, with no parallel edges, no self-loops and no negative edges (although the latter are permitted by some algorithms). Further it was assumed that we are only interested in finding elementary paths in the network.

Algorithms may calculate the KSPs from one node to another node (1–1), from one node to all nodes (1–all) or vice versa (all–1), or from all nodes to all nodes (all–all). In telecommunications we are interested in all of these, but of primary interest is the single node to node calculation and it is this that we shall concentrate on.

5 Literature Survey

A literature survey was conducted to find previous work on KSPs, revealing some 70 papers from many different branches of modern mathematics. Four algorithms [1, 2, 4, 5] from among these were chosen for implementation and speed comparison.

6 Algorithm Selection

The criteria for algorithm selection was based purely on the expected speed of operation of the algorithm in a network meeting the assumptions above. A strong indicator used during the literature survey to provide an insight into algorithm speed, was an algorithm's computational complexity. However, this could only be used as a rough guide, as our interest was in average (rather than worst case) performance on 'real' networks, and actual implementation would thus be the only way to determine the best algorithm.

The two original algorithms, Yen [1] and Lawler [2] were implemented to provide a reference to the expected speed and improvement available. Katoh [4] was included as it represented a comparatively recent update and modification to Yen [1]. The fourth algorithm, Hoffman [5], was implemented after further study, as it was felt that it had the potential to outperform the other algorithms.

Yen [1] is known to have a computational complexity of $O(kn^3)$, where $O(n^2)$ is due to the shortest-path calculation. Lawler [2] improves this by a constant factor, but the computational complexity remains the same. Katoh [4] (approximately) and Hoffman [5] both have a computational complexity of $O(kn^2)$, but use widely different means to derive the KSPs[1].

---

[1] Katoh himself [4] claims $O(kc(n, m)) \leq O(k \min[n^2, m \log n])$

7 Implementation

The algorithms were coded in 'C' on the department's UNIX workstations. Steps were taken to ensure that the coding of all the algorithms was done as efficiently as possible and such that none of the algorithms were particularly disadvantaged with respect to another. For simplicity, the network was represented as an adjacency matrix; arrays and pointers were used extensively; and particular care was taken in the management of dynamic memory.

7.1 Yen

The shortest path (containing $p \leq n$ nodes) is found using a standard shortest-path algorithm (*e.g.* Dijkstra's [6]) and placed in the results list (Yen's list A). Yen [1] takes every node in the shortest path, except the terminating node and calculates another shortest path (spur) from each selected node to the terminating node. For each such node, the path from the start node to the current node is the root path. Two restrictions are placed on the spur path: 1) It must not pass through any node on the root path (*i.e.* loopless) and 2) It must not branch from the current node on any edge used by a previously found [k-]shortest path with the same root. Node and edge marking is used to prevent the spur paths from looping or simply following the route of a previous [k-]shortest path. If a new spur path is found it is appended to the root path for that node, to form a complete path from start to end node, which is then a candidate for the next KSP. All such paths are stored (Yen's list B) and the shortest remaining unselected path is selected as the next KSP, and transferred to the results list (Yen's list A). The same process is repeated, calculating a spur path from each node in each new KSP, until the required number of KSP have been found.

Various improvements to Yen's original algorithm have been commented on in papers over the years. The most significant improvement can be made with the use of a heap to store Yen's list B, giving an improved performance although the overall computational complexity does not change. A further improvement is checking for non-existent spur paths (*i.e.* where the root path exists, but all spur paths from that root have been used previously).

If list B contains enough shortest paths all of the same next minimum length, to produce all the required paths, it is not necessary to extract each path in turn and perform the above calculations; only to extract the required number of paths and place them directly in list A, as no other shorter paths will be found.

7.2 Lawler

Lawler [2] presents a modification to Yen's algorithm, such that rather than calculating and then discarding any duplicate paths, they are simply never calculated.

The extra paths occur due to the calculation of spur paths from nodes in the root of the KSPs. Lawler points out that it is only necessary to calculate new paths from nodes that were on the spur of the previous KSP. Consider a node on the root of a path; Yen calculates a spur path from this node every time a new KSP is required. The path calculated each time will be the same provided that no extra edge marking has taken place at this node. Extra edge marking will only take place if a KSP has been chosen that branches from this node *i.e.* this node is on the spur. When one of these paths becomes a KSP, then all paths from the root of that KSP will have already been calculated and stored in the heap of prospective KSPs (Yen's list B).

It is therefore necessary to keep a record of the node where each path branched from its parent. This node marks the point from where the calculation of spur paths starts. The increased efficiency of Lawler can be explained therefore as, when finding more than two KSPs, the next KSP will on average branch from the middle of the path and so approximately a 50% improvement in speed is achieved over Yen.

7.3 Katoh

This algorithm [4] is claimed to be faster than Yen due to the way that the previous [k-]shortest path is broken down for the calculation of the next KSP. This path is broken into three sections rather than the $(p - 1 \leq n - 1)$ sections for Yen. Two shortest-path trees are calculated for each section, one from the start node and another to the end node.

The three sections used in the algorithm are derived from the previous KSP by recording (exactly as in Lawler's algorithm) the point that each path branches from its parent. The sections are: 1) all nodes after the branch, 2) the branch node and 3) all nodes before the branch (defined as P_a, P_b and P_c respectively in the paper). The algorithm then calculates one path from each of these sections. Restrictions are placed on this path such that it must deviate from its parent at some point and must not follow other previous KSPs. Each path is calculated by generating two shortest path trees: from the specially calculated start node in each section and from the given end node. The possible shortest paths are represented by paths across these two trees, where each path goes between the trees via a common node or a single edge.

7.4 Hoffman

Hoffman's algorithm [5] requires that the shortest path between the two nodes has been found and that the shortest-path tree from all nodes to the terminal node is known.

In Yen and Lawler, the spur paths are calculated from each node $O(n)$ on the previous [k-]shortest path in turn, using a shortest-path algorithm $O(n^2)$ with node marking, making a total time of $O(n^3)$. In contrast, Hoffman calculates the shortest-path tree from all nodes to the end node $O(n^2)$ at the start of execution. Then, to find the next shortest path, it is only necessary to search from each node $O(n)$ on the previous KSP spur to every other node $O(n)$, making a total time of $O(n^2)$. Each path is made up of three sections: 1) the start node to the selected (*i.e.* branching) node, 2) the edge from the selected node to the new node and 3) the branch of the shortest-path tree from the new node to the terminating node. Edge marking does not need to be used, as edges can be ignored when searching through those from the selected node. The shortest path from each selected node is placed in the heap. The next KSP is then the path with the shortest total length from the start node along the root, via the edge and along the tree to the end node. As the shortest-path tree is known beforehand, it is only a matter of searching at most $n \times n$ edges to determine each KSP.

An important point to note in this path generation, is that the paths evolve by the addition of exactly one edge to the existing root and spur tree paths. This gives rise to a complication, that is automatically excluded by Yen's and Lawler's algorithms, such that it is possible (and quite likely) that looping paths will be generated. These paths are essential to the operation of the algorithm and must be kept like any other previous or prospective KSP. The looping path obtained will follow a previous KSP, but with the addition of a loop making the path slightly longer. The importance is that elementary paths may evolve from the looping section, again by the addition of one edge. The way looping paths are generated is determined by the geometry of the network and whether it obeys the rules of Euclidean geometry. The generation of looping paths adds an extra overhead to this algorithm, which is not easily quantifiable, but will depend on the network features.

A further speed improvement is available with this algorithm, as it is not necessary to calculate the full path details (*i.e.* creating the route) until the path is finally selected as a KSP and removed from the heap. Additionally, the same shortest-path tree to the terminal node can be used for all–1 KSP calculations.

8   Results

In determining the performance of the algorithms, it is useful to summarise their main characteristics. The majority of time taken to calculate the KSPs occurs in the call to a shortest-path algorithm. This algorithm is used not only to find the first KSP (*i.e.* the shortest path), but also several times to calculate each KSP ($k > 1$), except in Hoffman. Katoh generates a maximum of three prospective KSPs from any previous [k-]shortest path, while Yen generates a number of paths according to the path length (*i.e.* the number of hops). Lawler generates some smaller number (approximately 50%) than Yen, depending where the path branched from its parent. For any realistically sized network therefore, Katoh will normally generate significantly fewer paths than either Yen or Lawler. However, in comparison Hoffman only generates one shortest-path tree for all KSP.

The number of paths calculated is not the only concern; the method of calculation and its speed is also important. Both Yen and Lawler calculate the shortest path, while Katoh calculates shortest-path trees. It can be shown that on average a shortest-path tree takes twice as long to calculate as a single shortest path. In order for Katoh to be faster than another algorithm, that algorithm must make at least twelve $(2 \times 2 \times 3)$ calls to the shortest-path algorithm for each KSP calculated (although Katoh may not always calculate P_a or P_c). This means that for Yen's algorithm the path lengths must be an average of at least 13 nodes long (12 hops), but for Lawler's algorithm the spur section of the paths must be at least 13 nodes long, and hence on average the path must be about 25 nodes long. For networks to have average path lengths this long, it is necessary for the network to be very large and have a low connectivity. Path hops this long are unlikely to exist in general telecommunications networks and so Katoh does not present a viable algorithm. A typical comparison of path (and tree) calculations is shown in Figure 2.

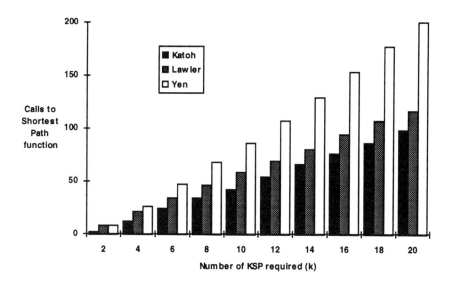

Figure 2: Number of KSP *shortest path* function calls made

However, while this is a better indication of the performance than the computational complexity, it does not demonstrate the required proof of overall best speed. This can only be achieved by comparing the run-times of the algorithms in tests on real networks. The results of one such series of runs on the COST 239 European Optical Network [7] are shown in Figure 3.

This figure demonstrates a staggering improvement in the performance of the algorithms, between Katoh and Hoffman; and yet Katoh is the later, more complicated algorithm. The improvement from Yen to Lawler is clearly indicated.

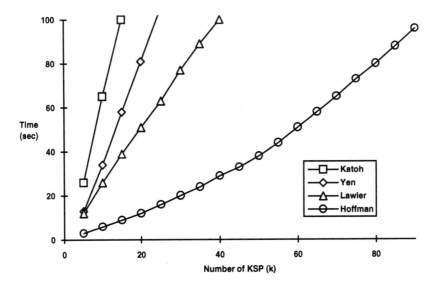

Figure 3: Times for the European Optical Network from London to all nodes

9 Conclusions and Further Work

Lawler's algorithm [2] has been shown to provide the expected speed improvement over Yen's algorithm [1].

However, the best overall speed for the network tested was achieved using Hoffman's algorithm [5]. The results indicate that the performance advantage of Hoffman's algorithm may decay with problem size, but this will require further investigation.

The KSP algorithms studied are very dependent on the efficient implementation of the shortest-path [tree] algorithm and any improvements to this may alter the performance balance.

In general the telecommunications networks of interest to us will be [well] connected. However, the algorithms will also work on unconnected networks, but their efficiency will be improved if the subnetworks are analysed separately.

With algorithms available from so many different branches of mathematics, it is difficult to compare them all without implementing them and as such, the following papers present alternative techniques from other fields that may yet perform faster than Hoffman's algorithm: Shier [8], Carraresi [9], and Boffey [10]. However, these do not all meet the restrictions of loopless and non-disjoint paths, so extra precautions will have to be taken.

10 Acknowledgements

The research that is described in this paper was undertaken by the first author (supervised by the second) as part of an M.Sc. in Telecommunications and Information Systems at the Department of Electronics Systems Engineering, University of Essex. The first author was supported by an EPSRC Advanced Studentship.

References

[1] Yen JY. Finding the k shortest loopless paths in a network. Management Science 1971; 17:712–716

[2] Lawler EL. A procedure for computing the k best solutions to discrete optimisation problems and its application to the shortest path problem. Management Science, Theory Series 1972; 18:401–405

[3] Ahuja RK, Magnanti TL, Orlin JB. Network Flows: Theory, Algorithms, and Applications. Prentice Hall, New Jersey, 1993

[4] Katoh N, Ibaraki T, Mine H. An efficient algorithm for k shortest simple paths. Networks 1982; 12:411–427

[5] Hoffman W, Pavley R. A method for the solution of the nth best path problem. Journal of the Association for Computing Machinery (ACM) 1959; 6:506–514

[6] Dijkstra EW. A note on two problems in connexion with graphs. Numerische Mathematik 1959; 1:269–271

[7] O'Mahony MJ, Sinclair MC, Mikac B. Ultra-high capacity optical transmission network: European research project COST 239. Information, Telecommunications, Automata Journal 1993; 12:33-45

[8] Shier DR. Iterative methods for determining the k shortest paths in a network. Networks 1976; 6:205–229.

[9] Carraresi P, Sodini C. A binary enumeration tree to find k shortest paths. Methods of Operations Research (Germany) 7th Symposium on Operations Research, St. Gallen, Switzerland 1983; 177–188

[10] Boffey B. The all-to-all alternative route problem. Operation Research – Rairo-Recherche Operationelle 1993; 27:375–387

Author Index